D0140283

Advanced Trigonometry

ADVANCED TRIGONOMETRY

C.V. DURELL
AND A. ROBSON

DOVER PUBLICATIONS, INC.
MINEOLA, NEW YORK

Bibliographical Note

This Dover edition, first published in 2003, is an unabridged republication of the work originally published by G. Bell and Sons, Ltd., London, in 1930.

Library of Congress Cataloging-in-Publication Data

Durell, Clement V. (Clement Vavasor), b. 1882.
Advanced trigonometry / C.V. Durell and A. Robson.
p. cm.
Originally published: London : G. Bell and Sons, Ltd., 1930.
Includes index.
ISBN 0-486-43229-7 (pbk.)
1. Trigonometry. I. Robson, A. (Alan), b. 1888. II. Title.

QA531.D786 2003
516.24—dc21

2003057276

Manufactured in the United States of America
Dover Publications, Inc., 31 East 2nd Street, Mineola, N.Y. 11501

PREFACE

MOST teachers will agree that at the present time the work of mathematical specialists in schools is heavily handicapped by the absence of suitable text-books. There have been such radical changes in method and outlook that it has become necessary to treat large sections of some of the standard books merely as (moderately) convenient collections of examples and to supply the bookwork in the form of notes; especially is this true of Algebra, Trigonometry, and the Calculus.

Dividing lines between these subjects tend nowadays to be obliterated. Methods of the Calculus are freely used in courses of Algebra and Trigonometry, while matter which used to find a place in the Algebra text-book is now included more conveniently elsewhere. Perhaps the most important example of this re-arrangement is the treatment of the logarithmic function. For many years past leading mathematicians have advocated a definition which transfers the chapter on the theory of logarithms from the Algebra to the Calculus text-book, and makes it the basis from which the exponential function is discussed, thus reversing the order commonly followed. The authors are convinced by their own experience that this is the best mode of approach. On general principles it would seem desirable also to follow the same order for the complex variable, but unfortunately in practice this point of view appears to be too difficult for school work. By tradition the theory of the exponential and logarithmic functions of a complex variable is included in books on Advanced Trigonometry and this is a very reasonable arrangement; it seems equally desirable to include also the theory of the corresponding functions of a real variable instead of relegating it to the Calculus book.

The interest and value of advanced trigonometry lies in regarding it as an introduction to modern analysis. The methods by which results are obtained are often more important—that is, educationally more valuable—than the results themselves. The character of the treatment in this book is shaped and controlled by that idea. Thus the methods for expanding functions in series focus attention on "remainders" and "limits"; the methods for factorizing functions turn on establishing possible forms and then using the fundamental factor-theorem; the discussion of complex numbers emphasises the fact that complex numbers are just as "real" as real numbers, etc. For the same reason no apology need be offered for the prevalence, in this book, of "inequalities." Their importance in higher mathematics can hardly be exaggerated, and they are invaluable too in elementary work. The "useful inequalities" of Chapter IV will, it is believed, be found fully worthy of their name.

The authors are planning text-books parallel to the present volume on Advanced Algebra and Calculus, written from a similar point of view. In all these subjects, it must be admitted, there are certain difficulties which the average student will never face, but which are all-important for the real mathematician; these include, for example, the purely arithmetical treatment of real number, limits, continuity, convergence, mean-value theorems, the analysis of area, length of a curve, etc. The authors propose to deal with these matters in a book which is cited as a "companion volume on Analysis," limiting the treatment, however, to what seems suitable for specialist work at schools. Although planned, no part of this book is yet written. The theory of Infinite Products has been left for this companion volume; it is not so easy to provide a satisfactory *ab initio* treatment for products as it is for series and the alternative of taking for granted everything that really matters is undesirable. Happily also Infinite Products are of small value in elementary work and they are not required for most examinations. See, however, pp. 223, 240.

As a text-book on Trigonometry, this volume is a continuation of Durell and Wright's *Elementary Trigonometry*, and Chapter I should be regarded partly as a revision course. The sole object of Chapter XIV is to give opportunity for practice in mechanical manipulation to those who require it. The course really closes with Chapter XIII, which deals with a difficult subject and one which should be done carefully if it is done at all.

A Key is published, for the convenience of teachers, in which solutions are given in considerable detail, and in some cases alternative methods of solution are supplied, so that to some extent the *Key* forms a supplementary teaching manual.

The authors gratefully acknowledge help with the proofs received from Mr. J. C. Manisty, whose numerous criticisms and suggestions have enabled them to effect many improvements.

CONTENTS

CHAPTER PAGE

I. PROPERTIES OF THE TRIANGLE - - - - - - 1

(Methods of solution, p. 1; circumcentre, incentre, orthocentre, nine-point centre, polar circle, pp. 4-6; centroid, p. 10; distance between special points, pp. 15-17; errors, p. 20.)

II. PROPERTIES OF THE QUADRILATERAL - - - 24

(Cyclic quadrilateral, p. 24; general quadrilateral, p. 27; circumscribable quadrilateral, p. 27.)

III. EQUATIONS, SUB-MULTIPLE ANGLES, INVERSE FUNCTIONS - - - - - - - - - - - 33

(General solutions, p. 34; sub-multiple angles, p. 41; inverse functions, p. 46.)

IV. A HYPERBOLIC FUNCTION AND LOGARITHMIC AND EXPONENTIAL FUNCTIONS - - - - - - 52

(Area-function for rect. hyperbola, p. 52; differentiation, p. 57; addition theorem, p. 60; properties of $\log x$ and e^x, pp. 63, 64; useful inequalities, p. 67; Euler's constant, pp. 69, 70.)

V. EXPANSIONS IN POWER-SERIES - - - - - - 77

(Convergence, p. 77; expansions of $\sin x$ and $\cos x$, p. 79; expansion of $\log(1+x)$, p. 84; expansion of $\tan^{-1}x$, p. 88; evaluation of π, p. 89; expansion of e^x, p. 90; $\lim_{n\to\infty}\left(1+\frac{x}{n}\right)^n$, p. 93.)

VI. THE SPECIAL HYPERBOLIC FUNCTIONS - - - 104

(Definitions, $\text{sh}\,x$, $\text{ch}\,x$, $\text{th}\,x$, p. 104; formulae, p. 105; calculus applications, p. 107; $\text{sh}^{-1}x$, $\text{ch}^{-1}x$, $\text{th}^{-1}x$, p. 110.)

VII. PROJECTION AND FINITE SERIES - - - - - 118

(Projections and general angles, p. 118; $\cos(A+B)$, $\sin(A+B)$, p. 123; $\Sigma\cos[\alpha+(r-1)\beta]$, etc., p. 125; difference series, p. 130.)

VIII. COMPLEX NUMBERS - - - - - - - - - 137

(Definitions, p. 138; notation and manipulation, p. 140; modulus and amplitude, p. 145; use of Argand Diagram, p. 148; products and quotients, pp. 150, 151; principal values, p. 155.)

CHAPTER PAGE

IX. DE MOIVRE'S THEOREM AND APPLICATIONS - - 162

(De Moivre's theorem, p. 162; principal values, p. 164; values of $z^{\frac{p}{q}}$, p. 165; powers and roots in Argand Diagram, p. 165; expansions of $\cos^n\theta$, $\sin^n\theta$, p. 169; expansions of $\cos n\theta$, $\sin n\theta$, $\tan n\theta$, p. 172; $\Sigma x^r \cos r\theta$, etc., p. 174; $\cos n\theta$, $\frac{\sin n\theta}{\sin\theta}$ as polynomials in $\cos\theta$, etc., p. 178.)

X. ONE-VALUED FUNCTIONS OF A COMPLEX VARIABLE - 189

(Absolute convergence, p. 189; series of complex terms, p. 190; exponential series and exponential function, p. 191; modulus and amplitude of exp (z), p. 194; $\sin z$, $\cos z$, $\tan z$, p. 197; sh z, ch z, th z, p. 198.)

XI. ROOTS OF EQUATIONS - - - - - - - - 204

(Formation of equations, p. 204; symmetric functions of the roots, p. 206; essentially distinct roots, p. 212.)

XII. FACTORS - - - - - - - - - - - 219

(Algebraic functions, p. 219; trigonometric functions, $\sin n\theta$, etc., p. 222; $x^{2n} - 2x^n \cos n\alpha + 1$, p. 226; comparison of series and products, p. 228; partial fractions, p. 231.)

XIII. MANY-VALUED FUNCTIONS OF A COMPLEX VARIABLE - 241

(Log w, p. 241; expansion of $\log(1+w)$, p. 245; circle of convergence, p. 247; z^w, p. 252; binomial series, p. 253; logarithms to any base, p. 253; inverse functions, p. 256.)

XIV. MISCELLANEOUS RELATIONS - - - - - - 263

(General identities, p. 263; conditional identities, p. 265; miscellaneous transformations, p. 268; elimination, p. 270; inequalities, p. 274.)

MISCELLANEOUS EXAMPLES ON CHAPTERS I-XIV - - - 278

ANSWERS - - - - - - - - - - - - 285

INDEX - - - - - - - - - - - - 333

SYMBOLS - - - - - - - - - - - - 336

CHAPTER I

PROPERTIES OF THE TRIANGLE

A LIST of the fundamental formulae connecting the elements of a triangle, proofs of which have been given in *Durell and Wright's Elementary Trigonometry*, will be found in Section D of the formulae at the beginning of that book; references to these proofs will be indicated by the prefix *E.T.*

For geometrical proofs of theorems on the triangle, the reader is referred to some geometrical text-book. When these theorems are quoted or illustrated in this chapter, references, indicated by the prefix *M.G.*, are given to *Durell's Modern Geometry.*

Revision. Examples for the revision of ordinary methods of solving a triangle are given in Exercise I. a, below.

It is sometimes convenient to modify the process of solution. If, for example, the numerical values of b, c, A are given and if the value of a *only* is required, we may proceed as follows:

$$a^2 = b^2 + c^2 - 2bc \cos \mathsf{A};$$

$$\therefore\ a^2 = (b+c)^2 - 2bc(1+\cos \mathsf{A}) = (b+c)^2 - 4bc\cos^2 \tfrac{1}{2}\mathsf{A};$$

$$\therefore\ a^2 = (b+c)^2 - (b+c)^2 \cos^2\theta, \text{ where } \cos^2\theta = \frac{4bc\cos^2 \frac{1}{2}\mathsf{A}}{(b+c)^2};$$

$$\therefore\ a = (b+c)\sin\theta, \quad \ldots\ldots\ldots\ldots\ldots\ldots(1)$$

where

$$\cos\theta = \frac{2\sqrt{(bc)}\cos\frac{1}{2}\mathsf{A}}{b+c}. \quad \ldots\ldots\ldots\ldots\ldots\ldots(2)$$

θ is first found from (2) and then a is obtained from (1), both equations being adapted to logarithmic work.

An angle θ, used in this way, is called a **subsidiary angle.** For other examples of the use of subsidiary angles, see Ex. I. **a, Nos. 21** to 25.

EXERCISE I. a.

[*Solution of Triangles*]

1. What are the comparative merits of the formulae for cos A, $\cos\frac{\mathsf{A}}{2}$, $\sin\frac{\mathsf{A}}{2}$, $\tan\frac{\mathsf{A}}{2}$, when finding the angles of a triangle from given numerical values of a, b, c?

2. Given $a = 100$, $b = 80$, $c = 50$, find A.

3. Given $a=37$, $b=61$, $c=37$, find B.

4. Given $a=11{\cdot}42$, $b=13{\cdot}75$, $c=18{\cdot}43$, find A, B, C.

5. Given $A=17^\circ\ 55'$, $B=32^\circ\ 50'$, $c=251$, find a from the formula

$$a=c \sin A \operatorname{cosec} C.$$

6. Given $B=86^\circ$, $C=17^\circ\ 42'$, $b=23$, solve the triangle.

7. Given $b=16{\cdot}9$, $c=24{\cdot}3$, $A=154^\circ\ 18'$, find $\frac{1}{2}(B-C)$ from the formula $\tan \frac{1}{2}(B-C)=\frac{b-c}{b+c}\cot\frac{A}{2}$, and complete the solution of the triangle.

8. Given $b=27$, $c=36$, $A=62^\circ\ 35'$, find a.

Solve the triangles in Nos. 9-13:

9. $A=39^\circ\ 42'$, $B=81^\circ\ 12'$, $c=47{\cdot}6$.

10. $b=6{\cdot}32$, $c=8{\cdot}47$, $B=43^\circ$.

11. $a=110$, $b=183$, $c=152$.

12. $a=6{\cdot}81$, $c=9{\cdot}06$, $B=119^\circ\ 45'$.

13. $b=16{\cdot}9$, $c=12{\cdot}3$, $C=51^\circ$.

[*The Ambiguous Case*]

14. Given $A=20^\circ\ 36'$, $c=14{\cdot}5$, find the range of values of a such that the number of possible triangles is 0, 1, 2. Complete the solution if a equals (i) $8{\cdot}3$, (ii) $16{\cdot}2$, (iii) $3{\cdot}2$, (iv) $5{\cdot}1$.

15. Given b, c, and B, write down the quadratic for a, and the sum and product of its roots, a_1 and a_2. Verify the results geometrically.

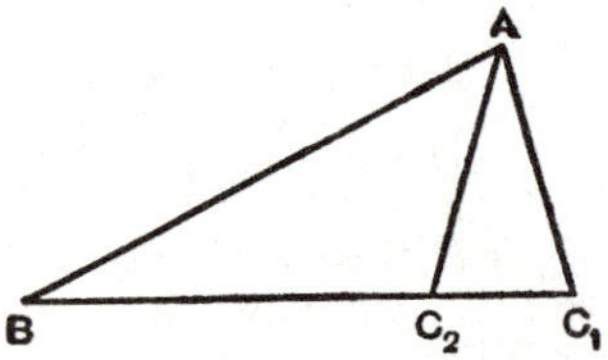

If A_1, C_1 and A_2, C_2 are the remaining angles of the two triangles which satisfy the data, find C_1+C_2 and A_1+A_2.

16. With the data of No. 15, prove that

(i) $a_1-a_2=\pm 2\sqrt{(b^2-c^2\sin^2 B)}$; (ii) $\sin\frac{1}{2}(A_1-A_2)=\frac{a_1-a_2}{2b}$.

17. With the data of No. 15, prove that

$$(a_1-a_2)^2+(a_1+a_2)^2\tan^2 B=4b^2.$$

18. (i) With the data of No. 15, if $a_1=3a_2$, prove that

$$2b=c\sqrt{(1+3\sin^2 B)}.$$

(ii) With the data of No. 15, if $C_2=2C_1$, prove that

$$2c\sin B=b\sqrt{3}.$$

19. If the two triangles derived from given values of c, b, B have areas in the ratio 3 : 2, prove that $25(c^2 - b^2) = 24c^2 \cos^2 B$.

20. With the data of No. 15, if $A_1 = 2A_2$, prove that

$$4c^3 \sin^2 B = b^2(b + 3c).$$

[*Subsidiary Angles*]

21. Given $b = 16{\cdot}9$, $c = 24{\cdot}3$, $A = 154° \, 18'$, find a from formulae (1) and (2), p. 1.

22. Show that the formula $c = b \cos A \pm \sqrt{(a^2 - b^2 \sin^2 A)}$ may be written in the form $c = a \sin(\theta \pm A) \operatorname{cosec} A$, where $\sin\theta = \dfrac{b}{a} \sin A$.

23. Show how to apply the method of the subsidiary angle to $a^2 = (b - c)^2 + 2bc(1 - \cos A)$.

24. In any triangle, prove that $\tan \frac{1}{2}(B - C) = \tan(45° - \theta) \cot \frac{1}{2}A$, where $\tan\theta = \dfrac{c}{b}$.

Hence find $\frac{1}{2}(B - C)$ if $b = 321$, $c = 436$, $A = 119° \, 15'$.

25. Express $a \cos\theta - b \sin\theta$ in a form suitable for logarithmic work.

[*Miscellaneous Relations*]

26. If $a = 4$, $b = 5$, $c = 6$, prove that $C = 2A$.

27. Express in a symmetrical form $\dfrac{a}{bc} + \dfrac{\cos A}{a}$.

28. Prove that $b^2(\cot A + \cot B) = c^2(\cot A + \cot C)$.

29. Simplify $\operatorname{cosec}(A - B) \,.\, (a \cos B - b \cos A)$.

30. Prove that $a^2 \sin(B - C) = (b^2 - c^2) \sin A$.

31. Prove that $\dfrac{b \sec B + c \sec C}{\tan B + \tan C} = \dfrac{c \sec C + a \sec A}{\tan C + \tan A}$.

32. If $b \cos B = c \cos C$, prove that either $b = c$ or $A = 90°$.

33. Prove that $\sin^2 A + \sin B \sin C \cos A = \dfrac{2\Delta^2(a^2 + b^2 + c^2)}{a^2 b^2 c^2}$.

34. Prove that $\dfrac{1 + \cos(A - B)\cos C}{1 + \cos(A - C)\cos B} = \dfrac{a^2 + b^2}{a^2 + c^2}$.

35. Prove that

$$a \cos B \cos C + b \cos C \cos A + c \cos A \cos B = \frac{2\Delta \sin A}{a}.$$

36. Express $\cos \frac{1}{2}(A - B) \,.\, \operatorname{cosec} \dfrac{C}{2}$ in terms of a, b, c.

37. If $b + c = 2a$, prove that $4\Delta = 3a^2 \tan \dfrac{A}{2}$.

38. If $a^2 = b(b+c)$, prove that $A = 2B$.

39. Prove that $c^2 = a^2 \cos 2B + b^2 \cos 2A + 2ab \cos (A - B)$.

40. Prove that $\dfrac{b-c}{b+c} \cot \dfrac{A}{2} + \dfrac{b+c}{b-c} \tan \dfrac{A}{2} = 2 \operatorname{cosec} (B - C)$.

41. Prove that

$$a(1 + 2\cos 2A)\cos 3B + b(1 + 2\cos 2B)\cos 3A = c(1 + 2\cos 2C).$$

42. If $\cos A \cos B + \sin A \sin B \sin C = 1$, prove that $A = 45° = B$.

The Circumcentre. The centre O of the circle through A, B, C is found by bisecting the sides of the triangle at right angles, and the radius is given by the formulae

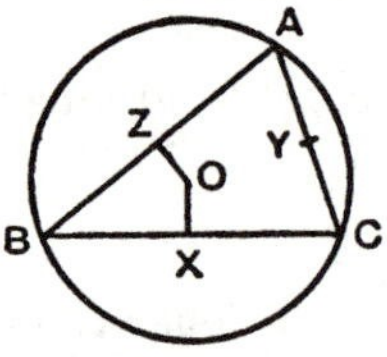

FIG. 1.

$$R = BX \operatorname{cosec} BOX = \frac{a}{2 \sin A}; \quad \text{.........(3)}$$

$$\therefore \quad R = \frac{abc}{2bc \sin A} = \frac{abc}{4\Delta}. \quad \text{.....................(4)}$$

The reader should prove that these formulae hold also when $\angle BAC$ is obtuse.

The in-centre and e-centres. The centres I, I_1, I_2, I_3 of the circles which touch the sides are found by bisecting the angles of the triangle, internally and externally.

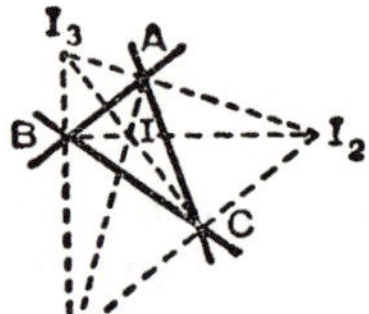

FIG. 2.

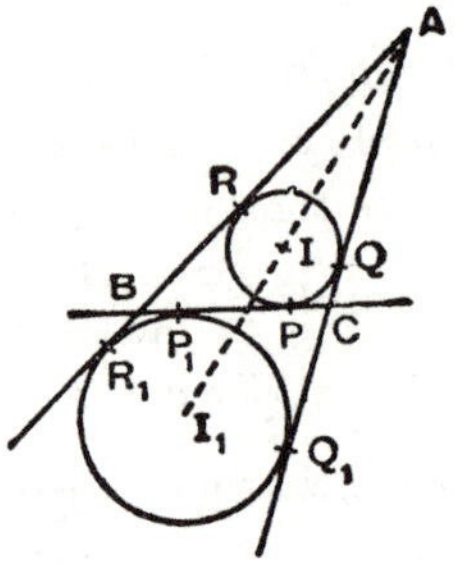

FIG. 3.

The radii of these circles are given by

$$r = \frac{\Delta}{s}; \quad r_1 = \frac{\Delta}{s-a}, \quad \text{etc. ....(5)}$$

$$r = 4R \sin \frac{A}{2} \sin \frac{B}{2} \sin \frac{C}{2}; \quad r_1 = 4R \sin \frac{A}{2} \cos \frac{B}{2} \cos \frac{C}{2}, \quad \text{etc. ......(6)}$$

Also in Fig. 3, we have

$$AR = s - a; \quad AR_1 = s; \quad BP_1 = s - c; \quad \ldots\ldots\ldots\ldots\ldots\ldots(7)$$

$$\therefore\ r = (s - a)\tan\frac{A}{2}; \quad r_1 = s\tan\frac{A}{2}. \quad \ldots\ldots\ldots\ldots\ldots\ldots(8)$$

For proofs of these formulae and further details, see *E.T.*, pp. 184-186, 277, 278 and *M.G.*, pp. 11, 24, 25.

The Orthocentre and Pedal Triangle. The perpendiculars AD, BE, CF from the vertices of a triangle to the opposite sides meet at a point H, called the **orthocentre**; the triangle DEF is called the **pedal triangle** (*M.G.*, p. 20).

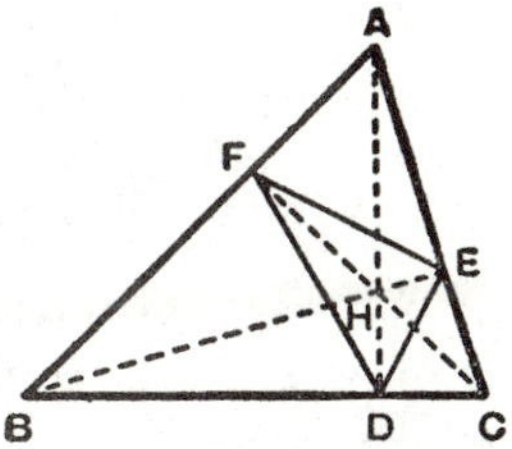

FIG. 4.

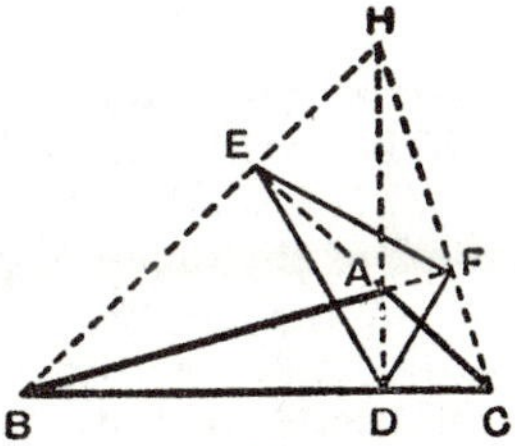

FIG. 5.

If $\triangle ABC$ is *acute-angled*, (Fig. 4), H lies inside the triangle.

Since BFEC is a cyclic quadrilateral, AFE and ACB are similar triangles;

$$\therefore\ \frac{EF}{BC} = \frac{AF}{AC} = \cos A;$$

$$\therefore\ \mathbf{EF = a\cos A.} \quad \ldots\ldots\ldots\ldots\ldots\ldots\ldots\ldots\ldots(9)$$

Since HECD is a cyclic quadrilateral, $\angle HDE = \angle HCE = 90° - A$; similarly $\angle HDF = 90° - A$;

$$\therefore\ \angle EDF = 180° - 2A. \quad \ldots\ldots\ldots\ldots\ldots\ldots\ldots(10)$$

Further, HD bisects $\angle EDF$ and similarly HE bisects $\angle DEF$; $\therefore$ H is the in-centre of $\triangle DEF$. Also since BC is perpendicular to AD, it is the external bisector of $\angle EDF$; hence A, B, C are the e-centres of the pedal triangle.

We have also

$$\mathbf{AH} = AE \operatorname{cosec} AHE = c\cos A \operatorname{cosec} C = \mathbf{2R\cos A}, \quad \ldots\ldots(11)$$

and $$\mathbf{DH} = BH\cos BHD = \mathbf{2R\cos B\cos C}. \quad \ldots\ldots\ldots\ldots\ldots\ldots\ldots(12)$$

The reader should work out the corresponding results for Fig. 5, where the triangle is *obtuse-angled*.

If $\angle BAC$ is obtuse, $\angle EDF = 2A - 180°$ and other results are modified by writing $-\cos A$ for $\cos A$. [See Ex. I. b, No. 27 and note the difference of form in No. 36. See also *Example* 3.]

The Nine-Point Circle. The circle which passes through the mid-points X, Y, Z of the sides BC, CA, AB passes also through D, E, F and

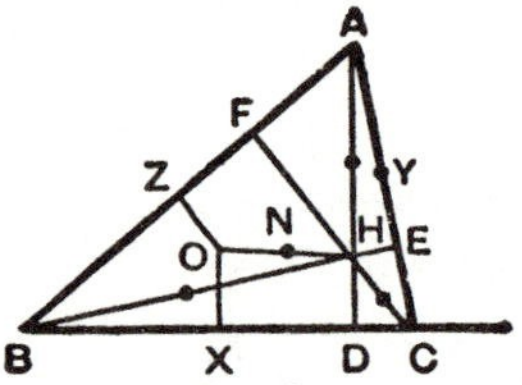

FIG. 6.

through the mid-points of HA, HB, HC; it is therefore called the nine-point circle and its centre N is the mid-point of OH (*M.G.*, p. 27).

Since $\triangle XYZ$ is similar to $\triangle ABC$ and of half its linear dimensions, the radius of the nine-point circle is $\frac{1}{2}R$.

Since each of the points H, A, B, C is the orthocentre of the triangle formed by the other three, the circumcircle of $\triangle DEF$ is the common nine-point circle of the four triangles ABC, BCH, CHA, HAB.

Also, since $\triangle ABC$ is the pedal $\triangle$ of $\triangle I_1I_2I_3$ and of $\triangle II_2I_3$, etc., the circumradius of each of these triangles is 2R.

The Polar Circle. In Fig. 6 and Fig. 7 we have, by cyclic quadrilaterals,

$$HA \,.\, HD = HB \,.\, HE = HC \,.\, HF.$$

In Fig. 7, where $\angle BAC$ is *obtuse*, A and D are on the same side of H, and so also are B, E and C, F. In this case, if $HA \,.\, HD = \rho^2$, it follows that the polars of A, B, C w.r.t. the circle, centre H, radius ρ, are BC, CA, AB.

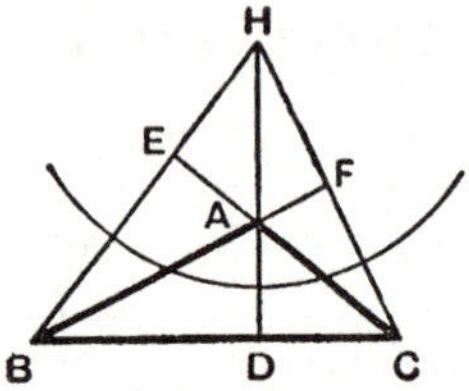

FIG. 7.

The triangle ABC is therefore self polar w.r.t. this circle; and the circle is called the polar circle of $\triangle ABC$.

We have

$$\rho^2 = HA \,.\, HD = (-2R\cos A)(2R\cos B\cos C);$$

$$\therefore\ \rho^2 = -4R^2 \cos A \cos B \cos C. \quad \ldots(13)$$

An acute-angled triangle in real geometry has no polar circle.

Notation. The lettering already adopted for special points connected with the triangle will be employed throughout the Chapter. This will shorten the statement of many of the examples.

We add some illustrative examples.

Example 1. Prove $s^2=\Delta(\cot\frac{1}{2}A+\cot\frac{1}{2}B+\cot\frac{1}{2}C)$.

Since $\frac{1}{2}(A+B+C)=90°$,

$$\Sigma\cot\tfrac{1}{2}A=\cot\tfrac{1}{2}A\cot\tfrac{1}{2}B\cot\tfrac{1}{2}C,\text{ (see } E.T.\text{, p. 272, Ex. V.)}$$

$$\therefore\ \Sigma\cot\tfrac{1}{2}A=\sqrt{\left\{\frac{s(s-a)}{(s-b)(s-c)}\cdot\frac{s(s-b)}{(s-c)(s-a)}\cdot\frac{s(s-c)}{(s-a)(s-b)}\right\}}$$

$$=\frac{s^2}{\Delta}.$$

Example 2. Express $\dfrac{(ab-r_1r_2)}{r_3}$ in a symmetrical form.

Since $$r_1r_2=\frac{\Delta^2}{(s-a)(s-b)}=s(s-c),$$

$$4ab-4r_1r_2=4ab-(a+b+c)(a+b-c)$$
$$=c^2-(a-b)^2$$
$$=4(s-a)(s-b);$$

$$\therefore\ \frac{(ab-r_1r_2)}{r_3}=\frac{(s-a)(s-b)}{r_3}$$

$$=\frac{(s-a)(s-b)(s-c)}{\Delta}=\frac{\Delta}{s}.$$

Example 3. If J is the in-centre of BHC, express the radius of the circle BJC in terms of R and A.

By equation (3) the radius is $\frac{1}{2}$BC cosec BJC, but

$$\angle BJC=90°+\tfrac{1}{2}\angle BHC=180°-\tfrac{1}{2}A,\text{ if B and C are acute angles;}$$

$$\therefore\ \text{the radius}=\frac{a}{2\sin\frac{1}{2}A}=\frac{2R\sin A}{2\sin\frac{1}{2}A}=2R\cos\tfrac{1}{2}A.$$

If either B or C is obtuse, $\angle BJC=90°+\frac{1}{2}A$, and then the radius $=2R\sin\frac{1}{2}A$.

EXERCISE I. b.

1. If $a=15{\cdot}1$, $A=24°\ 36'$, find R.
2. If $a=3$, $b=5$, $c=7$, find R and r.
3. If $a=13$, $b=14$, $c=15$, find r_1, r_2, r_3.
4. If $a=23{\cdot}5$, $A=62°$, and $b=c$, find R and r.
5. Prove that

$$\text{(i) } \angle BAI_3=90°-\frac{A}{2}=\angle I_3I_1I_2;\quad \text{(ii) } II_1=4R\sin\frac{A}{2};$$

$$\text{(iii) } I_2I_3=a\operatorname{cosec}\frac{A}{2}=4R\cos\frac{A}{2}.$$

6. Verify Equation (6), p. 4, by using the formulae for $\sin\frac{A}{2}$, $\cos\frac{A}{2}$, etc., in terms of the sides.

7. Express $a(\cos A+\cos B\cos C)$ in a symmetrical form.

Prove the following relations :

8. $s=4R\cos\frac{A}{2}\cos\frac{B}{2}\cos\frac{C}{2}$.

9. $s-a=4R\cos\frac{A}{2}\sin\frac{B}{2}\sin\frac{C}{2}$.

10. $r_2r_3\tan\frac{A}{2}=\Delta$.

11. $r_2r_3+r_3r_1+r_1r_2=s^2$.

12. $r_2+r_3=4R\cos^2\frac{A}{2}$.

13. $r-r_1+r_2+r_3=2a\cot A$.

14. $AI\,.\,AI_1=bc$.

15. $IA\,.\,IB=4Rr\sin\frac{C}{2}$.

16. $IA\,.\,IB\,.\,IC=\frac{abc\Delta}{s^2}$.

17. $II_1\,.\,II_2\,.\,II_3=16R^2r$.

18. $\triangle ABI:\triangle ACI=c:b$.

19. $AD^2(\cot B+\cot C)=2\Delta$.

20. $AD=2r\operatorname{cosec}\frac{A}{2}\cos\frac{B}{2}\cos\frac{C}{2}$.

21. $\triangle OI_2I_3:\triangle OI_3I_1=(b+c):(a+c)$.

22. $AH=a\cot A=2OX$.

23. $AH+BH+CH=2(R+r)$.

24. If $a=14$, $b=13$, $c=15$, prove that $AD=12$.

25. Given $B=37°$, $C=46°$, $BE=9{\cdot}3$, find b.

26. If $BP\,.\,PC=\Delta$, (see Fig. 3), prove that $A=90°$.

27. In Fig. 5, where $\angle BAC$ is obtuse, prove that

(i) $EF=-a\cos A$, $FD=b\cos B$, $DE=c\cos C$;

(ii) $\angle FDE=2A-180°$, $\angle DEF=2B$, $\angle EFD=2C$;

(iii) $AH=-2R\cos A$, $BH=2R\cos B$, $CH=2R\cos C$;

(iv) $HD=2R\cos B\cos C$, $HE=-2R\cos C\cos A$,
$HF=-2R\cos A\cos B$.

28. If $a=13$, $b=9$, $c=5$, find ρ (see p. 6).

29. Find an expression for the radius of the polar circle of $\triangle II_2I_3$ in terms of R, r_1.

30. Prove that the circumradius of $\triangle HBC$ equals R.

31. Prove that the circumradius of $\triangle OBC$ is $>\frac{1}{2}R$.

32. Prove that the in-radius of $\triangle AEF$ is $r\cos A$.

33. Prove that the area of $\triangle DEF$ is $\pm 2\Delta\cos A\cos B\cos C$.

34. Given b, c, B, prove that the product of the in-radii of the two possible triangles is $c(c-b)\sin^2\frac{1}{2}B$.

35. Prove that the in-radius of $\Delta I_1I_2I_3$ is $2R\left\{\Sigma\left(\sin\frac{A}{2}\right)-1\right\}$.

36. If ΔABC is acute-angled, prove that the perimeter of ΔDEF is $4R \sin A \sin B \sin C$. If $\angle BAC$ is obtuse, prove that the perimeter is $4R \sin A \cos B \cos C$.

37. Find in terms of A, B, C, R the in-radius of ΔDEF (i) if ΔABC is acute-angled, (ii) if $\angle BAC$ is obtuse.

38. Prove that $a \sin B \sin C + b \sin C \sin A + c \sin A \sin B = \frac{3\Delta}{R}$.

39. Express $\frac{\Delta}{a} + r \cos A - R \cos^2 A$ in a symmetrical form.

40. Prove that

(i) $a^2 \cos^2 A = b^2 \cos^2 B + c^2 \cos^2 C + 2bc \cos B \cos C \cos 2A$;

(ii) $a^2 \cos^2 A \cos^2 2A = b^2 \cos^2 B \cos^2 2B + c^2 \cos^2 C \cos^2 2C + 2bc \cos B \cos C \cos 2B \cos 2C \cos 4A$;

(iii) $a^2 \operatorname{cosec}^2 \frac{A}{2} = b^2 \operatorname{cosec}^2 \frac{B}{2} + c^2 \operatorname{cosec}^2 \frac{C}{2} - 2bc \operatorname{cosec} \frac{B}{2} \operatorname{cosec} \frac{C}{2} \sin \frac{A}{2}$.

Any Line through a Vertex. Suppose any line through A cuts BC at K. Denote $\frac{BK}{KC}$ by $\frac{z}{y}$, so that K is the centroid of masses y, z at B, C respectively.

Let $\angle BAK = \beta$, $\angle KAC = \gamma$, $\angle AKC = \theta$.

Draw BB′, CC′ perpendicular to AK.

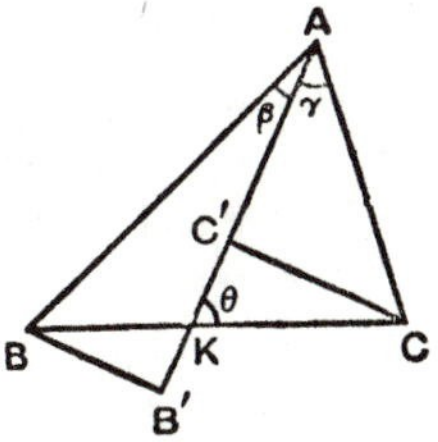

FIG. 8.

Then $$\frac{BK}{KC} = \frac{BB'}{CC'} = \frac{c \sin \beta}{b \sin \gamma}. \quad \text{............(14)}$$

This may be written

$$\frac{z}{y} = \frac{\sin C \sin(\theta - B)}{\sin B \sin(\theta + C)} = \frac{\sin(\theta - B)}{\sin B \sin \theta} \cdot \frac{\sin C \sin \theta}{\sin(\theta + C)};$$

$$\therefore \frac{z}{y} = \frac{\cot B - \cot \theta}{\cot C + \cot \theta};$$

$$\therefore (y + z) \cot \theta = y \cot B - z \cot C. \quad \text{....................(15)}$$

This relation, which determines θ for a given triangle and given position of K, is often useful in three-force problems in statics (cf. Ex. I. c, No. 11); an alternative method of proof is indicated in Ex. I. c, No. 8. Sometimes (cf. Ex. I. c, No. 12) it is convenient to have an expression for θ in terms of β, γ.

From (14),

$$\frac{z}{y}=\frac{\sin(\theta+\gamma)\sin\beta}{\sin(\theta-\beta)\sin\gamma}=\frac{\sin(\theta+\gamma)}{\sin\theta\sin\gamma}\cdot\frac{\sin\theta\sin\beta}{\sin(\theta-\beta)};$$

$$\therefore\ \frac{z}{y}=\frac{\cot\gamma+\cot\theta}{\cot\beta-\cot\theta};$$

$$\therefore\ (y+z)\cot\theta=z\cot\beta-y\cot\gamma. \qquad\qquad (16)$$

The Centroid. The centroid of k_1 at (x_1, y_1), k_2 at (x_2, y_2), k_3 at (x_3, y_3), etc., is the *centre of mass* of particles of masses proportional to k_1, k_2, k_3, etc., at these points. The centroid is also called the *centre of mean position.* The point may also be defined geometrically, and its coordinates are $\left(\frac{\Sigma(kx)}{\Sigma k}, \frac{\Sigma(ky)}{\Sigma k}\right)$; thus the idea of mass is not really involved. The values of the k's need not all be positive, but Σk must not be zero. (*M.G.*, pp. 58-64.)

If T is any point in AK, we have with the notation of Fig. 9,

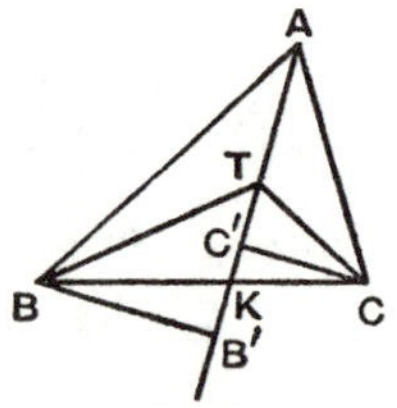

FIG. 9.

$$\frac{BK}{KC}=\frac{\frac{1}{2}BB'\,.\,AT}{\frac{1}{2}CC'\,.\,AT}=\frac{\Delta ATB}{\Delta ATC}; \qquad\qquad (17)$$

$\therefore$ K is the centroid of ΔCTA at B and ΔATB at C;

$\therefore$ the centroid of ΔBTC at A, ΔCTA at B, ΔATB at C lies on AK, that is on AT ; similarly it must lie on BT, and it is therefore at T.

Hence, if any point T is the centroid of masses λ, μ, ν at A, B, C, then $\lambda:\mu:\nu=\Delta TBC:\Delta TCA:\Delta TAB$.

If, with the same notation as before, K is the centroid of y at B and z at C, the length of AK is given by a theorem of Apollonius (*M.G.*, p. 61) :

$$y\,.\,AB^2+z\,.\,AC^2=(y+z)\,.\,AK^2+y\,.\,KB^2+z\,.\,KC^2, \qquad\qquad (18)$$

where
$$\frac{BK}{z}=\frac{KC}{y}=\frac{BC}{z+y}=\frac{a}{z+y}.$$

And more generally (*M.G.*, p. 62) if G is the centroid of k_1 at P_1, k_2 at P_2, etc., and if O is any point,

$$\Sigma(k\,.\,OP^2)=(\Sigma k)\,.\,OG^2+\Sigma(k\,.\,GP^2). \qquad\qquad (19)$$

Equation (19) is useful in dealing with expressions connected with ΔABC of the form $\lambda\,.\,TA^2+\mu\,.\,TB^2+\nu\,.\,TC^2$. (Cf. Ex. I. c, Nos. 39, 40 and Ex. I. d, Nos. 22-28.)

Medians. If $y = z$, AK is a median; we then have from (15) and (18)

$$2 \cot AXC = \cot B - \cot C, \qquad \text{.....................(20)}$$

$$b^2 + c^2 = 2AX^2 + \tfrac{1}{2}a^2. \qquad \text{.......................(21)}$$

The three medians of a triangle are concurrent at a point G, which is the centroid of *equal* masses at A, B, C or of *equal* masses at X, Y, Z.

Further $GX = \frac{1}{3}AX$ and in addition G is the point on OH such that $OG = \frac{1}{3}OH$. (*M.G.*, p. 28.)

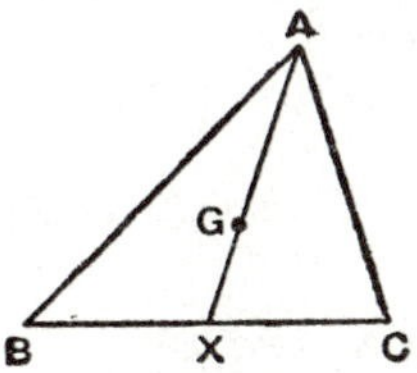

FIG. 10.

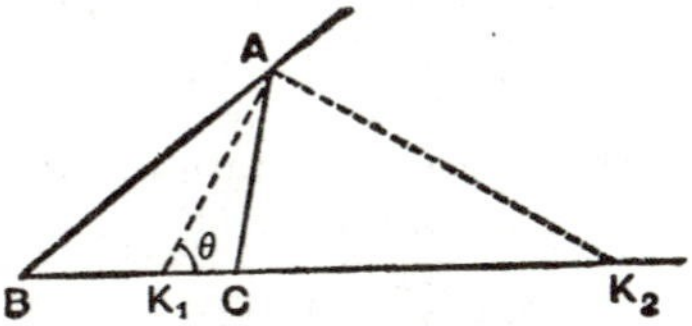

FIG. 11.

Angle-Bisectors. If AK_1 is the internal bisector of $\angle BAC$, $\beta = \gamma = \frac{1}{2}A$ and $\theta = B + \frac{1}{2}A = 90° - \frac{1}{2}(C - B)$.

Also $z : y = c : b$; $\therefore$ from (18), we have

$$bc^2 + cb^2 = (b + c)AK_1^2 + b\left(\frac{ac}{b+c}\right)^2 + c\left(\frac{ab}{b+c}\right)^2;$$

$$\therefore \text{ on reduction, } AK_1^2 = bc\left\{1 - \frac{a^2}{(b+c)^2}\right\}.$$

If AK_2 is the external bisector of $\angle BAC$,

$$\beta = 90° + \tfrac{1}{2}A; \quad \gamma = -(90° - \tfrac{1}{2}A); \quad \theta = B + \beta = 180° - \tfrac{1}{2}(C - B).$$

Also $z : y = c : -b$; $\therefore$ from (18) as before, we have

$$-bc^2 + cb^2 = (c - b)AK_2^2 - b\left(\frac{ac}{c-b}\right)^2 + c\left(\frac{ab}{c-b}\right)^2,$$

or $$AK_2^2 = bc\left\{\frac{a^2}{(c-b)^2} - 1\right\}.$$

Direct methods of proof are indicated in Ex. I. c, Nos. 15, 16.

Example 4. Show that $\Sigma \cot AXC = 0$

and $\Sigma \cot BAX = \Sigma \cot CAX$.

Equation (20) gives

$$2 \cot AXC = \cot B - \cot C,$$

and equation (16), with $y=z$, gives

$$2\cot \mathrm{AXC} = \cot \mathrm{BAX} - \cot \mathrm{CAX},$$

from which the required results follow.

Example 5. Find what masses at the vertices have their centroid at the circumcentre, and deduce that, if S is on the circle ABC,

$$\mathrm{SA}^2 \sin 2\mathrm{A} + \mathrm{SB}^2 \sin 2\mathrm{B} + \mathrm{SC}^2 \sin 2\mathrm{C} = 8\mathrm{R}^2 \sin \mathrm{A} \sin \mathrm{B} \sin \mathrm{C}.$$

The area BOC $= \frac{1}{2}\mathrm{R}^2 \sin 2\mathrm{A}$;
∴ the ratios of the areas BOC, COA, AOB are

$$\sin 2\mathrm{A} : \sin 2\mathrm{B} : \sin 2\mathrm{C};$$

∴ the masses are proportional to sin 2A, sin 2B, sin 2C (see p. 10).
Hence, by equation (19),

$$\begin{aligned} \Sigma[\mathrm{SA}^2 \sin 2\mathrm{A}] &= \Sigma[\mathrm{OA}^2 \sin 2\mathrm{A}] + [\Sigma \sin 2\mathrm{A}] \,.\, \mathrm{OS}^2 \\ &= \mathrm{R}^2 \,.\, [\Sigma \sin 2\mathrm{A}] + [\Sigma \sin 2\mathrm{A}] \,.\, \mathrm{R}^2 \\ &= 2\mathrm{R}^2 \,.\, (\Sigma \sin 2\mathrm{A}) \\ &= 8\mathrm{R}^2 \sin \mathrm{A} \sin \mathrm{B} \sin \mathrm{C}. \quad (\textit{E.T.}, \text{ p. } 271.) \end{aligned}$$

EXERCISE I. c.

1. If AO meets BC at K, prove that $\dfrac{\mathrm{BK}}{\mathrm{KC}} = \dfrac{\sin 2\mathrm{C}}{\sin 2\mathrm{B}}$.

2. If K is a point on the base BC of an equilateral triangle ABC and if $\angle \mathrm{BAK} = 15°$, calculate $\dfrac{\mathrm{BK}}{\mathrm{KC}}$.

3. If $\mathrm{B} = \mathrm{C} = 30°$ and if the perpendicular at A to AC cuts BC at K, prove that $\mathrm{BK} = \frac{1}{2}\mathrm{KC}$.

4. If $a = 13$, $b = 14$, $c = 15$, find cot B, cot C and cot AXC.

5. If $a = 61$, $b = 11$, $c = 60$ and if K divides BC internally as 3 : 2, find cot AKC.

6. If $a = 85$, $b = 13$, $c = 84$ and if K divides BC externally as 3 : 2, find cot AKC.

7. Prove that $\tan \mathrm{AXC} = \dfrac{4\Delta}{c^2 - b^2}$.

8. If B, K, D, C are any four collinear points, prove that

$$\mathrm{KD} \,.\, \mathrm{BC} = \mathrm{BD} \,.\, \mathrm{KC} - \mathrm{BK} \,.\, \mathrm{DC}.$$

From this relation, deduce equation (15) on p. 9.

9. Prove that $abc \cot \mathrm{AXB} = \mathrm{R}(b^2 - c^2)$.

10. If the trisectors of $\angle$BAC meet BC in K, K′, prove that

$$\frac{BK}{KC} \div \frac{BK'}{K'C} = \tfrac{1}{4}\sec^2\frac{A}{3}.$$

11. A uniform rod AB, 1 ft. long, is suspended from O by strings OA, OB of lengths 10 in., 7 in.; find the angle between AB and the vertical.

12. A uniform rod AB rests with its ends on two smooth planes, as shown; XOY is horizontal, find the angle between AB and the vertical.

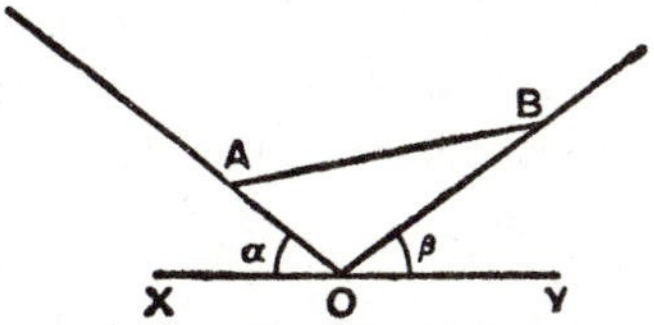

FIG. 12.

13. If $a=5$, $b=4$, $c=6$ and if K divides BC internally as 3 : 2, find AK.

14. If $\angle XAC = 90°$, prove that $\tan A + 2\tan C = 0$.

15. If the internal bisector of $\angle$BAC meets BC at K, prove that $\tfrac{1}{2}(b+c)\,.\,AK\sin\frac{A}{2} = \Delta$, and deduce that $AK = \frac{2bc}{b+c}\cos\tfrac{1}{2}A$, and that

$$AK^2 = bc\left\{1 - \frac{a^2}{(b+c)^2}\right\}.$$

16. If the external bisector of $\angle$BAC meets BC produced at K′, prove that $\tfrac{1}{2}(c-b)\,.\,AK'\cos\frac{A}{2} = \Delta$, and deduce that

$$AK'^2 = bc\left\{\frac{a^2}{(c-b)^2} - 1\right\}.$$

17. If the internal bisector of $\angle$BAC meets BC at K, prove that

(i) $AI : IK = (b+c) : a$; (ii) $a\,.\,PD = (c-b)(s-a)$;

(iii) $\tan APC = \frac{2r_1}{c-b}$.

18 If the internal bisector of $\angle$BAC meets BC at K and the circumcircle at L, prove that $AL = \tfrac{1}{2}(b+c)\sec\frac{A}{2}$. Find AK . AL and show that

$$AL : KL = (b+c)^2 : a^2.$$

19. Find the areas of ΔBOC, ΔBHC and deduce the area of ΔBNC. If AN meets BC at K, find $\frac{BK}{KC}$.

20. Show that I is the centroid of a at A, b at B, c at C.

21. What is the centroid of $-a$ at A, b at B, c at C?

22. If H is the centroid of x at A, y at B, z at C, find $x : y : z$.

23. Find the centroid of

(i) 1 at A, 1 at B, 1 at C, 1 at H ;
(ii) 3 at G, -2 at O.

24. Prove that $AX^2+BY^2+CZ^2=\frac{3}{4}(a^2+b^2+c^2)$.

25. If BY is perpendicular to CZ, prove that $b^2+c^2=5a^2$.

26. Prove that $\tan BGC=\dfrac{12\Delta}{b^2+c^2-5a^2}$.

27. If BY cuts AD at T, prove that $AT=\dfrac{2\Delta}{a+c\cos B}$.

28. If $B=55°$, $C=23°\ 30'$, $AX=40$, prove that $BY \simeq 60$.

29. If $A=90°$, and if BC is trisected at T_1, T_2, prove that

$$AT_1^2+AT_2^2=\frac{5a^2}{9}.$$

30. If $A=B=45°$ and if K is on AB, prove that $AK^2+BK^2=2CK^2$.

31. If $AX=m$, $AD=h$, prove that $\cot A=\dfrac{4m^2-a^2}{4ah}$.

32. If $\angle BAX=\beta$, $\angle CAX=\gamma$, prove that $\tan\dfrac{\beta-\gamma}{2}=\dfrac{b-c}{b+c}\tan\dfrac{A}{2}$.

33. If the internal bisectors make angles θ, ϕ, ψ with the opposite sides, prove that $a\sin 2\theta+b\sin 2\phi+c\sin 2\psi=0$.

34. Prove that $3\cot BGC=\cot A-\dfrac{a^2}{\Delta}$.

35. If $C=2B$ and if CB is divided externally at Q in the ratio 4 : 1, prove that $AQ-AC=\frac{1}{2}QC$.

36. If A, B, C, D are collinear and O is any point, prove that

$$\frac{AB\,.\,CD}{AD\,.\,CB}=\frac{\sin AOB\,.\,\sin COD}{\sin AOD\,.\,\sin COB}.$$

37. If AU, BV, CW are concurrent lines cutting BC, CA, AB at U, V, W, prove that

$$\sin BAU\,.\,\sin CBV\,.\,\sin ACW=\sin UAC\,.\,\sin VBA\,.\,\sin WCB.$$

38. If three segments AB, BC, CD of a straight line are of lengths α, β, γ and subtend equal angles θ at a point P, prove that

$$4\alpha\gamma\cos^2\theta=(\alpha+\beta)(\beta+\gamma).$$

39. (i) Use equation (19) and No. 20 to show that

$$a\,.\,TA^2+b\,.\,TB^2+c\,.\,TC^2$$

is least when T coincides with I.

(ii) For what position of T is $TA^2+TB^2+TC^2$ least ?

40. What is the locus of T, if

$$TA^2\,.\sin 2A+TB^2\,.\sin 2B+TC^2\,.\sin 2C$$

is constant ?

Distances between Special Points. With the usual notation,

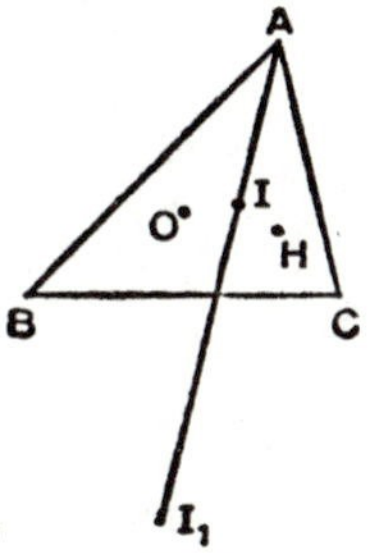

FIG. 13.

$$IA = r \operatorname{cosec}\frac{A}{2} = 4R\sin\frac{B}{2}\sin\frac{C}{2};$$

$$I_1A = r_1\operatorname{cosec}\frac{A}{2} = 4R\cos\frac{B}{2}\cos\frac{C}{2};$$

$$HA = 2R\cos A \text{ if } A < 90°,$$

and $$= -2R\cos A \text{ if } A > 90°.$$

To find OI^2 (*M.G.*, p. 35).

Let C be one of the acute angles of the triangle ABC.

$$\angle BOA = 2C; \quad \therefore \angle OAB = 90° - C;$$

$$\therefore \angle OAI = \frac{A}{2} \sim (90° - C) = \frac{C \sim B}{2}.$$

$$\therefore OI^2 = OA^2 + IA^2 - 2OA \,.\, IA\cos OAI$$

$$= R^2 + 16R^2\sin^2\frac{B}{2}\sin^2\frac{C}{2} - 8R^2\sin\frac{B}{2}\sin\frac{C}{2}\left(\cos\frac{B}{2}\cos\frac{C}{2} + \sin\frac{B}{2}\sin\frac{C}{2}\right)$$

$$= R^2 + 8R^2\sin\frac{B}{2}\sin\frac{C}{2}\left(\sin\frac{B}{2}\sin\frac{C}{2} - \cos\frac{B}{2}\cos\frac{C}{2}\right)$$

$$= R^2 - 8R^2\sin\frac{B}{2}\sin\frac{C}{2}\sin\frac{A}{2};$$

$$\therefore \mathbf{OI^2 = R^2 - 2Rr.} \quad \text{.........................(22)}$$

In the same way it can be proved that

$$\mathbf{OI_1{}^2 = R^2 + 2Rr_1.} \quad \text{.........................(23)}$$

To find OH^2.

For an *acute-angled* triangle ABC, $\angle OAB = 90° - C$, $\angle HAB = 90° - B$,

$$\therefore \angle OAH = C \sim B;$$

$$\therefore OH^2 = OA^2 + HA^2 - 2OA \,.\, HA\cos OAH$$

$$= R^2 + 4R^2\cos^2 A - 4R^2\cos A\cos(C - B)$$

$$= R^2 - 4R^2\cos A\,[\cos(C + B) + \cos(C - B)];$$

$$\therefore \mathbf{OH^2 = R^2 - 8R^2\cos A\cos B\cos C = R^2 + 2\rho^2.} \quad \text{.........(24)}$$

If $A > 90°$, $\angle HAB = 90° + B$; $\therefore \cos OAH = -\cos(C - B)$; also

$$HA = -2R\cos A;$$

$\therefore$ the final result is the same as before.

To find I_1H^2.

For an *acute-angled* triangle ABC, $\angle I_1AH = \frac{A}{2} \sim (90° - C) = \frac{C \sim B}{2}$;

$$\therefore\ I_1H^2 = I_1A^2 + HA^2 - 2I_1A \,.\, HA \cos\frac{C-B}{2}$$

$$= 16R^2\cos^2\frac{B}{2}\cos^2\frac{C}{2} + 4R^2\cos^2 A$$

$$- 16R^2\cos A\cos\frac{B}{2}\cos\frac{C}{2}\left(\cos\frac{B}{2}\cos\frac{C}{2} + \sin\frac{B}{2}\sin\frac{C}{2}\right)$$

$$= 16R^2\cos^2\frac{B}{2}\cos^2\frac{C}{2}(1 - \cos A) + 4R^2\cos A(\cos A - \sin B\sin C)$$

$$= 32R^2\cos^2\frac{B}{2}\cos^2\frac{C}{2}\sin^2\frac{A}{2} - 4R^2\cos A\cos B\cos C;$$

$$\therefore\ \mathbf{I_1H^2 = 2r_1^2 - 4R^2\cos A\cos B\cos C = 2r_1^2 + \rho^2.} \quad ......(25)$$

In the same way it can be proved that

$$\mathbf{IH^2 = 2r^2 - 4R^2\cos A\cos B\cos C = 2r^2 + \rho^2.} \quad ....(26)$$

The reader should verify that these results are also true for an obtuse-angled triangle.

A geometrical method of proof of (24) is indicated in Ex. I. d, No. 21.

The reciprocity of the relations (24) and (26) is explained by the following argument: since $\triangle$ABC circumscribes its own in-circle and is self-polar w.r.t. its own polar circle, there exists a triangle $\alpha\beta\gamma$ which is inscribed in this polar circle, and is self-polar w.r.t. this in-circle (*Durell's Projective Geometry*, p. 209). $\therefore$ H is the circumcentre, ρ is the circumradius, I is the orthocentre, r is the polar-radius of $\triangle\alpha\beta\gamma$.

$\therefore$ applying (24) to $\triangle\alpha\beta\gamma$, we have $HI^2 = \rho^2 + 2r^2$.

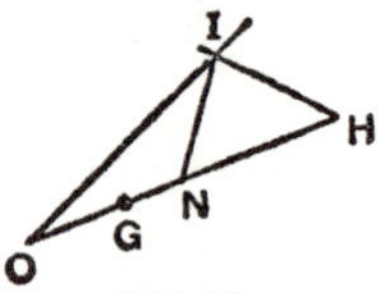

FIG. 14.

To find IN.

The nine-point centre N is the mid-point of OH;

$$\therefore\ OI^2 + IH^2 = 2IN^2 + 2ON^2;$$

$$\therefore\ (R^2 - 2Rr) + (2r^2 + \rho^2) = 2IN^2 + \tfrac{1}{2}(R^2 + 2\rho^2);$$

$$\therefore\ IN^2 = \tfrac{1}{4}R^2 - Rr + r^2 = (\tfrac{1}{2}R - r)^2.$$

But $OI^2 = R(R - 2r)$; $\therefore$ $R \geqslant 2r$;

$$\therefore\ \mathbf{IN = \tfrac{1}{2}R - r.} \quad ....(27$$

In the same way it can be proved that

$$\mathbf{I_1N = \tfrac{1}{2}R + r_1.} \quad ....(28)$$

Since the radius of the nine-point circle is $\frac{1}{2}$R, equations (27) and (28) prove that the nine-point circle touches the in-circle and the escribed circles [*Feuerbach's Theorem*, *M.G.*, p. 117].

Example 6. If IH is parallel to BC, find a relation between the cosines of the angles of the triangle.

I and H will be equidistant from BC, thus $r = 2R\cos B\cos C$, thus

$$4\sin\tfrac{1}{2}A \,.\, \sin\tfrac{1}{2}B \,.\, \sin\tfrac{1}{2}C = 2\cos B\cos C,$$
$$2\sin\tfrac{1}{2}A\,[\cos\tfrac{1}{2}(B-C) - \cos\tfrac{1}{2}(B+C)] = 2\cos B\cos C,$$
$$\cos B + \cos C - 2\sin^2\tfrac{1}{2}A = 2\cos B\cos C,$$
$$\Sigma\cos A = 1 + 2\cos B\cos C.$$

Example 7. Express IG in terms of the radii of the various circles connected with the triangle.

By equation (19) we have, since G is the centroid of 2 at O and 1 at H, (see Fig. 14)

$$2IO^2 + IH^2 = 2GO^2 + GH^2 + 3IG^2;$$
$$\therefore\ 3IG^2 = 2IO^2 + IH^2 - \tfrac{2}{9}OH^2 - \tfrac{4}{9}OH^2;$$
$$\therefore\ 9IG^2 = 6IO^2 + 3IH^2 - 2OH^2$$
$$= 6(R^2 - 2Rr) + 3(2r^2 + \rho^2) - 2(R^2 + 2\rho^2)$$
$$= 4R^2 - 12Rr + 6r^2 - \rho^2;$$
$$\therefore\ IG^2 = \tfrac{1}{9}(4R^2 - 12Rr + 6r^2 - \rho^2).$$

EXERCISE I. d.

1. If OI is parallel to BC, prove that $\cos B + \cos C = 1$.
2. If IG is parallel to BC, prove that $r_1 = 3r$.
3. Prove $OI^2 = R^2\,[3 - 2\Sigma(\cos A)]$.
4. Prove $II_1{}^2 = 4R(r_1 - r)$ and $I_2I_3{}^2 = 4R(r_2 + r_3)$.
5. Prove $II_2{}^2 + I_1I_3{}^2 = II_3{}^2 + I_1I_2{}^2$.
6. Prove $OH^2 = 9R^2 - a^2 - b^2 - c^2$.
7. If $A = 60°$, prove $OH^2 = (3R + 2r)(R - 2r)$.
8. Prove $\tan IAX = \tan^2\dfrac{A}{2}\tan\dfrac{B-C}{2}$.
9. If in a scalene triangle IG is perpendicular to BC, prove that
$$\sin\frac{A}{2} = \sin\frac{B}{2}\sin\frac{C}{2}.$$
10. If O lies on the in-circle, prove that $\cos A + \cos B + \cos C = \sqrt{2}$.
11. If OH makes an angle ϕ with BC, prove that
$$\tan\phi\,(\tan C - \tan B) = 3 - \tan B\tan C.$$
12. Prove that
$$4AN^2 = R^2 + b^2 + c^2 - a^2 = R^2(3 + 2\cos 2A - 2\cos 2B - 2\cos 2C).$$

13. Prove that $\rho^2 = (r + 2R)^2 - s^2$.

14. If the circumcircle cuts the nine-point circle orthogonally prove that $\cos A \cos B \cos C = -\frac{1}{2}$.

15. If $AH = r$, prove that the circumcircle cuts one escribed circle orthogonally.

16. If OI cuts AD at T, prove $OT = OI \cos \frac{B-C}{2} \operatorname{cosec} \frac{A}{2}$.

17. Prove that the area of ΔOIH is

$$\pm 2R^2 \sin \frac{B-C}{2} \sin \frac{C-A}{2} \sin \frac{A-B}{2}.$$

18. If S is the circumcentre of ΔBHC, prove that

$$SA^2 = R^2 (1 + 8 \cos A \sin B \sin C).$$

19. If $IO = IH$, prove that either $AO = AH$ or A, O, I, H are concyclic. Deduce that an angle of the triangle is $60°$.

20. Prove $NI + NI_1 + NI_2 + NI_3 = 6R$.

21. If $\angle BAC$ is obtuse and if HA cuts the circumcircle at T, prove that (i) $HT = 2HD$; (ii) $HA \cdot HT = 2\rho^2$. Hence show that

$$HO^2 = 2\rho^2 + R^2.$$

22. Prove that (i) $OI^2 + OI_1^2 + OI_2^2 + OI_3^2 = 12R^2$;

(ii) $NA^2 + NB^2 + NC^2 + NH^2 = 3R^2$.

23. Prove that (i) $DA^2 + DB^2 + DC^2 + DH^2 = 4R^2$;

(ii) $AI_1^2 + AI_2^2 + AI_3^2 + AI^2 = 16R^2$.

24. Prove that $HA^2 + HB^2 + HC^2 - HO^2 = 3R^2$.

25. Prove that $AG^2 + BG^2 + CG^2 = \frac{8}{3}R^2 (1 + \cos A \cos B \cos C)$.

26. Prove that $a \cdot IA^2 + b \cdot IB^2 + c \cdot IC^2 = 4Rrs$; find a similar expression for $a \cdot I_3A^2 + b \cdot I_3B^2 - c \cdot I_3C^2$.

27. If T is any point, prove that

$$TA^2 \cdot \sin 2A + TB^2 \cdot \sin 2B + TC^2 \cdot \sin 2C = 4 (R^2 + OT^2) \sin A \sin B \sin C.$$

28. If T is a point on the in-circle, prove that

$$a \cdot TA^2 + b \cdot TB^2 + c \cdot TC^2 = 2\Delta (r + 2R).$$

29. Prove that the common chord of the circumcircle and the escribed circle, centre I_1, is $\sqrt{\left\{\frac{r_1^3 (4R - r_1)}{R (R + 2r_1)}\right\}}$.

30. If t_1, t_2, t_3 are the lengths of the tangents from I_1, I_2, I_3 to the circumcircle, prove that (i) $\frac{t_1^2}{t_2^2} = \frac{r_1}{r_2}$; (ii) $t_1 t_2 t_3 = abc \sqrt{\left(\frac{R}{2r}\right)}$.

Solution of Triangles from Miscellaneous Data. No general rules can be given, but the following typical examples may be useful.

(i) **Given a, b − c, A.**

In Fig. 15, cut off $AK = AB$; then $KC = b - c$.

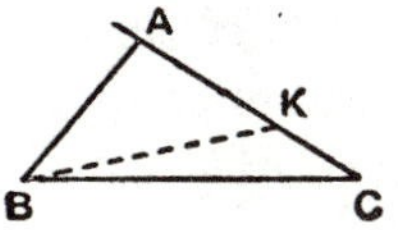

FIG. 15.

Also $\angle BKC = 90° + \frac{1}{2}A$ and $\angle KBC = \frac{1}{2}(B - C)$;

$$\therefore \text{ from } \triangle BKC, \quad \frac{a}{\cos\frac{1}{2}A} = \frac{b-c}{\sin\frac{1}{2}(B-C)}.$$

This determines $\frac{1}{2}(B - C)$ and therefore B, C.

(ii) **Given a, R, Δ.**

A is given by the relation $R = \dfrac{a}{2\sin A}$.

Also, $$\cos(B - C) + \cos A = 2\sin B\sin C = \frac{b\sin C}{R} = \frac{2\Delta}{aR}.$$

This determines $(B - C)$ and therefore B and C since A has been found.

(iii) **Given the altitudes p_1, p_2, p_3.**

The ratios $a : b : c$ are given by $2\Delta = ap_1 = bp_2 = cp_3$.

A is then given by $\tan\dfrac{A}{2} = \sqrt{\left\{\dfrac{(s-b)(s-c)}{s(s-a)}\right\}}$; also $c = p_2 \operatorname{cosec} A$.

(iv) **Given r_1, r_2, r_3.**

$$r_2r_3 = \frac{\Delta^2}{(s-b)(s-c)} = s(s-a);$$

$$\therefore r_2r_3 + r_3r_1 + r_1r_2 = \Sigma[s(s-a)] = s(3s - 2s) = s^2;$$

$$\therefore as = s^2 - r_2r_3 = r_1r_2 + r_1r_3 = r_1(r_2 + r_3);$$

$$\therefore a = \frac{r_1(r_2 + r_3)}{\sqrt{(r_2r_3 + r_3r_1 + r_1r_2)}}.$$

EXERCISE I. e.

1. Given $a - b = 19{\cdot}8$, $c = 22{\cdot}2$, $C = 29°\ 16'$, find B.

2. Given $b = 3{\cdot}36$, $a + c = 9{\cdot}28$, $B = 37°\ 25'$, find A.

3. Given $AD = 6$, $BE = 8$, $CF = 9$, find A and a.

4. Given $b=8$, $c=10$, AX $=7$, find A.
5. Given $A=45°$, $a=2(b-c)$, find B.
6. Given $r=5$, $r_1=12$, $r_2=20$, find a, c.
7. Given b, B and that $a+c=2b$, show how to solve the triangle.
8. Express bc and b^2+c^2 in terms of a, R, Δ.
9 Express $\sin\frac{1}{2}A$ in terms of r_1, r_2, r_3.
10. Express $\tan B$ in terms of b, c, A.
11. Express c in terms of a, b, Δ.
12. Given B, b, c^2-a^2, show how to find C.
13. Given $A=53°$ and BE $=2$CF, find B.
14. Given $A=42°$, $r=3{\cdot}5$, find the least possible value of a.
15. Given $a=\sqrt{57}$, $A=60°$, $\Delta=2\sqrt{3}$, find b, B.
16. Given $A=60°$, $b-c=4$, AD $=11$, find a and $\sin\dfrac{B-C}{2}$.
17. Given $\cot(B-C)=7$ and BC $=5$AD, find $\cot B$.
18. Given a, s, A, show how to find B.
19. Express a in terms of r, A, p where $p=$ AD.

Errors. If u is a known function, $f(x)$, of x, we have, using differentials: $du=f'(x)\,.\,dx$. If the value of u is calculated from a measured or observed value of x, the resulting error δu in u, due to an error δx in x is given by

$$\delta u \simeq f'(x)\,.\,\delta x.$$

If u is a known function of several independent variables x, y, z, etc., then the error in u due to errors in the values of x, y, z, etc., is given by

$$\delta u \simeq \frac{\partial u}{\partial x}\,.\,\delta x + \frac{\partial u}{\partial y}\,.\,\delta y + \frac{\partial u}{\partial z}\,.\,\delta z + \ldots .$$

Example 8. The area of $\triangle$ABC is calculated from measurements of B, C, a; find the error in the calculated value of Δ due to an error δB in the measured value of B.

$$\Delta=\frac{1}{2}ab\sin C=\frac{1}{2}a^2\frac{\sin B\sin C}{\sin A}=\frac{1}{2}a^2\sin C\,.\,\frac{\sin B}{\sin(B+C)};$$

$$\therefore\ \delta\Delta \simeq \frac{1}{2}a^2\sin C\,.\,\frac{\sin(B+C)\cos B-\sin B\cos(B+C)}{\sin^2(B+C)}\,.\,\delta B$$

$$=\frac{1}{2}\frac{a^2\sin^2 C}{\sin^2 A}\,.\,\delta B=\frac{1}{2}c^2\,.\,\delta B.$$

Here, δB is measured in radians; thus an error of $1'$ in B causes an error of approximately $\frac{1}{2}c^2 \cdot \frac{1}{60} \cdot \frac{\pi}{180}$ in Δ.

If there are errors δB, δC, δa in B, C, a, the resulting error in Δ is given by

$$\delta\Delta \simeq \frac{\sin B \sin C}{\sin(B+C)} \cdot a\delta a + \tfrac{1}{2}c^2 \cdot \delta B + \tfrac{1}{2}b^2 \cdot \delta C;$$

$$\therefore\ \delta\Delta \simeq b \sin C \cdot \delta a + \tfrac{1}{2}c^2 \cdot \delta B + \tfrac{1}{2}b^2 \cdot \delta C.$$

EXERCISE I. f.

1. If Δ is expressed as a function of a, B, C, prove that $\frac{\partial \Delta}{\partial a} = \frac{2\Delta}{a}$.

2. If Δ is expressed as a function of a, b, c, prove that

$$\frac{\partial \Delta}{\partial a} = R \cos A.$$

3. If A is calculated from measurements of a, b, c, prove that the error due to a small error y in b is about $\frac{y \cot C}{b}$ radians.

4. If R is calculated from measurements of a, b, c, prove that the error due to a small error x in a is about $\frac{1}{2}x \operatorname{cosec} A \cot B \cot C$.

5. The base AB of a triangle ABC is fixed and the vertex C moves along the arc of a circle of which AB is a chord, prove that

$$\cos B \cdot da + \cos A \cdot db = 0.$$

6. An observer, on the ground, 50 ft. from a vertical tower, observes the angles of elevation of two marks on the tower to be 45° and 30°. Find the approximate error to which the calculated distance between them is liable if there may be an error of $1'$ in each observed angle.

7. If c is calculated from measurements of a, b, C, prove that the error due to small errors x, y, γ in a, b, C is about

$$x \cos B + y \cos A + a \sin B \cdot \gamma.$$

8. With the data of No. 7, prove that the relative error $\frac{\delta\Delta}{\Delta}$ in the calculated area of the triangle is approximately $\frac{x}{a} + \frac{y}{b} + \gamma \cot C$.

9. If C is calculated from measurements of a, b, c in which there are small errors x, y, z, prove that the error in C is approximately $\frac{z}{a} \operatorname{cosec} B - \frac{x}{a} \cot B - \frac{y}{b} \cot A$.

10. If c is calculated from measurements of a, b, R and if there is a small error x in a, prove that the error in c is approximately

$$-\frac{x \cos C}{\cos A}.$$

11. The area Δ of a triangle on a given base c is expressed in terms of c, A, B. Prove that

$$\frac{\partial\Delta}{\partial A}=\tfrac{1}{2}b^2\,;\ \frac{\partial\Delta}{\partial B}=\tfrac{1}{2}a^2\,;\ \text{and}\ \frac{\partial^2\Delta}{\partial A\,.\,\partial B}=2\Delta\,.\,\text{cosec}^2\,C.$$

In finding the vertex when the base is accurately known and the base angles are subject to small errors $\pm\alpha$, $\pm\beta$, show that the area of the small region within which the vertex must lie is approximately

$$4\alpha\beta\frac{\partial^2\Delta}{\partial A\,.\,\partial B}.$$

12. The area ABC was calculated from the measured values a, b, 90° of BC, CA, ACB and it was found that the calculated area was too great by z and that $a-x$, $b-y$ were the true lengths of BC, CA. Show that the error in C was about $\frac{180}{\pi}\sqrt{\left\{\frac{2(2z-ay-bx)}{ab}\right\}}$ degrees, if z, x, y were small.

MISCELLANEOUS EXAMPLES

EXERCISE I. g.

1. If $a+b=2c$, prove that $\cot\frac{A}{2}+\cot\frac{B}{2}=2\cot\frac{C}{2}$.

2. Prove that $\Sigma(ab\sin^2 C)=2s\,.\,\Sigma(a\cos B\cos C)$.

3. Prove that $2aR\sin(B-C)=b^2-c^2$.

4. Prove that $\Delta=r^2\cot\frac{A}{2}\cot\frac{B}{2}\cot\frac{C}{2}$.

5. Prove that $(r_1-r)\cot^2\frac{A}{2}=r_2+r_3$.

6. Prove that $a^2=(r_1-r)(4R-r_1+r)$.

7. Prove that $IA\,.\,II_1=4Rr$.

8. Prove that $\triangle I_1I_2I_3=\frac{abc}{2r}$.

9. Prove that the circumradius of $\triangle IBC$ is $2R\sin\frac{A}{2}$ and find that of $\triangle I_1BC$.

10. Prove that $PD=4R\sin\frac{B}{2}\sin\frac{C}{2}\sin\frac{B\sim C}{2}$.

11. Prove that $\triangle OAI=\frac{R^2r}{a}(\cos B\sim\cos C)$.

12. Prove that AD cuts the in-circle at an angle

$$\cos^{-1}\left(\sin\frac{B-C}{2}\,\text{cosec}\,\frac{A}{2}\right).$$

13. Given $\frac{r}{2}=\frac{r_1}{12}=\frac{R}{5}$, prove that the triangle is right-angled.

14. If $B = 18°$, $C = 36°$, prove that $a - b = R$.

15. If $\cos A + \cos B = \frac{3}{2}$, prove that $2r - R = 2R \cos C$.

16. Given a, b, B, find the difference between the in-radii of the two triangles.

17. Prove that $\Sigma(a^3 \cos A) = abc(1 + 4 \cos A \cos B \cos C)$.

18. Prove that r_1, r_2, r_3 are the roots of

$$r^2x^3 - r^2(r + 4R)x^2 + \Delta^2 x - \Delta^2 r = 0.$$

19. Prove that $\dfrac{2\triangle ABC}{\triangle I_1I_2I_3} = \dfrac{a \cos A + b \cos B + c \cos C}{a + b + c}$.

20. If T is the mid-point of EF, prove that $XT = \frac{1}{2} a \sin A$.

21. If $a^2 \cos^2 A + b^2 \cos^2 B = c^2 \cos^2 C$, prove that one of the angles A, B, C is determinate, and find it.

22. If $\angle CAX = 90°$, prove that $3ac \cos A \cos C = 2(c^2 - a^2)$.

23. If a, b, c are in A.P., prove that

$$\cos A + \cos C - \cos A \cos C + \tfrac{1}{3} \sin A \sin C = 1.$$

24. ABC, ABD are equilateral triangles in perpendicular planes; calculate $\angle CAD$.

25. If $B = C = 2A$, prove that $IB = \dfrac{c^2}{2s}$.

26. If DT, DT′ are perpendicular to AB, AC, prove that $TT' = \dfrac{\Delta}{R}$.

27. Prove that the tangents at A, B, C to the circumcircle form a triangle of area $\pm R^2 \tan A \tan B \tan C$.

28. Prove that the radii of the circles which touch AB, AC and the circumcircle are $r \sec^2 \dfrac{A}{2}$ and $r_1 \sec^2 \dfrac{A}{2}$. [Use Inversion.]

29. If l, m are the directed lengths of the perpendiculars from A, B to any line through C, prove that

$$a^2l^2 + b^2m^2 - 2ablm \cos C = 4\Delta^2.$$

[If A and B are on opposite sides of the line, l and m must be regarded as opposite in sign.]

30. Prove that, if $a > b > c$, the length of the shortest line which bisects the area of ABC is $\sqrt{(2\Delta \tan \frac{1}{2}C)}$.

31. If the angles of a triangle are calculated from measured values of the sides, show that the small errors satisfy

$$\delta A = \frac{a}{2\Delta}(\delta a - \cos C \,.\, \delta b - \cos B \,.\, \delta c).$$

32. If r is calculated from measured values of a, b, c, show that the error due to an error x in a is

$$\frac{x}{2s}(2R \cos A - r).$$

CHAPTER II

PROPERTIES OF THE QUADRILATERAL

Notation. In dealing with a quadrilateral ABCD, we shall denote the angles by A, B, C, D, and shall represent the other elements (see Fig. 16), as follows:

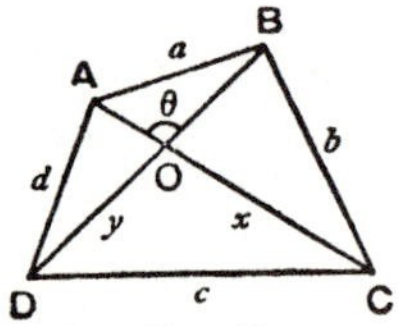

FIG. 16.

$$AB = a, \quad BC = b, \quad CD = c, \quad DA = d;$$

$$AC = x, \quad BD = y, \quad A\hat{O}B = \theta;$$

$$s = \tfrac{1}{2}(a + b + c + d); \quad S = \text{area ABCD}.$$

We assume the quadrilateral to be convex.

The Cyclic Quadrilateral. If a quadrilateral is known to be cyclic, and if the lengths of the sides, in order, are given, it is possible to calculate the other elements of the figure. Formulae for S, x, y, the circumradius R, and the angles, in terms of a, b, c, d may be obtained as follows:

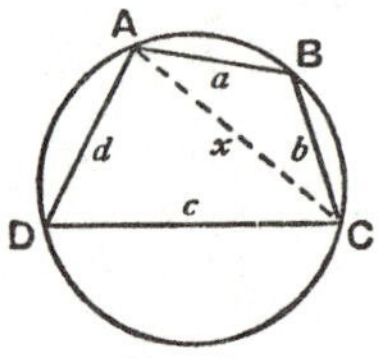

FIG. 17.

The **area** *of a cyclic quadrilateral is*

$$\sqrt{\{(s-a)(s-b)(s-c)(s-d)\}}.$$

We have $\quad S = \triangle ABC + \triangle ACD;$

$$\therefore \quad 4S = 2ab \sin B + 2cd \sin D. \quad \text{.....................(1)}$$

Also $a^2 + b^2 - 2ab \cos B = x^2 = c^2 + d^2 - 2cd \cos D;$

$$\therefore \quad a^2 + b^2 - c^2 - d^2 = 2ab \cos B - 2cd \cos D. \quad \text{...............(2)}$$

From (1) and (2), by squaring and adding,

$$16S^2 + (a^2 + b^2 - c^2 - d^2)^2 = 4a^2b^2 + 4c^2d^2 - 8abcd \cos(B + D) \quad \text{...(3)}$$

$$= (2ab + 2cd)^2, \text{ since } B + D = 180°;$$

$$\therefore \quad 16S^2 = (2ab + 2cd)^2 - (a^2 + b^2 - c^2 - d^2)^2$$

$$= \{(a+b)^2 - (c-d)^2\}\{(c+d)^2 - (a-b)^2\}$$

$$= (a+b+c-d)(a+b-c+d)(c+d+a-b)(c+d-a+b)$$

$$= (2s-2d)(2s-2c)(2s-2b)(2s-2a);$$

$$\therefore \quad S = \sqrt{\{(s-a)(s-b)(s-c)(s-d)\}}. \quad \text{..........................(4)}$$

This formula was first given by the Hindu mathematician *Brahmagupta* (630 A.D.), but he believed, wrongly, that it held good for any quadrilateral. The Greek mathematician *Hero* had, however, pointed out that the general quadrilateral is not determined by the four sides alone. An n-sided polygon is determined by $2n-3$ elements—a simple framework with n joints is 'just stiff' if it contains $2n-3$ rods.

The **diagonals** *of a cyclic quadrilateral are given in terms of the sides by the formulae*

$$\mathbf{x}^2 = \frac{(\mathbf{ac}+\mathbf{bd})(\mathbf{ad}+\mathbf{bc})}{\mathbf{ab}+\mathbf{cd}}, \quad \mathbf{y}^2 = \frac{(\mathbf{ac}+\mathbf{bd})(\mathbf{ab}+\mathbf{cd})}{\mathbf{ad}+\mathbf{bc}}.$$

Since $x^2 = a^2 + b^2 - 2ab\cos B,$

and $x^2 = c^2 + d^2 - 2cd\cos D = c^2 + d^2 + 2cd\cos B;$

$$\therefore (cd+ab)x^2 = cd(a^2+b^2) + ab(c^2+d^2)$$
$$= ac(ad+bc) + bd(ad+bc)$$
$$= (ac+bd)(ad+bc);$$

$$\therefore x^2 = \frac{(ac+bd)(ad+bc)}{ab+cd}. \quad \text{.............................(5)}$$

The formula for y is proved in the same way.

By multiplication, we have *Ptolemy's Theorem,*

$$\mathbf{xy} = \mathbf{ac} + \mathbf{bd}. \quad \text{..................................(6)}$$

By division, we have $\dfrac{x}{y} = \dfrac{ad+bc}{ab+cd}$.

The **circumradius, R,** *is given by*

$$4\mathbf{RS} = \sqrt{\{(\mathbf{ab}+\mathbf{cd})(\mathbf{ac}+\mathbf{bd})(\mathbf{ad}+\mathbf{bc})\}}.$$

Using the formula $\mathrm{R} = \dfrac{abc}{4\Delta}$ for a triangle, we have

$$4\mathrm{R}\,.\,\triangle \mathrm{ABC} = abx \quad \text{and} \quad 4\mathrm{R}\,.\,\triangle \mathrm{ACD} = cdx;$$

∴ adding,

$$4\mathrm{RS} = (ab+cd)x = (ab+cd)\sqrt{\left\{\frac{(ac+bd)(ad+bc)}{ab+cd}\right\}};$$

$$\therefore 4\mathrm{RS} = \sqrt{\{(ab+cd)(ac+bd)(ad+bc)\}}. \quad \text{.........(7)}$$

The **Angles** *of a cyclic quadrilateral* may be found from formulae like

$$\sin \mathbf{B} = \frac{2\mathbf{S}}{\mathbf{ab}+\mathbf{cd}}, \quad \cos \mathbf{B} = \frac{\mathbf{a}^2+\mathbf{b}^2-\mathbf{c}^2-\mathbf{d}^2}{2(\mathbf{ab}+\mathbf{cd})}, \quad \text{............(8)}$$

which follow from equations (1) and (2), p. 24, in virtue of $\mathrm{B}+\mathrm{D} = 180°$. From (8) it is easy to obtain

$$\tan^2 \frac{\mathbf{B}}{2} = \frac{(\mathbf{s}-\mathbf{a})(\mathbf{s}-\mathbf{b})}{(\mathbf{s}-\mathbf{c})(\mathbf{s}-\mathbf{d})}.$$

The expressions for x^2 and y^2 are easily remembered, if this is desired, by noting that the sides paired together in the *denominator* are on the *same* side of the required diagonal.

EXERCISE II. a.

[*The results in this Exercise refer to cyclic quadrilaterals.*]

1. Find the area of the cyclic quadrilateral whose sides in order are 4, 5, 6, 7.

2. With the data of No. 1, find the lengths of the diagonals x, y.

3. With the data of No. 1, find the length of the diameter of the circumcircle.

4. With the data of No. 1, find the interior angle between the sides of lengths 4, 5.

5. Prove that $\tan^2 \frac{1}{2}B = \dfrac{(s-a)(s-b)}{(s-c)(s-d)}$.

6. Prove that $(s-b)\tan\frac{1}{2}A = (s-d)\tan\frac{1}{2}B$.

7. Express $\tan\frac{1}{2}C \tan\frac{1}{2}D$ in terms of the sides.

8. Prove that $S = \frac{1}{4}\tan A(a^2 - b^2 - c^2 + d^2)$.

9. Interpret the results obtained from the formulae for S, cos B, and R by putting $d = 0$.

10. Simplify the expressions for S and $\tan^2\frac{1}{2}B$ when $a + c = b + d$. What is the geometrical meaning of this condition?

11. The sides of a quadrilateral taken in order are 7, 4, 4, 3, and the angle between the first two is 60°; prove that the quadrilateral is cyclic and find its circumradius.

12. From equations like $a^2 = AO^2 + OB^2 - 2AO \cdot OB \cos\theta$, prove that $2xy\cos\theta = b^2 + d^2 - a^2 - c^2$, and deduce that

$$\tan^2\frac{\theta}{2} = \frac{(s-b)(s-d)}{(s-a)(s-c)}.$$

13. Prove that $\sin\theta = \dfrac{2S}{ac+bd}$.

14. Prove that $\dfrac{OA}{ad} = \dfrac{OB}{ab} = \dfrac{OC}{bc} = \dfrac{OD}{cd} = \sqrt{\left\{\dfrac{ac+bd}{(ab+cd)(ad+bc)}\right\}}$.

15. Express $BO \cdot OD$ in terms of the sides.

16. If AB, DC are produced to meet at P, and DA, CB at Q, prove that $\dfrac{QA}{QB+b} = \dfrac{a}{c} = \dfrac{QB}{QA+d}$, and deduce expressions for QA, QB, QC, QD in terms of the sides. Write down similar expressions for PA, PB, PC, PD.

17. With the data of No. 16, prove that the other point of intersection K of the circles QAB, PBC lies on PQ, and deduce that

$$PQ^2 = PK \cdot PQ + QK \cdot QP = PB \cdot PA + QB \cdot QC$$
$$= (ab+cd)(ad+bc)\{ac(c^2-a^2)^{-2} + bd(d^2-b^2)^{-2}\}.$$

The General Quadrilateral. Equation (3), p. 24, is applicable to any quadrilateral, and it may be used to calculate the area. The equation may also be written

$$16S^2 = 4a^2b^2 + 4c^2d^2 - (a^2+b^2-c^2-d^2)^2 - 8abcd\cos(B+D)$$
$$= (2ab+2cd)^2 - (a^2+b^2-c^2-d^2)^2 - 8abcd\{1+\cos(B+D)\}$$
$$= 16(s-a)(s-b)(s-c)(s-d) - 8abcd\{1+\cos(B+D)\},$$

as on p. 24;

$$\therefore\ \mathbf{S^2 = (s-a)(s-b)(s-c)(s-d) - abcd\cos^2\frac{B+D}{2}}. \quad \ldots\ldots(9)$$

If the lengths of the sides of a quadrilateral are given, equation (9) shows that the area is greatest when $\cos\frac{1}{2}(B+D)=0$, *i.e.* when $B+D=180°$. Therefore *the area is greatest when the quadrilateral is cyclic.*

An extension of Ptolemy's Theorem. It was proved in *E.T.*, pp. 178, 179, that

$$2xy\sin\theta = 4S. \quad \ldots\ldots(10)$$

It may similarly be proved (see Ex. II. a, No. 12), that

$$2xy\cos\theta = b^2+d^2-a^2-c^2, \quad \ldots\ldots(11)$$

squaring and adding,

$$4x^2y^2 = 16S^2 + (b^2+d^2-a^2-c^2)^2,$$

substituting for $16S^2$ from equation (3),

$$4x^2y^2 = 4a^2b^2+4c^2d^2 - 8abcd\cos(B+D) - (a^2+b^2-c^2-d^2)^2 + (b^2+d^2-a^2-c^2)^2;$$

$$\therefore\ x^2y^2 = a^2b^2 + c^2d^2 - 2abcd\cos(B+D) + (b^2-c^2)(d^2-a^2);$$

$$\therefore\ \mathbf{x^2y^2 = a^2c^2 + b^2d^2 - 2abcd\cos(B+D)}. \quad \ldots\ldots(12)$$

Equations (10) and (11) lead to another expression for the area. By division,

$$\tan\theta = 4S/(b^2+d^2-a^2-c^2);$$

$$\therefore\ \mathbf{S = \tfrac{1}{4}(b^2+d^2-a^2-c^2)\tan\theta}. \quad \ldots\ldots(13)$$

The Circumscribable Quadrilateral. This is a quadrilateral in which a circle can be inscribed.

If P, Q, R, T are the points of contact of AB, BC, CD, DA, we have

$$AP = AT \quad \text{and} \quad BP = BQ;$$

$$\therefore\ a = AB = AT + BQ,$$

similarly

$$c = DT + CQ;$$

$$\therefore\ \mathbf{a + c = AD + BC = b + d}. \quad \ldots\ldots(14)$$

Conversely, it can be proved that, if $a+c=b+d$, a circle can be inscribed in the quadrilateral, see Ex. II. b, No. 8.

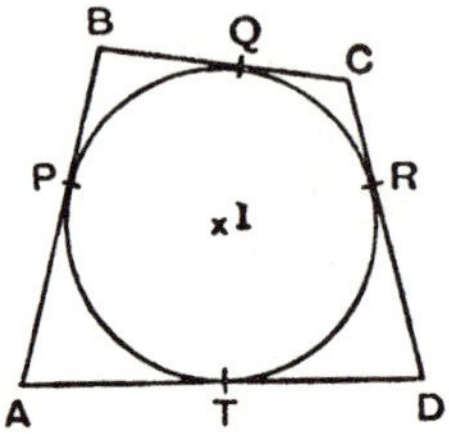

FIG. 18.

From (14), we have

$$a+c=b+d=\tfrac{1}{2}(a+b+c+d)=s;$$

$$\therefore\ s-a=c,\quad s-b=d,\quad s-c=a,\quad s-d=b;$$

thus equation (9), p. 27, becomes

$$S^2=abcd-abcd\cos^2\frac{B+D}{2}.$$

We have therefore for the *area of a circumscribable quadrilateral*

$$S=\sqrt{(abcd)}\,.\sin\tfrac{1}{2}(B+D). \quad \ldots\ldots\ldots\ldots\ldots\ldots(15)$$

If the quadrilateral is also *cyclic*, $\sin\frac{1}{2}(B+D)=\sin 90°=1$;

$$\therefore\ S=\sqrt{(abcd)}. \quad \ldots\ldots\ldots\ldots\ldots\ldots\ldots\ldots(16)$$

The in-radius of a circumscribable quadrilateral or polygon is $\dfrac{S}{s}$.

For, if I is the in-centre, $\triangle AIB=\frac{1}{2}ra$, etc.;

$\therefore$ by addition, $\quad S=\frac{1}{2}r(a+b+\ldots)=rs.$

For some properties of regular polygons, see *E.T.*, pp. 179, 180.

EXERCISE II. b.

1. Find the sum of two opposite angles of the quadrilateral in which $a=13$, $b=14$, $c=12$, $d=9$, $S=138$.

2. For the quadrilateral of No. 1, find the angle between the diagonals.

3. For the quadrilateral of No. 1, find xy.

4. If $a=7$, $b=8$, $c=9$, $d=11$, and $S=33$, find the angle between the diagonals.

5. If $a=7$, $b=8$, $c=9$, $d=11$, and $\theta=60°$, find S and xy.

6. The sides of a cyclic quadrilateral in order are 2, 4, 3, 5; calculate the cosine of the angle of intersection of the diagonals.

7. Explain the meaning of equation (12) when $B=D=0°$, and when $B=D=180°$.

8. In a quadrilateral for which $a+c=b+d$, where $a>d$, AX is cut off from AB equal to AD, and CY is cut off from CB equal to CD. Show that BX = BY and that the circumcentre of DXY is equidistant from the sides of the quadrilateral. (This proves the converse of relation (14) on p. 27.)

9. If $B + D = 90°$, prove that $x^2y^2 = a^2c^2 + b^2d^2$.

10. Show that the area of a circumscribable quadrilateral is
$\frac{1}{2}(ac - bd)\tan\theta$ or $\frac{1}{2}(b-a)(a-d)\tan\theta$ or $\frac{1}{2}\sqrt{\{x^2y^2 - (ac-bd)^2\}}$.

11. If a quadrilateral circumscribes a circle, show that the radius of the circle is $\dfrac{\sqrt{(abcd)} \,.\, \sin\frac{1}{2}(A+C)}{a+c}$.

12. If ABCD is circumscribable, prove that

$$\sqrt{(ab)} \,.\, \sin\tfrac{1}{2}B = \sqrt{(cd)} \,.\, \sin\tfrac{1}{2}D.$$

13. If ABCD is circumscribable, prove that

$$\tan\theta = \frac{2\sin\frac{1}{2}(A+C) \,.\, \sqrt{(abcd)}}{ac - bd}.$$

14. If $a + b = c + d$, deduce a formula for the area from the general formula.

15. If a circle can be drawn to touch the sides of a quadrilateral when produced, obtain a relation of the form $a + b = c + d$.

16. The sides, in order, of a cyclic quadrilateral are 4, 3, 5 and 6. Show that the quadrilateral is also circumscribable, and find (i) its area, (ii) its in-radius, (iii) the angle between the sides 4 and 6, (iv) the lengths of the diagonals, and (v) the circumradius.

17. The sides, taken in order, of a hexagon circumscribed about a circle are 13, 12, 8, 11, 9, and x. Find x, and if the area is 60, find the radius of the circle.

18. How many elements are required, in general, to determine a pentagon ? A cyclic pentagon has sides, in order, of lengths 39, 52, 39, 25, 33 and the longest diagonal is 65, find the area.

EASY MISCELLANEOUS EXAMPLES

EXERCISE II. c.

1. The sides of a cyclic quadrilateral taken in order are 1, 3, 4, 6 ; find the largest angle.

2. Three cyclic quadrilaterals have sides 8, 9, 10, 13, in different orders. Prove that their areas and circumradii are equal, and find the lengths of their diagonals.

3. Prove that there is a quadrilateral in which $a = b = y = 65$, $c = 50$, $d = 78$, $x = 112$, and find its area. Is the quadrilateral cyclic ?

4. If $a = 4$, $b = 1$, $c = 7$, $B = 120° = C$, find d.

5. If $a = 1$, $b = \sqrt{3}$, $c = 2$, $A = 60°$, $B = 150°$, find d and D.

6. When a, b, c, A, C are given, is the quadrilateral determined with or without ambiguity ?

7. If $a = 7$, $b = 12$, $c = 5$, $A = 60°$, $C = 90°$, find d.

8. If $a = 14$, $b = 12$, $c = 5$, $A = 60°$, $C = 90°$, find d.

9. If $A = 90°$, $B = 60°$, $C = 150°$, $a = 2$, $b = 1$, find c and d.

10. If $A = 120° = B$, $D = 90°$, $a = \sqrt{3}$, $c = 5$, find b and d.

11. If $a^2 + c^2 = b^2 + d^2$, prove that $\theta = 90°$.

12. If $A = 60°$ and $B = 90° = D$, prove that $3x^2 = 4y^2$.

13. If ABCD is cyclic and AD is a diameter, prove that

$$d(d^2 - a^2 - b^2 - c^2) = 2abc.$$

14. If ABCD is cyclic and $a - c = b - d$, prove that $S = bc \tan \frac{A}{2}$. Interpret the condition geometrically.

15. If ABCD is circumscribable, prove that

$$S = ab \sin \frac{B}{2} \operatorname{cosec} \frac{D}{2} \sin \frac{B+D}{2}.$$

16. If ABCD is both cyclic and circumscribable, prove that

$$\text{(i) } \sin \theta = \frac{2\sqrt{(abcd)}}{ac + bd}\text{; (ii) } \tan^2 \frac{\theta}{2} = \frac{bd}{ac}\text{; (iii) } \cos A = \frac{ad - bc}{ad + bc}.$$

17. Find the ratio of the areas of two regular polygons of n sides and $2n$ sides inscribed in the same circle.

18. Find the ratio of the areas of two regular polygons of n sides, inscribed in and circumscribed about a given circle.

19. The length of a side of a regular n-sided polygon is $2l$, and the areas of the polygon and of the inscribed and circumscribed circles are A, B, C; prove that $C - B = \pi l^2$ and $n^2 l^2 B = \pi A^2$.

20. Prove that the ratio of the areas of two regular polygons of n sides and $2n$ sides and of equal perimeters is

$$\cos \frac{\pi}{n} : \cos^2 \frac{\pi}{2n}.$$

21. If r_n and R_n denote the in-radius and circumradius of a regular n-sided polygon of given perimeter, prove that

$$\text{(i) } 2r_{2n} = r_n + R_n \quad \text{and} \quad \text{(ii) } R^2_{2n} = R_n \cdot r_{2n}.$$

22. If a square and a regular hexagon have equal perimeters, prove that the ratio of their areas $\simeq \frac{13}{15}$.

HARDER MISCELLANEOUS EXAMPLES

EXERCISE II. d.

1. Do equations (9) to (13) require any modification for the quadrilateral in Fig. 19?

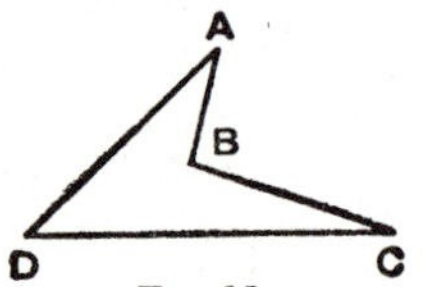

FIG. 19.

2. Two quadrilaterals ABCD, A′B′C′D′ of equal area, but not congruent, have their corresponding sides equal; also $B = D' = 90°$; prove that $ab = cd$.

3. If $a = 24$, $b = 7$, $c = 65$, $d = 60$, $x = 25$, find S, y, D.

4. If $a = 13$, $b = 14$, $c = 12$, $d = 9$, $S = 138$, show that

$$91 \cos B - 54 \cos D = 35 \quad \text{and} \quad 91 \sin B + 54 \sin D = 138.$$

Hence prove that $138 \sin D - 35 \cos D = 138$ and find D. Show also that $x = 15$ or $x \simeq 18{\cdot}1$.

5. Fig. 20 represents a "crossed" cyclic quadrilateral. What meaning must be given to S and what other conventions should be introduced to enable equations (1), (2), (3) on p. 24 to remain true? Obtain a result corresponding to equation (4).

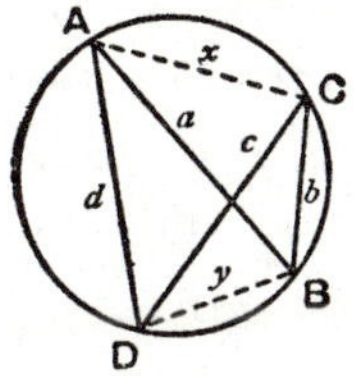

FIG. 20.

6. With the data of Fig. 20, find x and y in terms of a, b, c, d.

Show that 4 rods of lengths 8, 9, 10, 13 cannot be fitted together in any order to form a crossed cyclic quadrilateral.

7. If ABCD is cyclic and if AB, DC, when produced, cut at right angles, prove that $(ab + cd)^2 + (ad + bc)^2 = (b^2 - d^2)^2$.

8. If ABCD is cyclic and if $ac = bd$, prove that the tangents at A and C meet on BD. Conversely, if the tangents at A, C meet on BD (i.e. if ABCD is a harmonic system of points on a circle), prove that

$$ac = bd.$$

9. A quadrilateral is inscribed in a given circle of radius R, and one side subtends a given angle α at a point of the arc of the circle on the opposite side to the quadrilateral. Prove that the greatest possible area of the quadrilateral is $2R^2 \sin^3 \dfrac{2\alpha}{3}$.

10. If the sides of a cyclic quadrilateral are the roots of

$$x^4 - 2sx^3 + tx^2 - qx + 2p = 0,$$

express S in terms of p, q, s, t.

11. In a cyclic quadrilateral, prove that the productions of AB, DC meet at an angle ϕ, given by

$$\cos^2 \frac{\phi}{2} = \frac{(s-b)(s-d)(b+d)^2}{(ab+cd)(ad+bc)}.$$

12. Discuss the different ways in which a quadrilateral may be determined by five of the eight elements (4 sides, 4 angles), showing in which cases more than one quadrilateral may exist.

13. In any quadrilateral, prove that

$$x^2y^2(x^2+y^2-a^2-b^2-c^2-d^2)$$
$$+a^2c^2(a^2+c^2-b^2-d^2-x^2-y^2)+b^2d^2(b^2+d^2-a^2-c^2-x^2-y^2)$$
$$+x^2(a^2b^2+c^2d^2)+y^2(a^2d^2+b^2c^2)=0.$$

14. If ABCD is circumscribable, and α, β, γ, δ are the lengths of the tangents from A, B, C, D, prove that

$$abcd\cos^2\frac{A+C}{2}=(\alpha\gamma-\beta\delta)^2.$$

15. Prove that the distance between the circumcentres of ADC and BDC is $\frac{1}{2}a\sin(B+D)\operatorname{cosec}C\operatorname{cosec}D$.

16. If R and r are the circumradius and in-radius of a quadrilateral and if z is the distance between the centres of the circles, prove that $\frac{1}{(R+z)^2}+\frac{1}{(R-z)^2}=\frac{1}{r^2}$. State the corresponding result for a triangle.

[If I is the in-centre, let AI, CI meet the circumcircle in A′, C′; use $AI \,.\, IA'=R^2-z^2=CI \,.\, IC'$ to find IA′, IC′, and substitute in

$$IA'^2+IC'^2=2R^2+2z^2.]$$

17. Prove that the necessary and sufficient condition for the existence of a crossed cyclic quadrilateral with sides of lengths a, b, c, d in that order of magnitude is $b+c>a+d$. For $a=13$, $b=12$, $c=11$, $d=6$, find x and y.

18. A hexagon is inscribed in a circle of radius R; alternate sides are of lengths l, m; prove that $3R^2=l^2+lm+m^2$.

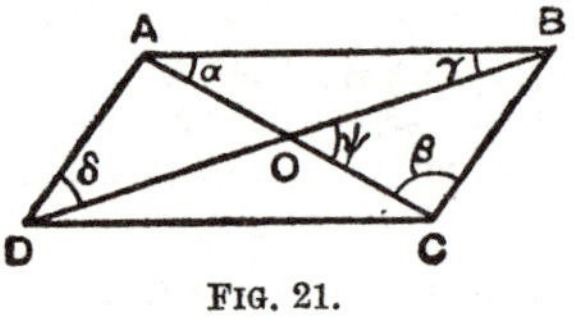

FIG. 21.

19. In Fig. 21, ABCD is a parallelogram. Prove that

(i) $\cot\alpha-\cot\gamma=\cot\beta-\cot\delta=2\cot(\alpha+\beta)$;

(ii) $\cot\alpha-\cot\beta=\cot\gamma-\cot\delta=2\cot\psi$.

20. If P_n and P'_n are the perimeters of regular n-gons inscribed and circumscribed to the same circle, prove that

$$\frac{P'_n-P'_{n+1}}{P_{n+1}-P_n}\to 2 \text{ as } n\to\infty.$$

21. The base of a pyramid is a horizontal regular n-gon, and its vertex is at height h vertically above the centre of the base. If a is the circumradius of the base and 2θ the angle between two adjacent sloping faces, prove that

$$(h^2+a^2)\cot^2\theta=h^2\tan^2\frac{\pi}{n}.$$

CHAPTER III

EQUATIONS AND SUB-MULTIPLE ANGLES

A FIRST discussion of Trigonometrical equations, for solutions from 0° to 360°, has been given in *E.T.*, Ch. XVIII, p. 258. We shall now obtain the general solutions of such equations, and shall usually work in radians.

Example 1. Solve $\sin\theta = \frac{1}{2}$.

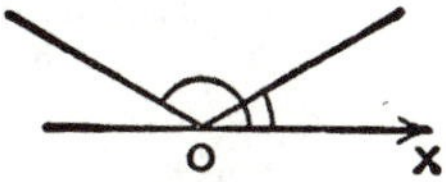

FIG. 22.

There are two, and only two, solutions between 0 and 2π, (see Fig. 22), $\theta = \frac{\pi}{6}$ or $\pi - \frac{\pi}{6}$.

But any angle differing from these by a multiple of 2π is also a solution; therefore the general solution is

$$\theta = 2n\pi + \frac{\pi}{6} \quad \text{or} \quad 2n\pi + \pi - \frac{\pi}{6};$$

$$\therefore\ \theta = 2n\pi + \frac{\pi}{6} \quad \text{or} \quad (2n+1)\pi - \frac{\pi}{6},$$

where n is any positive or negative integer or zero.

Example 2. Solve $\cos\theta = \frac{1}{2}$.

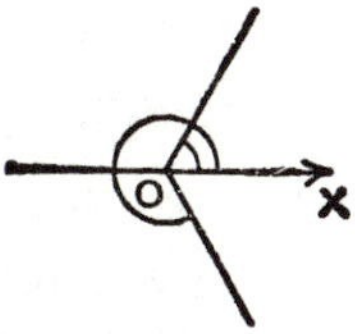

FIG. 23.

The solutions between 0 and 2π are

$$\theta = \frac{\pi}{3} \quad \text{or} \quad 2\pi - \frac{\pi}{3}$$

(see Fig. 23);

∴ the general solution is

$$\theta = 2r\pi + \frac{\pi}{3} \quad \text{or} \quad 2r\pi + 2\pi - \frac{\pi}{3};$$

$$\therefore \ \theta = 2n\pi \pm \frac{\pi}{3},$$

where n is any positive or negative integer or zero.

Example 3. Solve $\tan\theta = 1$.

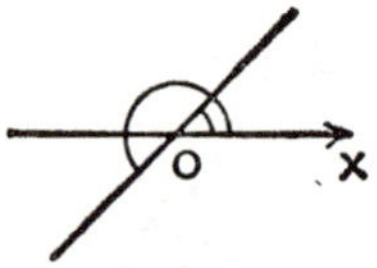

FIG. 24.

The solutions between 0 and 2π are

$$\theta = \frac{\pi}{4} \quad \text{or} \quad \pi + \frac{\pi}{4}$$

(see Fig. 24);

∴ the general solution is

$$\theta = 2r\pi + \frac{\pi}{4} \quad \text{or} \quad 2r\pi + \pi + \frac{\pi}{4};$$

$$\therefore \ \theta = n\pi + \frac{\pi}{4},$$

where n is any positive or negative integer or zero.

These examples illustrate the following general statements:

If $\sin\theta = \sin\alpha$, then $\theta = 2n\pi + \alpha$ or $(2n+1)\pi - \alpha$. ....(1)

If $\cos\theta = \cos\alpha$, then $\theta = 2n\pi \pm \alpha$. ....(2)

If $\tan\theta = \tan\alpha$, then $\theta = n\pi + \alpha$. ....(3)

where n is any positive or negative integer or zero.

The solution of $\sin\theta = \sin\alpha$ may also be written in the form, $\theta = m\pi + (-1)^m\alpha$, since, if m is even, this becomes $2n\pi + \alpha$ and, if m is odd, it becomes $(2n+1)\pi - \alpha$.

Note. There are certain specially simple equations which can be solved at sight by thinking of the figure without recourse to the general formulae. For example the values of θ for which $\cos\theta = 0$

are evidently odd numbers of right angles; this gives the solution in the form $\theta=(2n+1)\frac{\pi}{2}$, which is slightly better than the form $2n\pi\pm\frac{\pi}{2}$ given by the general formula; the two forms are, of course, equivalent.

EXERCISE III. a.

Use figures to write down the solutions of Nos. 1-6.

1. $\sin\theta=0$. 2. $\cos\theta=1$. 3. $\tan\theta=0$.

4. $\sin\theta=1$. 5. $\cos\theta=-1$. 6. $\sin\theta=-1$.

Apply the general formulae (1), (2), (3) to write down the solutions of Nos. 7-12.

7. $\sin\theta=\frac{\sqrt{3}}{2}$. 8. $\cos\theta=\frac{1}{\sqrt{2}}$. 9. $\tan\theta=\sqrt{3}$.

10. $\sin\theta=-\frac{1}{2}$. 11. $\cos\theta=-\frac{\sqrt{3}}{2}$. 12. $\tan\theta=-1$.

Solve the following:

13. $\cos 2\theta=1$. 14. $\sin 3\theta=0$. 15. $\tan 4\theta=0$.

16. $\sin 3\theta=-1$. 17. $\tan 3\theta=-1$. 18. $\cos 4\theta=0$.

19. $\sin\theta=\cos\theta$. 20. $\sin\theta+\cos\theta=0$.

21. $\sec\theta=2$. 22. $\operatorname{cosec}\theta=\operatorname{cosec}\alpha$.

Some useful methods of solving equations are illustrated in the following examples:

Example 4. Solve $4\sin^2\theta=1$.

This may be written

$$2(1-\cos 2\theta)=1,$$

$$\text{or}\qquad \cos 2\theta=\tfrac{1}{2};\quad\therefore\ 2\theta=2n\pi\pm\frac{\pi}{3};$$

$$\therefore\ \theta=n\pi\pm\frac{\pi}{6}.$$

The procedure just adopted is more convenient than using equation (1) to solve $\sin\theta=\pm\frac{1}{2}$.

Example 5. Solve $\sin 7\theta=\sin 5\theta$.

From equation (1),

$$7\theta=5\theta+2n\pi\quad\text{or}\quad 7\theta=\pi-5\theta+2n\pi;$$

$$\therefore\ \theta=n\pi\quad\text{or}\quad\frac{(2n+1)\pi}{12}.$$

Example 6. Solve $\tan A = -\cot 2A$.

This may be written

$$\tan A = \tan\left(\frac{\pi}{2} + 2A\right);$$

$$\therefore\ A = n\pi + \frac{\pi}{2} + 2A;$$

$$\therefore\ A = -n\pi - \frac{\pi}{2} = m\pi - \frac{\pi}{2},$$

where m is any integer or zero. This result is equivalent to

$$A = k\pi + \frac{\pi}{2}.$$

Example 7. Solve $\sin 2\theta = 1 + \cos 2\theta$.

First Method. $2 \sin\theta \cos\theta = 2\cos^2\theta$.

$$\therefore\ \cos\theta = 0 \quad \text{or} \quad \tan\theta = 1;$$

$$\therefore\ \theta = (2n+1)\frac{\pi}{2} \quad \text{or} \quad k\pi + \frac{\pi}{4}.$$

Second Method. This consists in applying the general process which is applicable to $A\cos\theta + B\sin\theta = C$. We have

$$\sin 2\theta - \cos 2\theta = 1;$$

$$\therefore\ \frac{1}{\sqrt{2}}\sin 2\theta - \frac{1}{\sqrt{2}}\cos 2\theta = \frac{1}{\sqrt{2}};$$

$$\therefore\ \sin\left(2\theta - \frac{\pi}{4}\right) = \sin\frac{\pi}{4};$$

$$\therefore\ 2\theta - \frac{\pi}{4} = 2n\pi + \frac{\pi}{4} \quad \text{or} \quad 2n\pi + \frac{3\pi}{4};$$

$$\therefore\ \theta = n\pi + \frac{\pi}{4} \quad \text{or} \quad (2n+1)\frac{\pi}{2}.$$

An alternative method is to express the sine and cosine each in terms of the tangent of half the angle (see *E.T.*, p. 263). Equations of this type should not be solved by methods involving the squaring of both sides; for since this process is not reversible the solutions obtained need not necessarily satisfy the equation, and testing becomes necessary.

Example 8. Solve $\sin 2\theta = \sin\theta$.

Using the same method as in Example 5, we find

$$\theta = 2n\pi \quad \text{or} \quad (2n+1)\frac{\pi}{3},$$

and this result follows also by using

$$\sin 2\theta - \sin\theta = 2\cos\tfrac{3}{2}\theta\sin\tfrac{1}{2}\theta.$$

We can, however, write the equation in the form $2\sin\theta\cos\theta = \sin\theta$;

$$\therefore\ \sin\theta = 0 \quad\text{or}\quad \cos\theta = \tfrac{1}{2};$$

$$\therefore\ \theta = n\pi \quad\text{or}\quad 2n\pi \pm \frac{\pi}{3}.$$

This illustrates the fact that different methods of solution sometimes give results of different forms; in the present example the aggregate of values of $(2n+1)\frac{\pi}{3}$ can be separated into two parts, those for which $2n+1$ is and is not a multiple of 3; the first of these taken with the values of $2n\pi$ just make up the values of $n\pi$, and the second part is the same as the values given by $2n\pi \pm \frac{\pi}{3}$.

EXERCISE III. b.

Solve:

1 $2\cos^2\theta = 1$.

2. $\cos 7\theta = \cos 5\theta$.

3. $\sin 3\theta = \sin 7\theta$.

4. $\sin 3\theta = \cos 2\theta$.

5. $\sin 6\theta + \sin\theta = 0$.

6. $\sin 5\theta + \cos 3\theta = 0$.

7. $\cot 5\theta = \cot 2\theta$.

8. $\tan 3\theta = \cot 4\theta$.

9. $\sin 2\theta = 1 - \cos 2\theta$.

10. $\sec\theta = \sec 2\theta$.

11. $\sin\theta + \sqrt{3}\,.\cos\theta = 1$.

12. $\cos\theta - \sin\theta = 1$.

13. $\sin 3\theta = 3\sin\theta$.

14. $3\sin\theta + 4\cos\theta = 2\tfrac{1}{2}$.

15. $4\cos\theta = \operatorname{cosec}\theta$.

16. $13\sin\theta - 84\cos\theta = 17$.

17. $\cos\theta + \cos 2\theta + \cos 3\theta = 0$.

18. $\sin 7\theta = \sin\theta + \sin 3\theta$.

19. $\cos 3\theta = \cos\theta\cos 2\theta$.

20. $\tan\theta + \tan 2\theta = \tan 3\theta$.

21. $4\cos\theta\cos 2\theta\cos 3\theta = 1$.

22. $\sec\theta + \operatorname{cosec}\theta = 2\sqrt{2}$.

23. $\cos\theta + \sin\theta = 1 + \sin 2\theta$.

24. $\tan\theta + \sec 2\theta = 1$.

25. $\cos x + \tan\alpha\sin x = \tfrac{1}{2}\sec\alpha$.

26. $\cos 9x\cos 7x = \cos 5x\cos 3x$.

27. $\cot x - \operatorname{cosec} 2x = 1$.

28. $\cos(x-\alpha)\cos(x-\beta) = \cos\alpha\cos\beta + \sin^2 x$.

29. $\cos^3 x - \cos x\sin x - \sin^3 x = 1$.

30. $\operatorname{cosec} 4\alpha - \operatorname{cosec} 4x = \cot 4\alpha - \cot 4x$.

31. $\tan(\cot\theta) = \cot(\tan\theta)$.

32. $\tan^{-1}\left(\frac{1-x}{1+x}\right) = \tfrac{1}{2}\tan^{-1}x$.

33. Discuss the solution of $\sin\theta+\cos\theta=k$ when (i) $k=1$, (ii) $k=2$, (iii) $k=\frac{1}{2}(1+\sqrt{3})$.

34. Find θ such that $\tan\theta=\sqrt{3}$ and $\sec\theta=-2$ simultaneously.

35. Find θ such that $\sin\theta+\sin 3\theta=\cos\theta$ and $\sin 4\theta=\frac{\sqrt{3}}{2}$ simultaneously.

36. Show that the aggregates of values given by

$$(2n-1)\frac{\pi}{2}+(-1)^n\frac{\pi}{3} \quad \text{and by} \quad 2n\pi\pm\frac{\pi}{6}$$

are identical, n being any integer or zero.

37. Are the aggregates of values of $n\pi+\frac{\pi}{2}$ and $n\pi\pm\frac{\pi}{2}$ identical, n being any integer or zero ?

MISCELLANEOUS EQUATIONS

Example 9. Find from graphical considerations the number of real roots of $x=3\pi(1-\sin x)$, x being measured in radians.

The equation may be written, $\sin x=1-\frac{x}{3\pi}$. Sketch the graphs of

$$y=\sin x \quad \text{and} \quad y=1-\frac{x}{3\pi}.$$

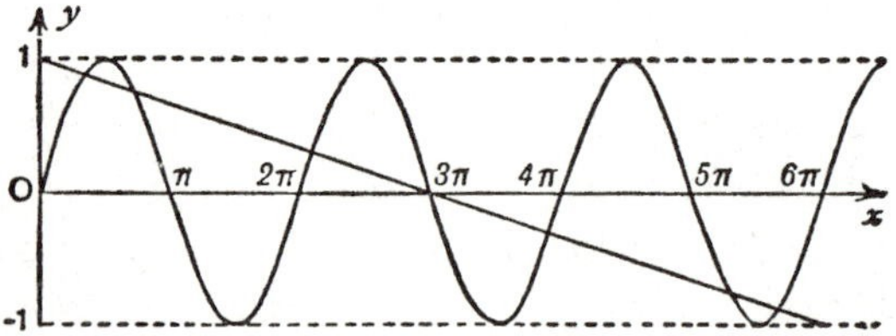

FIG. 25.

The graph of $y=\sin x$ lies between the lines $y=1$ and $y=-1$.

The graph of $y=1-\frac{x}{3\pi}$ is the straight line joining (0, 1) to $(6\pi, -1)$.

It is evident that the two graphs intersect at 7, and only 7, points.

$\therefore$ the equation $x=3\pi(1-\sin x)$ has 7 real roots.

Approximate values of these roots may be found by plotting the graphs carefully ; see also Chapter V, Example 4, p. 82.

Example 10. Solve $\sin x + \sin y = \sin c$; $\cos x + \cos y = \cos c$.

We have*
$$2 \sin \frac{x+y}{2} \cos \frac{x-y}{2} = \sin c, \quad \text{.......................(i)}$$

and
$$2 \cos \frac{x+y}{2} \cos \frac{x-y}{2} = \cos c; \quad \text{...................(ii)}$$

$\therefore$ by division,
$$\tan \frac{x+y}{2} = \tan c,$$

since $\cos \frac{x-y}{2} \neq 0$, as this would involve $\sin c = 0 = \cos c$;

$$\therefore \frac{x+y}{2} = m\pi + c; \quad \text{..........................(iii)}$$

$\therefore$ from (ii),
$$2 \cos(m\pi + c) \cos \frac{x-y}{2} = \cos c;$$

$$\therefore 2(-1)^m \cos c \cos \frac{x-y}{2} = \cos c;$$

$\therefore$ if $\cos c \neq 0$,
$$\cos \frac{x-y}{2} = \frac{1}{2}(-1)^m = \cos\left(m\pi + \frac{\pi}{3}\right);$$

$$\therefore \frac{x-y}{2} = 2n\pi \pm \left(m\pi + \frac{\pi}{3}\right); \quad \text{..............................(iv)}$$

$\therefore$ from (iii) and (iv),

$$x = 2(n+m)\pi + c + \frac{\pi}{3} \quad \text{or} \quad 2n\pi + c - \frac{\pi}{3},$$

$$y = -2n\pi + c - \frac{\pi}{3} \quad \text{or} \quad 2(m-n)\pi + c + \frac{\pi}{3}.$$

These solutions may be written

$$x = 2p\pi + c \pm \frac{\pi}{3}, \quad y = 2q\pi + c \mp \frac{\pi}{3};$$

where p, q are any integers and the upper signs are taken together, as also the lower signs.

If $\cos c = 0$, $\cos \frac{x+y}{2}$ must be zero; $\therefore x + y = (2n+1)\pi$;

$$\therefore \sin x + \sin x = \sin c = \pm 1;$$

$$\therefore \sin x = \pm \tfrac{1}{2};$$

$$\therefore x = m\pi \pm \frac{\pi}{6}, \quad y = (2n - m + 1)\pi \mp \frac{\pi}{6}.$$

* This solution illustrates some points of importance; other methods would be shorter for this particular example; e.g. the answers might be written down by geometrical considerations.

It happens that these solutions are contained in the previous general solutions, but as $\cos c \neq 0$ was assumed in obtaining the general solutions, it could not be anticipated that this would be the case.

EXERCISE III. c.

1. Solve graphically $x^2 = \cos x$.

2. Solve graphically $x = \cos^2 x$.

3. How many roots has $x = 10 \sin x$?

4. How many roots has $2x = 3\pi(1 - \cos x)$?

5. Solve graphically $x^2 = 4(1 - \sin x)$.

6. How many roots are there of $x + \tan 2x = \frac{\pi}{4}$, which lie between $x = 0$ and $x = \pi$.

7. Find the general expression for the range of values of θ if $x + \frac{1}{x} = 4 \cos \theta$ is satisfied by two values of x.

8. Show that the condition for $\sin x(\cos x + \sin x) = c$ to have roots is $\frac{1}{2}(1 + \sqrt{2}) \geqslant c \geqslant \frac{1}{2}(1 - \sqrt{2})$.

9. Prove that $\sec x + \operatorname{cosec} x = c$ has two roots between 0 and 2π if $c^2 < 8$, and four roots if $c^2 > 8$.

10. Solve graphically by a geometrical construction

$$\cos \theta + \cos \phi = a, \quad \sin \theta + \sin \phi = b$$

where a, b are given positive numbers. What limitation is there to the values of a and b to ensure that solutions exist ?

How can the cases when a or b is negative be dealt with ?

Solve the equations in Nos. 11-20 :

11. $\sin (x + y) = \frac{1}{2}$, $\cos (x - y) = -\frac{\sqrt{3}}{2}$.

12. $\cos x \cos y = \frac{1}{4}\sqrt{6}$, $\sin x \sin y = \frac{1}{4}\sqrt{2}$.

13. $\tan x = \sin 3y$, $\sin x = \tan 3y$.

14. $\sin^2 x - \sin x = \sin x \sin y = \sin^2 y + \sin y$.

15. $\cos x + \sqrt{3} \sin y = 2 \cos (x + y) = \sqrt{3} \cos x - \sin y$.

16. $\sin \theta + \sin \phi = \frac{3}{4}$; $\cos \theta + \cos \phi = \frac{2}{5}$, for values of θ, ϕ between 0° and 360°.

17. $5 \sin x - 2 \sin y = 1$; $5 \cos x - 2 \cos y = 4$, for values between 0° and 360°.

18. $\cos(x+3y)=\sin(2x+2y)$; $\sin(3x+y)=\cos(2x+2y)$ given that $x-y\neq 2n\pi+\frac{\pi}{2}$.

19. $\tan x=\tan 2y$; $\tan y=\tan 2z$; $\tan z=\tan 2x$ for values between 0 and π.

20. $\sin x+\sin 2y=\sin y+\sin 2z=\sin z+\sin 2x=0$ for values between 0 and π.

Submultiple Angles. *Given the value of* $\cos\theta$, *to find the values of* $\sin\frac{1}{2}\theta$ *and* $\cos\frac{1}{2}\theta$.

From the formulae,

$$1+\cos\theta=2\cos^2\frac{\theta}{2},\quad 1-\cos\theta=2\sin^2\frac{\theta}{2},$$

we have $\cos\frac{\theta}{2}=\pm\sqrt{\frac{1}{2}(1+\cos\theta)},\quad \sin\frac{\theta}{2}=\pm\sqrt{\frac{1}{2}(1-\cos\theta)}.$

The ambiguous signs are due to the fact that it does not follow from $\cos\theta=\cos\alpha$ that $\theta=\alpha$; the correct conclusion is $\theta=2n\pi\pm\alpha$, from which

$$\cos\frac{\theta}{2}=\cos\left(n\pi\pm\frac{\alpha}{2}\right)=\pm\cos\frac{\alpha}{2},$$

and

$$\sin\frac{\theta}{2}=\sin\left(n\pi\pm\frac{\alpha}{2}\right)=\pm\sin\frac{\alpha}{2}.$$

The ambiguity of sign can, of course, be removed if the actual value of θ is given, and this is done most easily by reference to a figure.

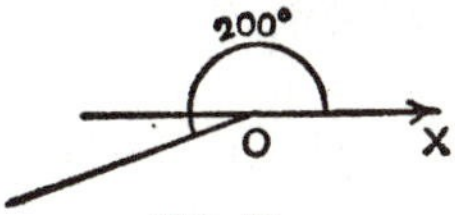

FIG. 26.

For example, if $\theta=400°$, $\cos\frac{\theta}{2}\left(=\cos 200°\right)$ is negative and $\sin\frac{\theta}{2}\left(=\sin 200°\right)$ is also negative, thus, in that case, the minus sign must be taken in both formulae. For the general results, see Ex. III. d, Nos. 2, 3.

Given the value of $\sin\theta$, *to find the values of* $\sin\frac{1}{2}\theta$ *and* $\cos\frac{1}{2}\theta$.

We begin by finding $\sin\frac{\theta}{2}+\cos\frac{\theta}{2}$ and $\sin\frac{\theta}{2}-\cos\frac{\theta}{2}$.

Since

$$\left(\sin\frac{\theta}{2}+\cos\frac{\theta}{2}\right)^2=\sin^2\frac{\theta}{2}+2\sin\frac{\theta}{2}\cos\frac{\theta}{2}+\cos^2\frac{\theta}{2}=1+\sin\theta,$$

$$\sin\frac{\theta}{2}+\cos\frac{\theta}{2}=\pm\sqrt{(1+\sin\theta)}, \quad\ldots\ldots\ldots\ldots(4)$$

and similarly

$$\sin\frac{\theta}{2}-\cos\frac{\theta}{2}=\pm\sqrt{(1-\sin\theta)}. \quad\ldots\ldots\ldots\ldots(5)$$

On account of the *two* ambiguous signs, there are *four* possibilities. From $\sin\theta=\sin\alpha$, the conclusion is

$$\frac{\theta}{2}=n\pi+\frac{\alpha}{2} \quad\text{or}\quad n\pi+\frac{\pi}{2}-\frac{\alpha}{2},$$

whence $$\sin\frac{\theta}{2}=\pm\sin\frac{\alpha}{2} \quad\text{or}\quad \pm\cos\frac{\alpha}{2},$$

and $$\cos\frac{\theta}{2}=\pm\cos\frac{\alpha}{2} \quad\text{or}\quad \pm\sin\frac{\alpha}{2}.$$

If the actual value of θ is given, it is possible to remove the ambiguities. For example if $\theta=320°$, then $\frac{\theta}{2}=160°$, so $\sin\frac{\theta}{2}$ is positive and $\cos\frac{\theta}{2}$ is negative; also the cosine is numerically the greater; thus, in that case,

$$\sin\frac{\theta}{2}+\cos\frac{\theta}{2}=-\sqrt{(1+\sin\theta)}$$

and $$\sin\frac{\theta}{2}-\cos\frac{\theta}{2}=+\sqrt{(1-\sin\theta)},$$

whence $$2\sin\frac{\theta}{2}=-\sqrt{(1+\sin\theta)}+\sqrt{(1-\sin\theta)},$$

$$2\cos\frac{\theta}{2}=-\sqrt{(1+\sin\theta)}-\sqrt{(1-\sin\theta)}.$$

For the general results, see Ex. III. d, Nos. 9, 10.

We can also determine for what values of θ a particular formula, such as

$$2\sin\frac{\theta}{2}=+\sqrt{(1+\sin\theta)}-\sqrt{(1-\sin\theta)}$$

will hold; for this requires

$$\sin\frac{\theta}{2}+\cos\frac{\theta}{2}=+\sqrt{(1+\sin\theta)},$$

$$\sin\frac{\theta}{2}-\cos\frac{\theta}{2}=-\sqrt{(1-\sin\theta)}.$$

Now $$\sin\frac{\theta}{2}+\cos\frac{\theta}{2}=\sqrt{2}\sin\left(\frac{\theta}{2}+\frac{\pi}{4}\right),$$

which is positive for $0<\frac{\theta}{2}+\frac{\pi}{4}<\pi$; therefore the first result holds if the angle $\frac{\theta}{2}(=\angle XOP)$ is such that its arm OP lies within the angle shown in Fig. 27.

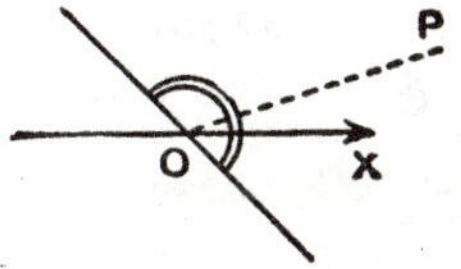

FIG. 27.

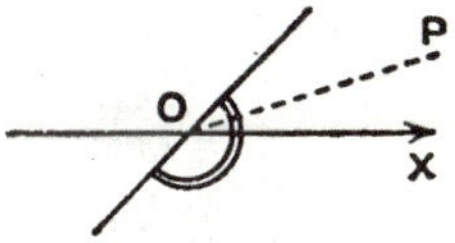

FIG. 28.

Similarly, $$\sin\frac{\theta}{2}-\cos\frac{\theta}{2}=\sqrt{2}\sin\left(\frac{\theta}{2}-\frac{\pi}{4}\right);$$

and so the second result holds if OP lies within the angle shown in Fig. 28. Thus for both results to be true the arm must lie within the angle shown in Fig. 29,

i.e. $$(8n-1)\frac{\pi}{4}<\frac{\theta}{2}<(8n+1)\frac{\pi}{4},$$

or $$(8n-1)\frac{\pi}{2}<\theta<(8n+1)\frac{\pi}{2}.$$

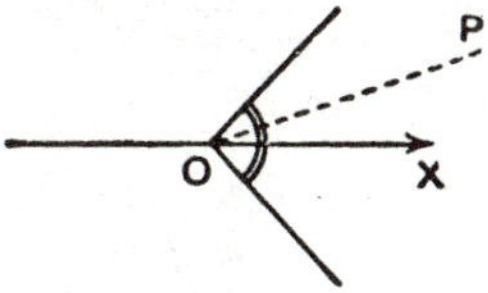

FIG. 29.

Given the value of $\cos\theta$, *to find the values of* $\cos\frac{\theta}{3}$.

There will be *three* values; for if $\cos\theta=\cos\alpha$, $\theta=2n\pi\pm\alpha$,

$$\therefore\ \frac{\theta}{3}=\frac{2n\pi}{3}\pm\frac{\alpha}{3},$$

and so $\cos\frac{\theta}{3}$ equals $\cos\frac{\alpha}{3}$, $\cos\left(\frac{2\pi}{3}+\frac{\alpha}{3}\right)$, or $\cos\left(\frac{4\pi}{3}+\frac{\alpha}{3}\right)$.

Hence from the identity $\cos\theta=4\cos^3\frac{\theta}{3}-3\cos\frac{\theta}{3}$, it follows that the values are the roots of the equation $4\cos^3\frac{\theta}{3} - 3\cos\frac{\theta}{3} = \cos\alpha$, which is a cubic for $\cos\frac{\theta}{3}$.

Cubic Equations. The general cubic equation

$$ax^3 + 3bx^2 + 3cx + d = 0$$

can be transformed by the substitution

$$y = ax + b \quad \text{into} \quad y^3 + 3\mathsf{H}y + \mathsf{G} = 0,$$

where $\mathsf{H} = ac - b^2$ and $\mathsf{G} = a^2d - 3abc + 2b^3$.

The further substitution $y = k\cos\theta$ gives

$$k^3\cos^3\theta + 3\mathsf{H}k\cos\theta = -\mathsf{G},$$

and if k is chosen so that $k^3 : 3\mathsf{H}k = 4 : -3$, this becomes

$$4\cos^3\theta - 3\cos\theta = \frac{\mathsf{G}}{2\mathsf{H}\sqrt{-\mathsf{H}}},$$

or

$$\cos 3\theta = \frac{\mathsf{G}}{2\mathsf{H}\sqrt{-\mathsf{H}}}$$

and

$$k^2 = -4\mathsf{H}.$$

From this, it may be possible to find 3θ, and three possible values of $\cos\theta$, giving three values of y and hence of x.

The conditions for the possibility are that H should be negative, and $\frac{\mathsf{G}}{2\mathsf{H}\sqrt{-\mathsf{H}}}$ numerically less than unity. Both conditions are included in

$$\mathsf{G}^2 + 4\mathsf{H}^3 < 0. \qquad \text{.......(6)}$$

This is precisely the condition for the cubic equation to have three real roots, and this method of solution is applicable therefore just to that case in which the usual algebraic solution breaks down.

EXERCISE III. d.

1. Give the signs to be used in the formulae

$$\cos\tfrac{1}{2}\theta = \pm\sqrt{\{\tfrac{1}{2}(1+\cos\theta)\}}, \quad \sin\tfrac{1}{2}\theta = \pm\sqrt{\{\tfrac{1}{2}(1-\cos\theta)\}}$$

when θ is (i) $70°$; (ii) $110°$; (iii) $200°$;
(iv) $-50°$; (v) $300°$; (vi) $3000°$.

2. Show that

(i) $\cos\frac{1}{2}\theta = +\sqrt{\{\frac{1}{2}(1+\cos\theta)\}}$, if $(4n-1)\pi < \theta < (4n+1\ \pi$;

(ii) $\cos\frac{1}{2}\theta = -\sqrt{\{\frac{1}{2}(1+\cos\theta)\}}$, if $(4n+1)\pi < \theta < (4n+3)\pi$.

3. Obtain results like those in No. 2 for $\sin \frac{1}{2}\theta$.

4. Determine the signs of $\sin \frac{1}{2}\theta + \cos \frac{1}{2}\theta$ and $\sin \frac{1}{2}\theta - \cos \frac{1}{2}\theta$ when θ is (i) 340°, (ii) 480°, (iii) 1360°.

5. If $\theta = \frac{11\pi}{6}$, prove $2 \sin \frac{1}{2}\theta = -\sqrt{(1+\sin\theta)} + \sqrt{(1-\sin\theta)}$, and obtain the formula for $2\sin\frac{1}{2}\theta$, when θ is

$$\text{(i) } \frac{11\pi}{8}; \quad \text{(ii) } \frac{8\pi}{3}; \quad \text{(iii) } \frac{23\pi}{6}.$$

6. Determine the signs in $2\cos\frac{1}{2}\theta = \pm\sqrt{(1+\sin\theta)} \pm \sqrt{(1-\sin\theta)}$, when θ is in the neighbourhood of 280°.

7. Determine the signs in

$$2\sin\theta = \pm\sqrt{(1+\sin 2\theta)} \pm \sqrt{(1-\sin 2\theta)},$$

when θ lies between 495° and 585°.

8. Determine the signs in

$$2\cos\theta = \pm\sqrt{(1+\sin 2\theta)} \pm \sqrt{(1-\sin 2\theta)},$$

when 2θ lies between $\frac{\pi}{2}$ and $\frac{3\pi}{2}$.

9. By writing $\sin\frac{\theta}{2} + \cos\frac{\theta}{2}$ in the form $\sqrt{2} . \sin\left(\frac{\theta}{2} + \frac{\pi}{4}\right)$, show that $\sin\frac{\theta}{2} + \cos\frac{\theta}{2} = +\sqrt{(1+\sin\theta)}$, if $(4n - \frac{1}{2})\pi < \theta < (4n + \frac{3}{2})\pi$, and that it $= -\sqrt{(1+\sin\theta)}$, if $(4n + \frac{3}{2})\pi < \theta < (4n + \frac{7}{2})\pi$.

10. Obtain results corresponding to those in No. 9 for $\sin\frac{\theta}{2} - \cos\frac{\theta}{2}$.

Determine the ranges of values of θ for which the following results (Nos. 11-14) hold:

11. $2\sin\frac{\theta}{2} = +\sqrt{(1+\sin\theta)} + \sqrt{(1-\sin\theta)}$.

12. $2\cos\frac{\theta}{2} = +\sqrt{(1+\sin\theta)} - \sqrt{(1-\sin\theta)}$.

13. $2\sin\theta = -\sqrt{(1+\sin 2\theta)} - \sqrt{(1-\sin 2\theta)}$.

14. $2\cos\theta = -\sqrt{(1+\sin 2\theta)} - \sqrt{(1-\sin 2\theta)}$.

15. Draw figures and use them to obtain the possible values of $\sin\frac{\theta}{2}$, when (i) $\cos\theta = \cos 120°$; (ii) $\cos\theta = \cos 300°$.

16. Draw figures and use them to obtain the possible values of $\cos\frac{\theta}{2}$, when (i) $\sin\theta = \sin 60°$; (ii) $\sin\theta = \sin 240°$.

17. Prove that $\cos\frac{\pi}{8} = \frac{1}{2}\sqrt{(2+\sqrt{2})}$, and find $\sin\frac{\pi}{16}$.

18. If $\cos\theta = \frac{1}{9}$ and $(4n-2)\pi < \theta < 4n\pi$, find $\sin\dfrac{\theta}{2}$.

19. If $\cos\theta = -\frac{7}{25}$ and $(4n-1)\pi < \theta < (4n+1)\pi$, find $\cos\dfrac{\theta}{2}$.

20. Express $\tan\dfrac{\theta}{2}$ in terms of $\tan\theta$. Determine the ambiguous sign for the cases when θ is between

$$\text{(i) } \frac{\pi}{2} \text{ and } \pi\,; \quad \text{(ii) } \pi \text{ and } \frac{3\pi}{2}\,; \quad \text{(iii) } 0 \text{ and } -\frac{\pi}{2}.$$

21. Prove $\tan x + \cot x = 2\operatorname{cosec} 2x$, and use it to express $\tan\dfrac{\theta}{2}$ in terms of $\sin\theta$.

22. Prove $\tan\frac{1}{2}\theta = \pm\sqrt{\dfrac{1-\cos\theta}{1+\cos\theta}}$ and show how to determine the sign.

23. If $\sin\theta$ has the given value $\sin\alpha$, find the possible values of $\sin\dfrac{\theta}{3}$.

24. Draw figures and use them to obtain the possible values of $\cos\dfrac{\theta}{3}$, when (i) $\cos\theta = \cos 60°$, (ii) $\cos\theta = \cos 210°$.

25.* Prove that $\sin\dfrac{\theta}{2} = (-1)^{\left[\frac{\theta}{2\pi}\right]}\sqrt{\{\frac{1}{2}(1-\cos\theta)\}}$.

26. Solve (i) $x^3 - 12x + 8 = 0$; (ii) $x^3 - 12x = 4$.

27. Solve (i) $x^3 - 27x - 27 = 0$; (ii) $x^3 - 27x + 40 = 0$.

28. Solve $x^3 + 3x^2 - 9x - 3 = 0$.

29. By putting $x = k\cos\theta$, reduce the equation $x^5 - 5x^3 + 5x + 1 = 0$ to the form $\cos 5\theta = c$ and hence solve it.

Inverse Functions. The equation $y = \sin x$, regarded as an equation for x in terms of a given number y (between -1 and $+1$) has an unlimited number of solutions.

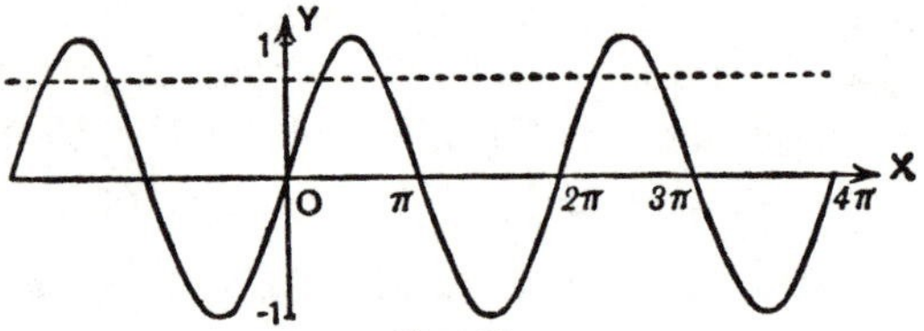

FIG. 30.

* $[x]$ denotes the greatest integer that is not greater than x; thus $[\frac{11}{3}] = 3$, $[5] = 5$, $[-4] = -4$, and $[-\frac{5}{2}] = -3$.

If $x = a$ is one solution, the others are

$$x = 2n\pi + a \quad \text{and} \quad x = 2n\pi + \pi - a.$$

The equation $x = \sin^{-1} y$ will be used, at present, to signify that x is equal to *one* of these values.

Similarly $\cos^{-1} k$ will denote any one of the angles whose cosine is k, and $\tan^{-1} k$ any one of the angles whose tangent is k.

Thus $\sin^{-1} k$, $\cos^{-1} k$, $\tan^{-1} k$ are *many-valued* functions of k.

Relations between Inverse Functions. Ordinary trigonometrical identities can often, with advantage, be expressed in terms of the inverse functions. For example, from

$$\tan(\theta - \theta') = \frac{\tan\theta - \tan\theta'}{1 + \tan\theta\tan\theta'}, \quad \text{......................(7)}$$

by putting $\tan\theta = m$, $\tan\theta' = m'$, we get

$$\tan(\theta - \theta') = \frac{m - m'}{1 + mm'},$$

which may be written

$$\theta - \theta' = \tan^{-1}\frac{m - m'}{1 + mm'}$$

or

$$\mathbf{\tan^{-1} m - \tan^{-1} m' = \tan^{-1}\frac{m - m'}{1 + mm'}}. \quad \text{........................(8)}$$

This form happens to be more convenient for certain purposes; it only implies that when any value of $\tan^{-1} m'$ is subtracted from any value of $\tan^{-1} m$ the result is *one* of the values of

$$\tan^{-1}\frac{m - m'}{1 + mm'}.$$

Similarly from the expansion of $\tan(\theta + \theta')$ we deduce the result

$$\mathbf{\tan^{-1} m + \tan^{-1} m' = \tan^{-1}\frac{m + m'}{1 - mm'}}. \quad \text{..................(9)}$$

Similarly from

$$\sin(\alpha + \beta) = \sin\alpha\cos\beta + \cos\alpha\sin\beta,$$

by putting $\sin\alpha = x$, $\sin\beta = y$, we get

$$\sin^{-1} x + \sin^{-1} y = \sin^{-1}\{x\sqrt{(1 - y^2)} + y\sqrt{(1 - x^2)}\}, \quad \text{......(10)}$$

and from $\cos(\alpha + \beta) = \cos\alpha\cos\beta - \sin\alpha\sin\beta$, we get

$$\cos^{-1} x + \cos^{-1} y = \cos^{-1}\{xy - \sqrt{(1 - x^2)}\,.\,\sqrt{(1 - y^2)}\}. \quad \text{......(11)}$$

Equations like (7) and (8) are really alternative statements of the same fact, and the reader should try to pass from one to the other without going through the process of substitution.

EXERCISE III. e.

1. (i) Prove that $\cos^{-1}x = \pm\sin^{-1}\sqrt{(1-x^2)}$.

(ii) If $\cos^{-1}x = \tan^{-1}p = \text{cosec}^{-1}q$, express p and q in terms of x.

2. If $\tan^{-1}x = \sin^{-1}p = \cos^{-1}q = \cot^{-1}r$, express p, q, r in terms of x.

3. How can you construct geometrically $\text{cosec}^{-1}1\frac{1}{4}$? Use the figure to express this angle in the forms, $\cos^{-1}p$, $\tan^{-1}q$.

4. Prove that $\sin(\cos^{-1}x) = \pm\sqrt{(1-x^2)}$.

5. Express in terms of x (i) $\cos(\sin^{-1}x)$; (ii) $\tan(\sin^{-1}x)$.

6. Prove that the general value of $2\cos^{-1}x$ equals

$$2n\pi \pm \cos^{-1}(2x^2-1).$$

Give two simple values of $2\cos^{-1}x + \cos^{-1}(2x^2-1)$.

7. Express $2\sin^{-1}x$ in the form $\sin^{-1}y$.

8. Prove that $\cos^{-1}x = \pm 2\tan^{-1}\sqrt{\dfrac{1-x}{1+x}}$.

9. Find the simplest value of $\tan^{-1}\frac{1}{2} + \tan^{-1}\frac{1}{3}$.

10. Find the simplest value of

$$4\tan^{-1}\tfrac{1}{5} - \tan^{-1}\tfrac{1}{70} + \tan^{-1}\tfrac{1}{99}.$$

11. Prove that $\tan^{-1}x = \pm\frac{1}{2}\cos^{-1}\dfrac{1-x^2}{1+x^2}$.

12. Simplify $\cos(2\sin^{-1}x)$.

13. Find a value of x, such that $\tan^{-1}\dfrac{x}{x+1} = 2\tan^{-1}\dfrac{1}{x}$.

14. Express $3\sin^{-1}x$ in the form $\sin^{-1}y$.

15. Find a simple value of $\text{cosec}^{-1}\sqrt{5} + \cot^{-1}3$.

16. Find the general value of $\tan^{-1}(\cot x) + \cot^{-1}(\tan x)$.

17. Evaluate $\cos 2\{\tan^{-1}x + \tan^{-1}y\}$.

18. Prove $\cos^{-1}\dfrac{b+a\cos x}{a+b\cos x} = 2\tan^{-1}\left(\dfrac{\sqrt{(a-b)}}{\sqrt{(a+b)}}\tan\dfrac{x}{2}\right)$.

19. Prove

$$\tan(\tan^{-1}x + \tan^{-1}y + \tan^{-1}z) = \cot(\cot^{-1}x + \cot^{-1}y + \cot^{-1}z).$$

20. Express $\sin^{-1}p = \cos^{-1}q$ as an algebraic relation between p and q.

21. If $\tan^{-1}p+\tan^{-1}q+\tan^{-1}r=\frac{\pi}{2}$, prove $qr+rp+pq=1$.

22. If $\cos^{-1}a+\cos^{-1}b+\cos^{-1}c=\pi$, prove that
$$a^2+b^2+c^2+2abc=1.$$

Find the simplest solutions of the following equations:

23. $\cot^{-1}2=\cot^{-1}x+\cot^{-1}7$.

24. $\tan^{-1}x+\tan^{-1}(1-x)=\tan^{-1}\frac{9}{7}$.

25. $\tan^{-1}x+\tan^{-1}\frac{2x}{1-x^2}=\frac{\pi}{3}$.

26. $6\cos^{-1}(2x^2-1)=\pi$.

27. $2\tan^{-1}\frac{3x-x^3}{1-3x^2}=\pi$.

MISCELLANEOUS EXAMPLES.

EXERCISE III. f.

Solve the equations : (Nos. 1-10).

1. $\tan x+\tan 2x=0$.

2. $\sin 3x+\cos 2x=0$.

3. $\sin x+\sin 3x=2\cos x$.

4. $\sin^2 x+\cos 3x\cos x=1$.

5. $\tan x+\tan(x+\alpha)+\tan(x+\beta)=\tan x\tan(x+\alpha)\tan(x+\beta)$.

6. $\tan x\cot(x+\alpha)=\tan\beta\cot(\beta+\alpha)$.

7. $\cos x\cos c+\sin a\sin b=\cos(x-a)\cos(x-b)$.

8. $\sin\left(3\theta-\frac{\pi}{8}\right)=2\sin\left(\theta+\frac{5\pi}{8}\right)$.

9. $70\cos\theta-24\sin\theta=37$.

10. $\sin\theta+\sin 2\theta+\sin 3\theta+\sin 4\theta=0$.

11. If $\sec\alpha\sec\theta+\tan\alpha\tan\theta=\sec\beta$, find $\tan\frac{1}{2}\theta$.

12. Find for what values of θ, $2\sin\theta-\tan\theta>0$.

13. Solve $x+y=2a$, $\cot x+\cot y=2\cot a$.

14. Solve $a^2\cos\theta+b^2\cos\phi=c^2\cos\alpha$; $a^2\sin\theta+b^2\sin\phi=c^2\sin\alpha$.

15. Find all values of x, y, such that
$$\cos^2 x\cos^2 y+\sin^2 x\sin^2 y=1.$$

16. Solve
$$\cos x+\cos y=\cos z,\ \cos 2x+\cos 2y=\cos 2z,$$
$$\cos 3x+\cos 3y=\cos 3z.$$

17. Show that the roots of

$$\cos\theta\cos(\theta-\alpha)\cos(\theta-\beta)\cos(\theta-\gamma)$$
$$-\sin\theta\sin(\theta-\alpha)\sin(\theta-\beta)\sin(\theta-\gamma)=\cos\alpha\cos\beta\cos\gamma$$

are $\theta=n\pi$ or $n\pi+\tan^{-1}\frac{1}{2}(\tan\alpha+\tan\beta+\tan\gamma+\tan\alpha\tan\beta\tan\gamma)$.

18. Prove that the roots of $\tan^3 x\tan\frac{x}{2}=1$ also satisfy

$$\cos 2x=2-\sqrt{5}.$$

19. Investigate how many values of $\sin\theta$ (between -1 and $+1$) satisfy $\sin^2\theta-2c\sin\theta+5c-6=0$ for various values of c.

20. Discuss the solution of $\cos^2 x-2m\cos x+4m^2+2m-1=0$ for various values of m.

21. Show that if $bh\cos\theta+ak\sin\theta=ab$ has roots for $\cos\theta$, they always determine values of θ.

22. Express $\sin\frac{\theta}{2}$ in terms of $\sin\theta$, when θ is in the neighbourhood of 420°. For what precise neighbourhood is the result valid?

23. Prove that $\tan\frac{\theta}{4}$ is one of the values of $\frac{1\pm\sqrt{(1-\sin\theta)}}{1\pm\sqrt{(1+\sin\theta)}}$, and find the other values.

24. Prove $\cos\frac{\theta}{2}=(-1)^{\left[\frac{\theta+\pi}{2\pi}\right]}\sqrt{\{\frac{1}{2}(1+\cos\theta)\}}$. (See footnote, p. 46.)

25. If p is an integer and $-1<q<1$, find the number of possible values of $\sin x$, such that (i) $\sin 2px=q$, (ii) $\sin(2p+1)x=q$.

26. Solve $x^5-5k^2x^3+5k^4x=2k^5\cos\alpha$, for x in terms of α and k.

27. Simplify $\tan^{-1}\frac{p-q}{1+pq}+\tan^{-1}\frac{q-r}{1+qr}$.

28. Prove that $\tan^{-1}\frac{1}{p}=\tan^{-1}\frac{1}{p+q}+\tan^{-1}\frac{q}{p^2+pq+1}$.

29. Use the result of No. 28 to express $\frac{\pi}{4}$ in the form $\tan^{-1}\frac{1}{2}+\tan^{-1}\frac{1}{3}$.

Also express $\tan^{-1}\frac{1}{2}$ and $\tan^{-1}\frac{1}{3}$ each in the form $\tan^{-1}\frac{1}{m}+\tan^{-1}\frac{1}{n}$ where m and n are positive integers.

30. Prove that $\frac{\pi}{4}=2\tan^{-1}\frac{1}{4}+\tan^{-1}\frac{1}{7}+2\tan^{-1}\frac{1}{13}$.

31. Prove that $\frac{\pi}{4}=2\cot^{-1}5+\cot^{-1}7+2\cot^{-1}8$.

32. Prove that $\cos^{-1}\dfrac{\cos\theta+\cos\phi}{1+\cos\theta\cos\phi}=2\tan^{-1}\left(\tan\dfrac{\theta}{2}\tan\dfrac{\phi}{2}\right)$.

33. Find x if one value of $\cos^{-1}x+\cos^{-1}2x$ equals $\dfrac{\pi}{3}$.

34. Find a value of x between 0 and $\dfrac{\pi}{2}$, such that

$$\sqrt{(\pi^2-4x^2)}=\sin^{-1}(\cos x).$$

CHAPTER IV

HYPERBOLIC, LOGARITHMIC, AND EXPONENTIAL FUNCTIONS

THE trigonometrical functions are called *circular* functions, because they arise naturally in connection with the geometry of the circle. There are other functions which are associated with the geometry of the hyperbola and may therefore be classified as *hyperbolic* functions; this name is, however, usually restricted to certain special functions of this group. We use it, in this Chapter, in a general sense.

In developing the argument, we shall make use of geometrical ideas, and especially that of an area bounded by a *curve*. At a first reading, the reasoning is more easily understood if this method is followed. But it is important to realise that the functions themselves can be regarded as purely analytical and that their properties can be obtained by purely arithmetical arguments. The more abstract line of approach will be followed in the companion volume on Analysis.

The Area-function for the Rectangular Hyperbola. Fig. 31 shows part of the graph of the function $y=\frac{1}{x}$; this equation represents a rectangular hyperbola. We shall confine attention to that part of the curve for which $x>0$.

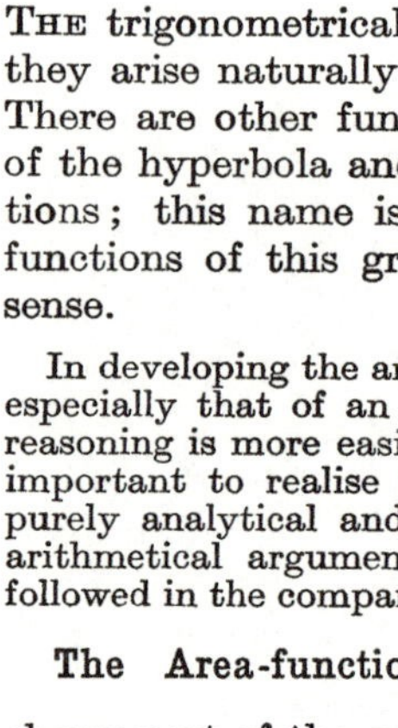

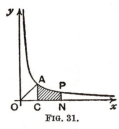

FIG. 31.

To every positive value of x there corresponds one, and only one, value of y; and, as x increases, y steadily decreases and tends to zero as x increases indefinitely. Further, if x tends down to zero from above, y increases without limit.

The point (1, 1) lies on the curve, and the curve is symmetrical about the line $y=x$, since, corresponding to any point P, $\left(t, \frac{1}{t}\right)$, on the curve, there is the point P′, $\left(\frac{1}{t}, t\right)$, also on the curve.

The curve is therefore shaped as in Fig. 31.

Consider the area bounded by the fixed ordinate CA, $x=1$, the variable ordinate NP, $x=t$, the curve and the x-axis. *This area* CNPA, *shaded in Fig.* 31, *is a function of* **t** *and will be denoted by the symbol* **hyp (t)**.

Expressed as a definite integral,

$$\text{area CNPA} = \int_1^t y\,dx = \int_1^t \frac{1}{x}\,dx;$$

$$\therefore\ \mathbf{hyp\ (t)} = \int_1^t \frac{1}{x}\,dx. \quad \ldots\ldots(1)$$

Approximate values of hyp (t) can be found for given values of t by the ordinary methods of practical geometry, such as counting squares or Simpson's rule.

It is best to adopt the usual sign conventions of the Integral Calculus for areas, as follows :

If $t > 1$, that is, if N is to the right of C, the area CNPA is represented by a positive number.

If $0 < t < 1$, that is if N lies between O and C, the area CNPA is represented by a negative number.

$$\therefore\ \text{hyp (t)} > 0 \text{ if } t > 1 \quad \text{and} \quad \text{hyp (t)} < 0 \text{ if } 0 < t < 1. \ \ldots\ldots(2)$$

If $t \leqslant 0$, we shall not discuss or even define hyp (t).

If $t = 1$, the shaded area CNPA vanishes ;

$$\therefore\ \text{hyp (1)} = 0. \quad \ldots\ldots(3)$$

Behaviour of hyp (t) when t increases indefinitely. If, in Fig. 32, P_1N_1, P_2N_2, ... are the ordinates $x = 2$, $x = 4$, ..., $x = 2^k$, then

$$\text{hyp } (2^k) = \text{area } ACN_kP_k;$$

but this is the sum of the areas

$$ACN_1P_1,\ P_1N_1N_2P_2,\ \ldots,\ P_{k-1}N_{k-1}N_kP_k,$$

and is therefore greater than the sum of the areas of the rectangles

$$CN_1P_1R_1,\ N_1N_2P_2R_2,\ \ldots,\ N_{k-1}N_kP_kR_k;$$

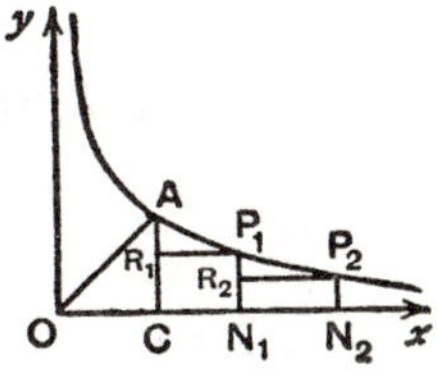

FIG. 32.

$$\therefore\ \text{hyp}(2^k) > (2-1)\cdot\tfrac{1}{2} + (4-2)\cdot\tfrac{1}{4}$$

$$+ (8-4)\cdot\tfrac{1}{8} + \ldots + (2^k - 2^{k-1})\cdot\frac{1}{2^k};$$

$$\therefore\ \text{hyp } (2^k) > \tfrac{1}{2} + \tfrac{1}{2} + \tfrac{1}{2} + \ldots \text{ to } k \text{ terms} = \frac{k}{2};$$

$$\therefore\ \text{if } t > 2^k,\ \text{hyp } (t) > \text{hyp } (2^k) > \frac{k}{2};$$

$$\therefore\ \text{hyp (t)} \to +\infty, \quad \text{when} \quad t \to +\infty.$$

This fact cannot be assumed without proof. For example, if $a(t)$ denotes the area under the curve $y=\frac{1}{x^2}$ from $x=1$ to $x=t$, we have

$$a(t)=\int_1^t \frac{1}{x^2}\,dx=\left[-\frac{1}{x}\right]_1^t=1-\frac{1}{t}.$$

$\therefore$ $a(t)$ tends to the *finite* limit 1, when $t\to\infty$.

Behaviour of hyp (t) when t tends to 0 from above. Using the same method as before, suppose, in Fig. 33, Q_1M_1, Q_2M_2, ... are the ordinates $x=\frac{1}{2}$, $x=\frac{1}{4}$, ..., $x=\frac{1}{2^k}$; and construct the rectangles CAS_1M_1, $M_1Q_1S_2M_2$, $M_2Q_2S_3M_3$, etc.

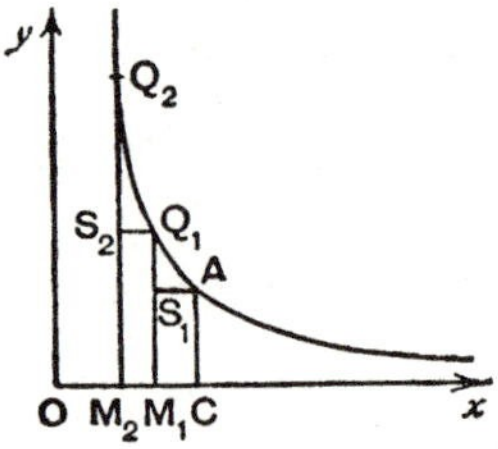

FIG. 33.

Then $$\text{hyp}\left(\frac{1}{2^k}\right)<-\left(1-\frac{1}{2}\right)1-\left(\frac{1}{2}-\frac{1}{4}\right).2-\left(\frac{1}{4}-\frac{1}{8}\right).4-\dots$$
$$\dots-\left(\frac{1}{2^{k-1}}-\frac{1}{2^k}\right).2^{k-1};$$

$$\therefore\ \text{hyp}\left(\frac{1}{2^k}\right)<-\frac{1}{2}-\frac{1}{2}-\frac{1}{2}-\dots\text{ to } k \text{ terms}=-\frac{k}{2};$$

$$\therefore\ \text{if } 0<t<\frac{1}{2^k},\ \text{hyp}(t)<\text{hyp}\left(\frac{1}{2^k}\right)<-\frac{k}{2};$$

$\therefore$ **hyp (t) $\to -\infty$, when t tends to 0 from above.**

EXERCISE IV. a.

1. Find from the graph of $y=\frac{1}{x}$, by some method of practical geometry, approximate values of hyp (2), hyp (3), hyp (4), hyp ($\frac{1}{2}$).

Draw a rough graph of $y=\text{hyp}(x)$ from $x=\frac{1}{2}$ to $x=4$.

2. Use Fig. 34 to show that $0\cdot5<\text{hyp}(2)<1$.

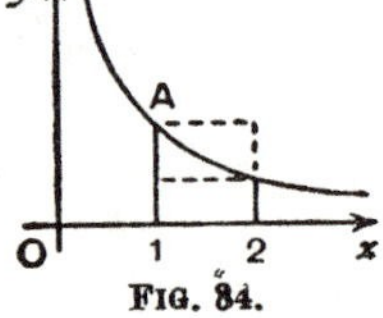

FIG. 34.

3. By drawing the ordinate $x = \frac{1}{2}$, show as in No. 2 that

$$-1 < \text{hyp}\left(\tfrac{1}{2}\right) < -0 \cdot 5.$$

4. Use Fig. 35 to show that hyp (2) lies between $\frac{7}{12}$ and $\frac{5}{6}$.

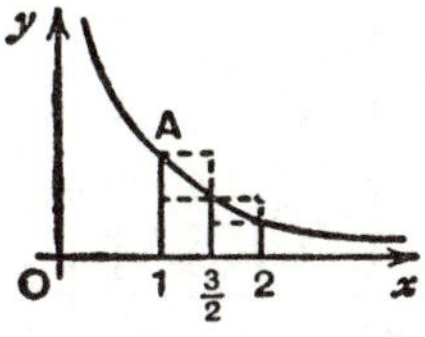

FIG. 35.

5. By drawing the ordinates $x = 1$, $x = \frac{2}{3}$, $x = \frac{1}{2}$, show as in No. 4 that $-\frac{5}{6} < \text{hyp}\left(\frac{1}{2}\right) < -\frac{7}{12}$.

6. By taking the ordinates $x = 1, 1 \cdot 1, 1 \cdot 2, \ldots 1 \cdot 9, 2$, show that hyp (2) lies between

$$0 \cdot 1\left(1 + \frac{1}{1 \cdot 1} + \frac{1}{1 \cdot 2} + \ldots + \frac{1}{1 \cdot 9}\right) \text{ and } 0 \cdot 1\left(\frac{1}{1 \cdot 1} + \frac{1}{1 \cdot 2} + \ldots + \frac{1}{1 \cdot 9} + \frac{1}{2}\right).$$

Deduce that $0 \cdot 66 < \text{hyp}(2) < 0 \cdot 72$. [Actually, hyp $(2) = 0 \cdot 693\ldots$.]

7. Show from a figure that $\text{hyp}(t_1) < \text{hyp}(t_2)$, if $0 < t_1 < t_2$.

8. Prove, as on p. 53, that $\text{hyp}(2^k) < k$, if k is a positive integer.

9. In Fig. 36, PN is the ordinate $x = t$ of any point on the line $y = 2x$. If the area of $\triangle$ONP is denoted by $\text{sq}(t)$, prove geometrically that (i) $\text{sq}(2t) = 4\text{sq}(t)$; (ii) $\text{sq}(t + t') - \text{sq}(t - t') = 4tt'$.

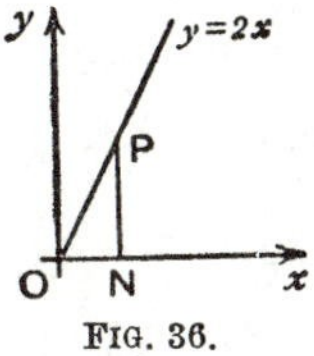

FIG. 36.

What does (ii) become if $t = t'$, and if $t = 0$?
Interpret geometrically $\text{sq}(-t)$.

10. Use geometrical methods to prove that

(i) $\frac{1}{3} < \text{hyp}\left(1\frac{1}{2}\right) < \frac{1}{2}$; (ii) $\frac{1}{5} < \text{hyp}\left(1\frac{1}{4}\right) < \frac{1}{4}$;

(iii) $\frac{1}{3} + \frac{1}{4} + \frac{1}{5} < \text{hyp}\left(2\frac{1}{2}\right) < 1$; (iv) $\text{hyp}(3) > 1$.

11. In Fig. 37, PN is the ordinate, $x = t$, where $t > 1$. Use the indicated construction to show that $1 - \dfrac{1}{t} < \text{hyp}(t) < t - 1$.

12. Draw, in Fig. 37, the ordinate $x=s$, where $s<1$, and by the method of No. 11, show that $1-\frac{1}{s}<\text{hyp}(s)<s-1$.

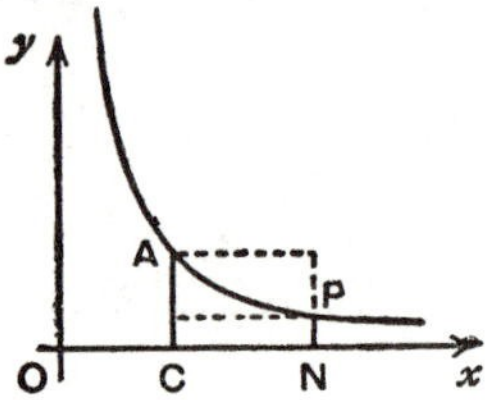

FIG. 37.

13. In Fig. 38, take PN and QM as the ordinates $x=t$, $x=t+h$ and hence show that $\frac{\text{hyp}(t+h)-\text{hyp}(t)}{h}$ lies between $\frac{1}{t}$ and $\frac{1}{t+h}$.

Is this result true if h is negative, $t+h$ being positive? What result is obtained by making h tend to 0?

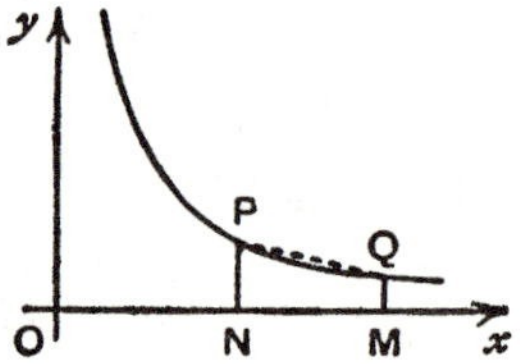

FIG. 38.

14. In Fig. 38, PN, QM are the ordinates, $x=p$, $x=q$, for the curve $y=\frac{1}{x}$. Prove that the area of the trapezium PNMQ is $\frac{1}{2}\left(\frac{q}{p}-\frac{p}{q}\right)$.

If P′N′, Q′M′ are the ordinates, $x=\frac{1}{p}$, $x=\frac{1}{q}$, prove that the area of the trapezium P′N′M′Q′ is $\frac{1}{2}\left(\frac{p}{q}-\frac{q}{p}\right)$.

What relation between the values of hyp (t) and hyp $\left(\frac{1}{t}\right)$ can be deduced from these results?

15. If, in Fig. 38, PN and QM are the ordinates, $x=\lambda p$, $x=\lambda q$, for the curve $y=\frac{1}{x}$, prove that the area of the trapezium PNMQ does not depend on the value of λ.

Use this fact to prove that the value of hyp $(\lambda t)-$ hyp (λ) does not depend on the value of λ, and so obtain its value in terms of t.

16. Draw a rough graph of $y=\frac{1}{1+x^2}$, marking the lengths of the ordinates at the points P_0, P_1, P_2, P_3 whose abscissae are 0, 1, 2, 3. By considering the area under the curve $P_0P_1P_2P_3$, prove that

$$\frac{4}{5}<\int_0^3 \frac{1}{1+x^2}\,dx<\frac{17}{10}.$$

17. Show that

$$\text{(i)} \int_0^1 \frac{1}{1+x^2}\,dx<1; \quad \text{(ii)} \int_1^t \frac{1}{1+x^2}\,dx<1-\frac{1}{t}, \text{ for } t>1.$$

Deduce that if $\tan^{-1}t$ is defined to be $\int_0^t \frac{1}{1+x^2}\,dx$, the function $\tan^{-1}t$ increases with t, but remains always less than 2.

18. By considering the area under the parabola $y=x^2$, show that $1^2+2^2+3^2+\ldots+n^2$ lies between $\frac{1}{3}n^3$ and $\frac{1}{3}\{(n+1)^3-1\}$.

19. Prove that

$$\tfrac{2}{3}n\sqrt{n}<\sqrt{1}+\sqrt{2}+\sqrt{3}+\ldots+\sqrt{n}<\tfrac{2}{3}\{(n+1)\sqrt{(n+1)}-1\}.$$

20. Prove that, if $0<\theta<\frac{\pi}{2n}$, $\sin\theta+\sin 2\theta+\ldots+\sin n\theta$ lies between $\frac{2}{\theta}\sin^2\frac{1}{2}n\theta$ and $\frac{2}{\theta}\sin\frac{n\theta}{2}\sin(\frac{1}{2}n+1)\theta$.

Differentiation of hyp (t). In Fig. 39, if $\mathsf{ON}=t$ and $\mathsf{ON}'=t+h$, we have

$$\text{hyp}(t+h)-\text{hyp}(t)=\text{area CN}'\text{P}'\text{A}-\text{area CNPA}$$
$$=\text{area NN}'\text{P}'\text{P}.$$

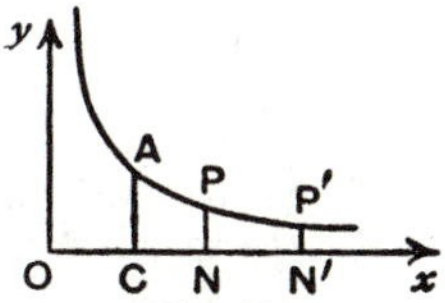

FIG. 39.

But area NN′P′P lies between NN′.NP and NN′.N′P′, i.e. between $\frac{h}{t}$ and $\frac{h}{t+h}$.

$$\therefore \frac{\text{hyp}(t+h)-\text{hyp}(t)}{h} \text{ lies between } \frac{1}{t} \text{ and } \frac{1}{t+h}.$$

But $\frac{1}{t+h}$ tends to the value $\frac{1}{t}$ as h tends to 0;

$$\therefore \frac{\text{hyp}(t+h)-\text{hyp}(t)}{h}\to\frac{1}{t} \text{ when } h\to 0;$$

$$\therefore \frac{d}{dt}\,\text{hyp}(t)=\frac{1}{t}. \qquad \text{(4)}$$

If h is negative, $t+h$ being positive, $\text{hyp}(t+h)-\text{hyp}(t)$ is negative.
$\therefore \frac{\text{hyp}(t+h)-\text{hyp}(t)}{h}$ is positive and still lies between $\frac{1}{t}$ and $\frac{1}{t+h}$.

Therefore it tends to the limit $\frac{1}{t}$ as h tends to 0 *in any manner.*

EXERCISE IV. b.

1. Differentiate with respect to x:

(i) $\text{hyp}\,(2x)$; (ii) $\text{hyp}\left(\frac{x}{3}\right)$.

(iii) $\text{hyp}\,(ax)$; (iv) $\text{hyp}\,(ax) - \text{hyp}\,(x)$.

What inference can be drawn from the last result ?

2. Differentiate with respect to x:

(i) $\text{hyp}\,(x^2)$; (ii) $\text{hyp}\,(x^3)$;

(iii) $\text{hyp}\left(\frac{1}{x}\right)$; (iv) $\text{hyp}\,(x^n)$.

What inference can be drawn from the last result ?

3. Differentiate with respect to x:

(i) $\text{hyp}\,(ax+b)^n$; (ii) $\text{hyp}\left(\frac{x+1}{x+2}\right)$.

4. Differentiate with respect to x:

(i) $\text{hyp}\,(\sin x)$; (ii) $\text{hyp}\,(\tan x)$; (iii) $\text{hyp}\,(\cot x)$.

5. Integrate the following with respect to x, giving the answers as hyp functions.

(i) $\frac{1}{x}$; (ii) $\frac{1}{x+1}$; (iii) $\frac{1}{x-1}$; (iv) $\frac{1}{1-x}$;

(v) $\frac{1}{2x+3}$; (vi) $\frac{1}{4-5x}$; (vii) $1-\frac{2}{x+2}$; (viii) $\frac{x}{x-3}$.

6. Write down an expression for $\frac{d}{dx}\{\text{hyp}\,[f(x)]\}$. Use the result to integrate with respect to x:

(i) $\frac{\cos x}{\sin x}$; (ii) $\tan x$; (iii) $\cot 2x$;

(iv) $\frac{ax+b}{ax^2+2bx+c}$; (v) $\frac{x^{n-1}}{1+x^n}$; (vi) $\frac{\cos x}{1+\sin x}$.

7. What is $\int \frac{\sec^2 x}{\tan x}\,dx$? Hence find $\int \text{cosec}\, 2x\, dx$.

8. What is $\frac{d}{dx}\{x\,\text{hyp}\,(x)\}$? Hence find $\int \text{hyp}\,(x)\,dx$.

9. What is the sign of $\frac{d}{dx}\{\text{hyp}\,(x)\}$? What inference can be drawn from the result ?

10. What is the sign of $\frac{d}{dx}\{x-1-\text{hyp}\,(x)\}$, (i) if $x>1$, (ii) if $0<x<1$. Use the result to prove that $\text{hyp}\,(x)<x-1$ for $x>0$, $x \neq 1$.

11. If $\tan^{-1} t \equiv \int_0^t \frac{1}{1+x^2}\,dx$, show by the method of p. 57, that

$$\frac{d\tan^{-1} t}{dt} = \frac{1}{1+t^2}.$$

12. By the substitution $x = \frac{1}{z}$, show that $\tan^{-1}\frac{1}{t} = \int_t^{\infty} \frac{1}{1+x^2}\,dx$, and deduce that $\tan^{-1} t + \tan^{-1}\frac{1}{t}$ is independent of t.

Other Properties of hyp (t). The following properties have all been illustrated in the previous examples.

To prove $$\text{hyp}\left(\frac{1}{t}\right) = -\text{hyp}(t). \qquad (5)$$

By definition, $$\text{hyp}\left(\frac{1}{t}\right) = \int_1^{\frac{1}{t}} \frac{1}{x}\,dx.$$

Put $x = \frac{1}{z}$, so that $z = t$ when $x = \frac{1}{t}$, and $z = 1$ when $x = 1$; also

$$dx = -\frac{1}{z^2}\,dz;$$

$$\therefore \text{hyp}\left(\frac{1}{t}\right) = \int_1^t z\left(-\frac{1}{z^2}\right)dz = -\int_1^t \frac{1}{z}\,dz;$$

$$\therefore \text{hyp}\left(\frac{1}{t}\right) = -\text{hyp}(t).$$

This result may be illustrated geometrically.

In Fig. 40, OC = OC′ = 1, ON = ON′ = t, NP, N′P′ are perpendiculars to Ox, Oy and P′K is perpendicular to Ox, so that $\text{OK} = \frac{1}{\text{P'K}} = \frac{1}{t}$.

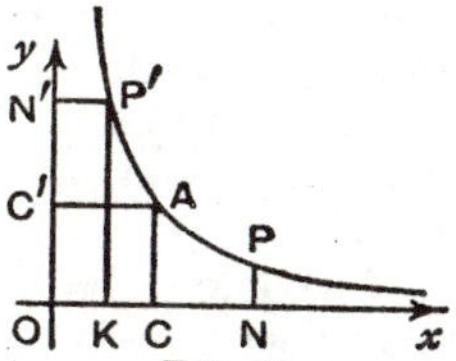

FIG. 40.

Since the curve is symmetrical about OA, the areas bounded by ON′P′AC and ONPAC′ are equal.

But OKP′N′ and OCAC′ are each of unit area; therefore the remainders, the areas KCAP′ and CNPA, are equal.

But these areas are $-\text{hyp}\left(\frac{1}{t}\right)$ and hyp (t) respectively.

An alternative geometrical method is indicated in Ex. IV. a, No. 14.

To prove $$\mathbf{hyp}\,(\mathrm{ab}) = \mathbf{hyp}\,(\mathrm{a}) + \mathbf{hyp}\,(\mathrm{b}), \quad \text{......................(6)}$$

$$\mathbf{hyp}\left(\frac{\mathrm{a}}{\mathrm{b}}\right) = \mathbf{hyp}\,(\mathrm{a}) - \mathbf{hyp}\,(\mathrm{b}). \quad \text{......................(7)}$$

Since $\frac{d}{dt}\{\text{hyp}\,(t)\} = \frac{1}{t}$; $\therefore \frac{d}{dt}\{\text{hyp}\,(ct)\} = \frac{1}{ct}\cdot c = \frac{1}{t}$,

where c is a constant;

$$\therefore \frac{d}{dt}\{\text{hyp}\,(ct) - \text{hyp}\,(t)\} = 0;$$

$\therefore$ the value of hyp (ct) − hyp (t) does not depend on the value of t and is therefore equal to the value obtained by putting $t=1$;

$$\therefore \text{hyp}\,(ct) - \text{hyp}\,(t) = \text{hyp}\,(c). \quad \text{......................(8)}$$

Putting $c=a$, $t=b$, we have hyp (ab) = hyp (a) + hyp (b).

Putting $c=\frac{a}{b}$, $t=b$, we have hyp $\left(\frac{a}{b}\right)$ = hyp (a) − hyp (b).

It should be noted that the result in (8) really contains those in (5), (6), and (7).

Relation (6) may be illustrated geometrically.

If, in Fig. 41, NP, MQ are the ordinates $x=p$, $x=q$, and if N′P′, M′Q′ are the ordinates $x=bp$, $x=bq$, then the trapeziums P′N′M′Q′, PNMQ are equal in area.

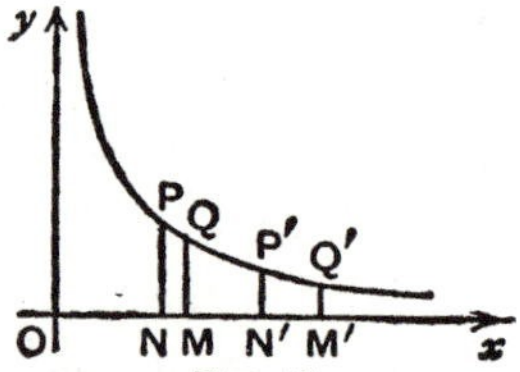

FIG. 41.

For, trapezium P′N′M′Q′

$$=\frac{1}{2}\left(\frac{1}{bp}+\frac{1}{bq}\right)(bq-bp)$$

$$=\frac{q-p}{2}\cdot\left(\frac{1}{p}+\frac{1}{q}\right)$$

= trapezium PNMQ.

Now draw a large number of ordinates between $x=1$ and $x=a$, and take, as above, the corresponding ordinates between $x=b$ and $x=ba$, (see Fig. 42). We then obtain corresponding pairs of trapeziums of equal area.

If we allow the number of ordinates to increase indefinitely and the width of each trapezium to tend to zero, we see that, in the limit, the area under the curve from $x=1$ to $x=a$ is equal to the area under the curve from $x=b$ to $x=ba$;

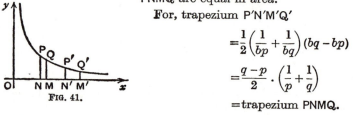

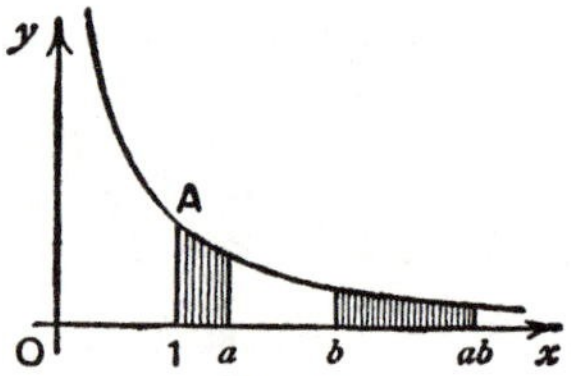

FIG. 42.

$$\therefore \text{hyp}\,(a) = \text{hyp}\,(ba) - \text{hyp}\,(b).$$

This geometrical illustration corresponds to the analytical method indicated in Ex. IV. c, No. 7.

To prove $$\text{hyp}(a^n) = n\,\text{hyp}(a), \qquad \text{(9)}$$

where n is any rational number.

If n is any positive integer, we have, by repeated applications of (6), $$\text{hyp}(a_1) + \text{hyp}(a_2) + \ldots + \text{hyp}(a_n) = \text{hyp}(a_1 a_2 \ldots a_n).$$

Putting $a_1 = a_2 = \ldots = a_n = a$, this becomes

$$\text{hyp}(a^n) = n\,\text{hyp}(a).$$

It is now possible to show by a similar argument that this result is true if n is any rational number; see Ex. IV. c, No. 9. Relation (9) may, however, be proved in a different way, as follows:

By definition, $$\text{hyp}(a^n) = \int_1^{a^n} \frac{1}{x}\,dx.$$

Put $x = y^n$, so that $y = a$ when $x = a^n$ and $y = 1$ when $x = 1$; also

$$dx = ny^{n-1}\,dy;$$

$$\therefore\ \text{hyp}(a^n) = \int_1^a \frac{ny^{n-1}}{y^n}\,dy = n\int_1^a \frac{1}{y}\,dy;$$

$$\therefore\ \text{hyp}(a^n) = n\,\text{hyp}(a).$$

EXERCISE IV. c.

1. Given that $\text{hyp}(2) \simeq 0{\cdot}693$ and $\text{hyp}(3) \simeq 1{\cdot}099$, find approximate values of $\text{hyp}(x)$ for $x = 4, 6, 8, 9, \frac{1}{2}, \frac{1}{4}, 1\frac{1}{2}, 2\frac{1}{4}, 2\frac{2}{3}$.

Draw on squared paper the graph of $\text{hyp}(x)$ from $x = 1$ to $x = 3$, and use it to solve $\text{hyp}(x) = 1$.

2. Use the data of No. 1 to evaluate:

(i) $\displaystyle\int_2^3 \frac{1}{x}\,dx$; (ii) $\displaystyle\int_{20}^{30} \frac{1}{x}\,dx$;

(iii) $\displaystyle\int_1^3 \frac{1}{x+1}\,dx$; (iv) $\displaystyle\int_0^{1\frac{1}{2}} \frac{1}{2x+1}\,dx$.

3. Evaluate $\displaystyle\int_a^{ab} \frac{1}{x}\,dx$ by putting $x = ay$.

What relation can be deduced from $\displaystyle\int_1^{ab} \frac{1}{x}\,dx = \int_1^a \frac{1}{x}\,dx + \int_a^{ab} \frac{1}{x}\,dx$?

4. Use the method of No. 3 to prove that

$$\text{hyp}\left(\frac{a}{b}\right) = \text{hyp}(a) - \text{hyp}(b).$$

5. Prove that $\text{hyp}(2^n) > \frac{n}{2}$ by applying the substitution $x = y^n$ to $\int \frac{1}{x}\,dx$.

6. Use the method of No. 5 to prove that $\text{hyp}(2^n) < n$.

7. By using a suitable substitution, prove that

$$\int_{ta}^{tb} \frac{1}{x}\,dx = \int_a^b \frac{1}{x}\,dx.$$

What is the geometrical meaning of this relation ? What property of hyp (t) is obtained from the relation ?

8. Prove that $\text{hyp}(x) > \frac{1}{2} + \frac{1}{3} + \frac{1}{4} + \ldots + \frac{1}{[x]}$, where $x > 1$ and $[x]$ denotes the greatest integer not greater than x.

Deduce that hyp (x) tends to $+\infty$ when x tends to $+\infty$.

9. Use the fact that $\text{hyp}(a^n) = n\,\text{hyp}(a)$ if n is a positive integer, to prove that $\text{hyp}\left(a^{\frac{p}{q}}\right) = \frac{p}{q}\text{hyp}(a)$, where p, q are positive integers [put $a = b^q$]. Prove also that $\text{hyp}\left(a^{-\frac{p}{q}}\right) = -\frac{p}{q}\text{hyp}(a)$.

10. Use the relation

$$\int_1^{2^n} \frac{1}{x}\,dx = \int_1^2 \frac{1}{x}\,dx + \int_2^4 \frac{1}{x}\,dx + \int_4^8 \frac{1}{x}\,dx + \ldots + \int_{2^{n-1}}^{2^n} \frac{1}{x}\,dx$$

to prove that $\frac{n}{2} < \text{hyp}(2^n) < n$.

11. By considering the area of the trapezium ACNP in Fig. 40, p. 59, show that $\text{hyp}(1+k) < k - \frac{k^2}{2(1+k)}$, where $k > 0$.

12. By using the method of No. 11, prove that

$$\text{hyp}(1-k) > -k - \frac{k^2}{2(1-k)}$$

where $0 < k < 1$.

The Function hyp (x). We have shown that, as x increases from zero to $+\infty$, hyp (x) increases *steadily* from $-\infty$ to $+\infty$ and is zero when $x = 1$. The graph is shown in Fig. 43.

Further, since $\text{hyp}(x+h) - \text{hyp}(x)$ lies between $\frac{h}{x}$ and $\frac{h}{x+h}$, (see p. 57), it follows that hyp (x) is a continuous function of x and, since everywhere it increases with x, we may conclude that it assumes once, and only once, any given value, as x passes from

0 to $+\infty$. In particular, there exists a *unique* value of x, such that $\text{hyp}(x)=1$. This value is always denoted by e, so that

$$\text{hyp}(e)=1. \quad \ldots\ldots\ldots\ldots\ldots(10)$$

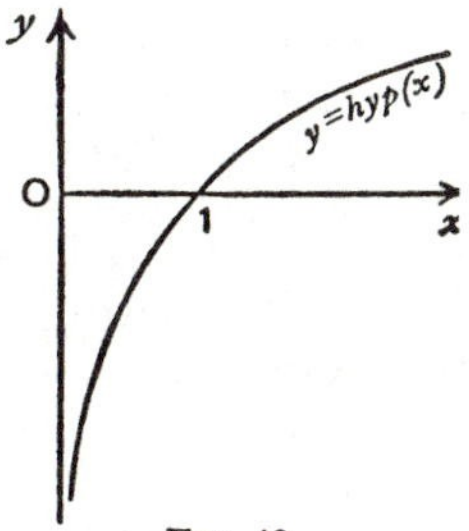

FIG. 43.

The number e is irrational and, moreover, like π, is not algebraic; that is to say, there is no algebraic equation of any degree, having rational coefficients, which has e (or π) for a root. It will be shown later how the value of e can be calculated to any number of places of decimals. For the present, we shall merely point out some limits between which e must lie. Thus from Ex. IV. a, No. 2, $\text{hyp}(2) < 1$ and $\text{hyp}(4) = 2\,\text{hyp}(2) > 1$; therefore $2 < e < 4$.

Again, from Ex. IV. a, No. 10, $\text{hyp}(2\tfrac{1}{2}) < 1 < \text{hyp}(3)$, therefore e lies between $2\tfrac{1}{2}$ and 3. See also Ex. IV. c, No. 1. Actually, $e = 2{\cdot}71828\ldots$.

The reader has probably solved by this time the mystery of the function $\text{hyp}(x)$.

From equations (9), (10), we have, if y is any rational number,

$$\text{hyp}(e^y) = y\,\text{hyp}(e) = y.$$

Therefore, if $e^y = x$, $\text{hyp}(x) = y$; and so $e^{\text{hyp}(x)} = x$.

In other words, the "hyp" of a number is the power to which e must be raised to make that number.

Therefore the "hyp" function is the logarithm of the number to the base e. Logarithms to base e are called natural logarithms or Napierian logarithms.

In mathematical work (as distinct from mere computation) the logarithms which occur are nearly always natural logarithms; and so the symbol $\log x$ is generally understood to mean the natural logarithm and is used as an abbreviation for $\log_e x$.

The argument, given above, therefore shows that

$$\text{hyp} \equiv \log,$$

and equations (1)-(10) of this chapter may now be re-written in this sense. The most important of these results are

$$\log x_1 + \log x_2 = \log (x_1 x_2) \ldots\ldots\ldots\ldots\ldots\ldots(11)$$

and

$$\frac{d}{dx}\{\log x\} = \frac{1}{x} \quad \text{or} \quad \int \frac{1}{x}\,dx = \log x. \ldots\ldots\ldots\ldots\ldots(12)$$

It should be noted that the function $\log x$ has been defined *for positive values of x only.*

The Exponential Function. If $y = \log x$, not only is y determined uniquely when x is given, but for any assigned value of y there is one and only one value of x, and that value of x is positive, since it has been shown that $\log x$ *increases steadily* from $-\infty$ to $+\infty$, as x increases from 0 to $+\infty$.

Therefore, if $y = \log x$, we may regard x as a function of y and this function is single-valued and everywhere positive. This function might be denoted by $\text{hyp}^{-1}(y)$ or by antilog (y), but it is in fact denoted by **exp (y)**, and is called the *exponential function* of y. We therefore write

$$\textbf{If} \quad \mathbf{y = \log x}, \quad \textbf{then} \quad \mathbf{x = \exp(y)} \quad \text{..................(13)}$$

and equation (11) may be expressed in the form

$$\mathbf{\exp(y_1 + y_2) = \exp(y_1) \times \exp(y_2).} \quad \text{..................(14)}$$

The graph of $x = \exp(y)$ is of course the same as that of $y = \log x$. We therefore obtain the graph of $y = \exp(x)$ by interchanging the axes of x and y in Fig. 43, or, equally well, by taking the image of $y = \log x$ in the line $y = x$.

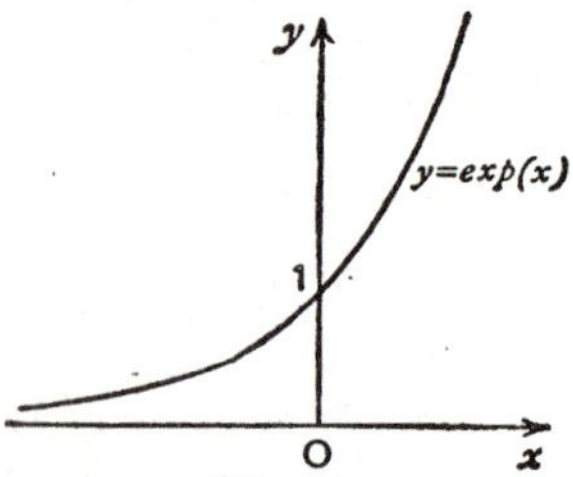

FIG. 44.

This gives the curve in Fig. 44.

If y is any rational number, we have from (9),

$$\log(e^y) = y \log e = y;$$

$$\therefore \quad \mathbf{\exp(y) = e^y.} \quad \text{..................(15)}$$

A discussion of the theory of irrational numbers is beyond the scope of this volume; we shall not, therefore, at this stage define the function e^y for irrational values of y. But it will be found that, when this function has been defined, equation (15) is true also when y is irrational.

Differentiation and Integration. If $y = \exp(x)$, then $x = \log y$;

$$\therefore \quad \frac{dx}{dy} = \frac{1}{y}; \quad \therefore \quad \frac{dy}{dx} = y;$$

$$\therefore \quad \frac{d}{dx}\exp(x) = \exp(x).$$

Thus
$$\mathbf{\frac{d}{dx}(e^x) = e^x} \quad \text{or} \quad \mathbf{\int e^x dx = e^x.} \quad \text{..................(16)}$$

The function a^x, where $a > 0$, may also be called an exponential function of x, but it is easily expressed as a power of e.

If $y=a^x$, then $\log y=\log(a^x)=x\log a$, a being supposed positive;

$$\therefore\ y=\exp(x\log a)=e^{x\log a}.$$

It follows that

$$\frac{\mathrm{d}}{\mathrm{dx}}(\mathrm{a}^{\mathrm{x}})=\frac{d}{dx}(e^{x\log a})=e^{x\log a}\,.\,\log a=\mathbf{a^x\,.\,\log a}. \quad \text{.........(17)}$$

Applications to the Calculus. The results of (12) and (16) may be used in conjunction with the ordinary processes of differentiation and integration; and the scope of the Calculus is thus extended to include many functions involving logarithms and exponentials. The most important applications are those of

$$\int x^{-1}\,dx=\log x.$$

Example 1. Integrate $\tan x$.

$$\frac{d}{dx}[\log(\cos x)]=\frac{d}{d(\cos x)}[\log(\cos x)]\,.\,\frac{d}{dx}(\cos x)=\frac{1}{\cos x}(-\sin x);$$

therefore it follows that

$$\int\tan x\,dx=-\log(\cos x)+c.$$

Whenever a function can be written in the form $\mathrm{A}\,.\,\dfrac{f'(x)}{f(x)}$, where the numerator is the differential coefficient of the denominator, the integral can be written down in the form $\mathrm{A}\,.\log[f(x)]$.

Example 2. Evaluate $\displaystyle\int\frac{dx}{1-x^2}$.

$$\frac{2}{1-x^2}\equiv\frac{1}{1+x}+\frac{1}{1-x};\quad\therefore\ 2\int\frac{dx}{1-x^2}=\log(1+x)-\log(1-x)+c;$$

$$\therefore\ \int\frac{dx}{1-x^2}=\tfrac{1}{2}\log\frac{1+x}{1-x}+c.$$

A very large class of functions can be integrated by the method of the last example, which consists in expressing the integrand as the sum of partial fractions; for the general method of doing this, see p. 231.

The formula for integration by parts will be required in some of the examples in the next exercise. It is

$$\int(uv)\,dx=uw-\int w\,\frac{du}{dx}\,dx,\ \text{where}\ w=\int v\,dx.$$

EXERCISE IV. d.

1. What is the connection between the graphs of

(i) $y = x^2$ and $y = \sqrt{x}$; (ii) $y = \sin x$ and $y = \sin^{-1} x$?

2. Sketch the graphs of 2^x and e^{2x}.

3. Sketch the graphs of

(i) $\log x$; (ii) $\log(2x)$; (iii) $\log(x^2)$; (iv) $\log\left(\frac{1}{x}\right)$.

4. What is the value of x, when

(i) $\log x = 1 + \log a$; (ii) $\log x = 1 - \log b$;

(iii) $\log(\log x) = 0$; (iv) $\log(\log x) = 1$?

5. Simplify (i) $e^{2 \log x}$; (ii) $e^{x \log 2}$; (iii) $\log(e^2 x)$.

6. Prove that if $x_1 < x_2$ then $e^{x_1} < e^{x_2}$.

7. Sketch the graphs of

(i) e^{-x}; (ii) e^{x^2}; (iii) e^{-x^2}; (iv) $e^{\frac{1}{x}}$.

Show graphically that $e^x = x + a$ has two roots if $a > 1$ and no roots if $a < 1$.

8. Differentiate $x \log x - x$ and write down the value of

$$\int_1^t \log x \, dx.$$

9. Differentiate $(x-1)e^x$ and write down the value of $\int_0^t xe^x \, dx$.

Differentiate with respect to x:

10. $x^2 \log x$. **11.** $\dfrac{\log x}{x}$. **12.** $\log\left(\dfrac{1}{x}\right)$. **13.** e^{ax}.

14. e^{x^3}. **15.** $e^{\log x}$. **16.** $\log(\cos x)$. **17.** $\log(\sec x)$.

18. $e^{\sin x}$. **19.** $e^{\tan^2 x}$. **20.** $\exp(x \sec x)$. **21.** $\cos(e^x)$.

22. $\operatorname{cosec}(\log x)$. **23.** $\log(a + bx)^n$. **24.** $\log(e^x + e^{-x})$.

25. (i) $\log\left(\tan\dfrac{x}{2}\right)$; (ii) $\log\left(\tan\overline{\dfrac{\pi}{4} + \dfrac{x}{2}}\right)$; (iii) $\log\{x + \sqrt{(a^2 + x^2)}\}$.

Integrate with respect to x:

26. $\dfrac{1}{3x+4}$. **27.** $\dfrac{2x+4}{2x+3}$. **28.** $\dfrac{x-1}{2x+3}$. **29.** e^{3x}.

30. $e^x - e^{-x}$. **31.** $\dfrac{\sin x}{2 + 3\cos x}$. **32.** $\dfrac{2x+3}{x^2 + 3x + 4}$. **33.** $\dfrac{\log x}{x}$.

34. $\dfrac{\sec^2 \frac{x}{2}}{\tan \frac{x}{2}}$. **35.** xe^{x^2}. **36.** $\cot 3x$. **37.** $\dfrac{1}{1+e^x}$.

38. (i) $\sec x$; (ii) $\operatorname{cosec} x$; (iii) $\dfrac{1}{\sqrt{a^2+x^2}}$. [See No. 25.]

Use the method of partial fractions to find :

39. $\displaystyle\int \frac{3x+1}{x^2-1}\,dx.$

40. $\displaystyle\int \frac{3x+7}{(x+1)(x+2)(x+3)}\,dx.$

41. $\displaystyle\int \frac{x-1}{x(x-2)^2}\,dx.$

42. $\displaystyle\int \frac{3x+1}{(x-3)(x^2+1)}\,dx.$

43. $\displaystyle\int \frac{x^3}{x-1}\,dx.$

44. $\displaystyle\int \frac{x^2}{(x-a)(x-b)}\,dx.$

Use the method of " integration by parts " to find :

45. $\displaystyle\int \log x\,dx.$

46. $\displaystyle\int x\log x\,dx.$

47. $\displaystyle\int x^n \log x\,dx.$

48. $\displaystyle\int xe^x\,dx.$

49. $\displaystyle\int x^2e^x\,dx.$

50. $\displaystyle\int \log(\sqrt{x})\,dx.$

51. Differentiate $e^{ax}\sin bx$ and $e^{ax}\cos bx$, and hence integrate the same two expressions.

52. Find the value of x for which $x^2\log x$ is a minimum.

53. Find the maximum value of $\dfrac{\log x}{x}$, and discuss the number of roots of the equation $\log x = \mathrm{A}x$ for different values of A.

Useful Inequalities. The results given in Ex. IV. a, Nos. 11, 12 should be noted. They may be deduced directly from the definition

$$\log t = \int_1^t \frac{1}{x}\,dx.$$

First, suppose $t>1$. Then throughout the range of values $1<x<t$, the integrand $\dfrac{1}{x}$ is <1 and is $>\dfrac{1}{t}$;

$$\therefore\ \int_1^t \frac{1}{t}\,dx < \int_1^t \frac{1}{x}\,dx < \int_1^t dx;$$

$$\therefore\ 1-\frac{1}{t} < \log t < t-1.$$

Next, suppose $0<t<1$ and put $t=\dfrac{1}{s}$, so that $s>1$;

$$\therefore\ 1-\frac{1}{s} < \log s < s-1;$$

$$\therefore\ 1-t<-\log t<\frac{1}{t}-1;$$

$$\therefore\ t-1>\log t>1-\frac{1}{t};$$

$\therefore$ for all positive values of t, except $t=1$, we have

$$1-\frac{1}{t}<\log t<t-1. \qquad \text{......................(18)}$$

Example 3. Prove that $\lim\limits_{x\to 1}\dfrac{\log x}{x-1}=1$.

From equation (18), $\dfrac{t-1}{t}<\log t<t-1$, if $t>0$, $t\neq 1$;

$\therefore$ if $t>1$, $\qquad \dfrac{1}{t}<\dfrac{\log t}{t-1}<1$,

and if $0<t<1$, $\qquad \dfrac{1}{t}>\dfrac{\log t}{t-1}>1$;

$\therefore$ if $x>0$, $x\neq 1$, $\dfrac{\log x}{x-1}$ lies between $\dfrac{1}{x}$ and 1;

$$\therefore\ \text{when } x\to 1,\ \frac{\log x}{x-1}\to 1.$$

Example 4. Prove that

(i) $\dfrac{\log x}{x}\to 0$, when $x\to\infty$;

(ii) $x\log x\to 0$, when $x\to 0$ through positive values.

(i) From equation (18), $\log t<t-1$ if $t>0$, $t\neq 1$.

Put $t=\sqrt{x}$; $\therefore$ $\log(\sqrt{x})<\sqrt{x}-1<\sqrt{x}$, if $x>0$, $x\neq 1$;

$$\therefore\ \tfrac{1}{2}\log x<\sqrt{x};$$

$\therefore$ if $x>1$, $\qquad 0<\dfrac{\log x}{x}<\dfrac{2\sqrt{x}}{x}=\dfrac{2}{\sqrt{x}}$.

But when $x\to\infty$, $\dfrac{2}{\sqrt{x}}\to 0$; $\therefore$ $\lim\limits_{x\to\infty}\dfrac{\log x}{x}=0$.

(ii) $\lim\limits_{x\to 0}x\log x=\lim\limits_{y\to\infty}\dfrac{1}{y}\log\dfrac{1}{y}=-\lim\limits_{y\to\infty}\dfrac{\log y}{y}=0$, from (i).

Example 5. Prove that

$$\text{(i) } \frac{e^t - 1}{t} \to 1 \text{ when } t \to 0;$$

$$\text{(ii) } n(\sqrt[n]{a} - 1) \to \log a \text{ when } n \to \infty, \text{ for } a > 0.$$

(i) $\lim\limits_{t \to 0} \dfrac{e^t - 1}{t}$ is the value of $\dfrac{d}{dx}(e^x)$ at $x=0$ and is therefore e^0, $=1$. This result can be obtained directly as follows: from equation (18), with e^t instead of t,

$$1 - e^{-t} < t < e^t - 1, \text{ if } t \neq 0;$$

$$\therefore\ t < e^t - 1 < t \,.\, e^t;$$

$$\therefore\ 1 < \frac{e^t - 1}{t} < e^t, \text{ if } 0 < t,$$

and

$$1 > \frac{e^t - 1}{t} > e^t, \text{ if } t < 0;$$

$\therefore$ when t tends to 0 in any manner, $\dfrac{e^t - 1}{t} \to 1.$

(ii) In (i), put $t = \dfrac{1}{n}\log a$, where a is any positive constant.

Then $e^t = e^{\frac{1}{n}\log a} = a^{\frac{1}{n}}$. Also when $t \to 0$, $n \to \infty$;

$$\therefore\ \lim_{n \to \infty} \frac{a^{\frac{1}{n}} - 1}{\frac{1}{n}\log a} = 1;$$

$$\therefore\ \lim_{n \to \infty} n(\sqrt[n]{a} - 1) = \log a, \text{ for } a > 0.$$

Example 6. Prove that the function

$$u_n \equiv 1 + \frac{1}{2} + \frac{1}{3} + \dots + \frac{1}{n} - \log n$$

decreases when n increases, but that it remains positive.

$$u_{n+1} - u_n = \frac{1}{n+1} - \log(n+1) + \log n = \frac{1}{n+1} - \log\left(1 + \frac{1}{n}\right),$$

but equation (18) with $1 + \dfrac{1}{n}$ instead of t, proves that

$$\log\left(1 + \frac{1}{n}\right) > \frac{1}{n+1};$$

thus $u_{n+1} < u_n$.

Equation (18) also gives

$$\frac{1}{n} > \log\left(1+\frac{1}{n}\right) = \log(n+1) - \log n;$$

$$\therefore \quad \sum_1^n \frac{1}{r} > \log(n+1) > \log n;$$

$\therefore$ u_n is positive.

Since u_n decreases but remains positive, it follows that $\lim_{n\to\infty} u_n$ exists and is not negative; but the theorem on which this depends is bound up with the theory of irrational numbers, and the discussion of it must be left to the companion volume on Analysis. The limit in the present example is called Euler's Constant, and it is denoted by γ. Since $u_1 = 1$, it follows that $\gamma < 1$; from Ex. IV. e, No. 14, it follows that $\gamma > \cdot 3$; the actual value is $\cdot 577\ldots$.

Example 7. Prove that $\lim_{n\to\infty}\left(\cos\frac{x}{n}\right)^n = 1$.

Since $\cos\frac{-x}{n} = \cos\frac{x}{n}$, we may assume that x is positive; also if $n > \frac{2x}{\pi}$, $\sec\frac{x}{n} > 1$.

Thus, from equation (18), putting $\sec\frac{x}{n}$ for t,

$$0 < \log\left(\sec\frac{x}{n}\right) < \sec\frac{x}{n} - 1 = 2\sec\frac{x}{n}\sin^2\frac{x}{2n},$$

$$\therefore \quad 0 < \log\left(\sec\frac{x}{n}\right)^n < 2n\sec\frac{x}{n}\sin^2\frac{x}{2n} < x\sec\frac{x}{n}\sin\frac{x}{2n}$$

since $\sin\theta < \theta$ (see *E.T.*, p. 162).

But when $n\to\infty$, $\sec\frac{x}{n}\to 1$, and $\sin\frac{x}{2n}\to 0$,

$$\therefore \quad \lim_{n\to\infty}\log\left(\sec\frac{x}{n}\right)^n = 0;$$

thus $$\lim_{n\to\infty}\left(\sec\frac{x}{n}\right)^n = 1, \quad\text{and}\quad \lim_{n\to\infty}\left(\cos\frac{x}{n}\right)^n = 1.$$

EXAMPLES IV. e.

1. If $1 < t_1 < t_2$, prove that $\log t_2 - \log t_1 < \frac{t_2 - t_1}{t_1}$.

2. If a and b are positive, prove that $\log(a+b) - \log a > \frac{b}{a+b}$.

3. If $1+x>0$, and $x \neq 0$, prove that

$$\frac{x}{1+x} < \log(1+x) < x.$$

4. If $0<x<1$, prove that $x < \log \frac{1}{1-x} < \frac{x}{1-x}$.

5. Prove that $e^x \geqslant 1+x$.

6. Prove that $e^x \leqslant \frac{1}{1-x}$ if $x<1$. What happens if $x>1$?

7. If p is positive, prove that $\left(\frac{e}{p}\right)^p \leqslant e$.

8. Show that $\frac{\log x}{x}$ steadily decreases as x increases from e upwards.

9. If $n>e$, prove that $n^{n+1}>(n+1)^n$.

10. Prove that $\lim_{x \to 0} \frac{\log(1+x)}{x} = 1$.

11. Prove that $\lim_{x \to \infty} \frac{\log x}{x^p} = 0$, where p is positive.

12. Prove that $\lim_{x \to 0} x^p \log x = 0$, where p and x are positive.

13. Prove that $1+\frac{1}{2}+\frac{1}{3}+\ldots+\frac{1}{n-1}-\log n$ increases with n, and that it is always less than unity.

14. Assuming $\log 2 = \cdot 69\ldots$, deduce from No. 13 that $\gamma > \cdot 3$.

15. Find, in terms of γ, the limit of $1+\frac{1}{2}+\frac{1}{3}+\ldots+\frac{1}{2n}-\log n$ when $n \to \infty$.

16. Evaluate $\lim_{n \to \infty}\left(\frac{1}{n+1}+\frac{1}{n+2}+\ldots+\frac{1}{2n}\right)$.

17. Evaluate $\lim_{n \to \infty}\left(1-\frac{1}{2}+\frac{1}{3}-\frac{1}{4}+\ldots+\frac{1}{2n-1}-\frac{1}{2n}\right)$.

18. Prove: $\lim_{n \to \infty}\left(1+\frac{1}{3}+\frac{1}{5}+\frac{1}{7}+\ldots+\frac{1}{2n+1}-\frac{1}{2}\log n\right) = \log 2+\frac{1}{2}\gamma$.

19. Prove that $1+\frac{1}{4}+\frac{1}{7}+\frac{1}{10}+\ldots+\frac{1}{3n+1}-\frac{1}{3}\log n$ tends to a limit when $n \to \infty$.

20. If p and $n-1$ are positive integers, prove that

$$\log\frac{n+p}{n-1}>\frac{1}{n}+\frac{1}{n+1}+\frac{1}{n+2}+\ldots+\frac{1}{n+p}>\log\frac{n+p+1}{n},$$

and deduce that

$$\lim_{n\to\infty}\left(\frac{1}{n}+\frac{1}{n+1}+\ldots+\frac{1}{nq}\right)=\log q,$$

where q is a given positive integer.

EASY MISCELLANEOUS EXAMPLES

EXERCISE IV. f.

1. Differentiate $\log\{x+\sqrt{(x^2-a^2)}\}$ with respect to x.

2. Show that $\displaystyle\int_{1\cdot 5}^{2\cdot 5}\frac{1}{\sqrt{(4x^2-9)}}\,dx=\tfrac{1}{2}\log 3$.

3. Evaluate $\displaystyle\int\frac{x^3+1}{(x-1)^3}\,dx$.

4. Obtain a relation between $\int x^n e^x\,dx$ and $\int x^{n+1}e^x\,dx$.

5. Find the maximum value of $x^{\frac{1}{x}}$.

6. If $y=a\cos(\log x)+b\sin(\log x)$, prove that

$$x^2\frac{d^2y}{dx^2}+x\frac{dy}{dx}+y=0.$$

7. If $y=e^{kx}$ satisfies $\dfrac{d^2y}{dx^2}-5\dfrac{dy}{dx}+6y=0$, find k.

8. Prove that $y=ae^{-mx}\sin(nx+\alpha)$ satisfies

$$\frac{d^2y}{dx^2}+2m\frac{dy}{dx}+(m^2+n^2)y=0.$$

9. If $y=e^x\sin x$, prove that $\dfrac{d^4y}{dx^4}=-4y$.

10. If $y=x^x$, prove that $\dfrac{dy}{dx}=x^x\log(ex)$.

11. Evaluate $\int e^{ax}\sin^2 bx\,dx$.

12. Compare the graphs of $\log x$ and $\log(\log x)$.

13. Prove that $\log(e^2t)\leqslant 2\sqrt{t}$ where $t>0$.

14. Napier used the approximation $\log \frac{a}{b} \simeq \frac{1}{2}(a-b)\left(\frac{1}{a}+\frac{1}{b}\right)$, where $\frac{a}{b}-1$ is small, for calculating logarithms. Express this as an approximate formula for $\log(1+x)$ if x^4 is negligible. [The error $\simeq \frac{1}{6}x^3$.]

15. Prove that $\lim_{x\to\infty} x^{\frac{1}{x}}=1$.

16. Prove that $\lim x^x=1$ if $x\to 0$ through positive values.

17. What results can be deduced from $\log t > 1-\frac{1}{t}$ by changing t into t^2 and into $\sqrt{t}$, where $t>1$. Which of the three inequalities gives most information?

18. If $x>1$, prove that $\log x < 2(\sqrt{x}-1)$.

19. Use the relation $\log(1+t)=\int_0^t \frac{1}{1+x}\,dx=\int_0^t\left(1-\frac{x}{1+x}\right)dx$ to show that if t is positive, $\log(1+t)$ lies between

$$t-\frac{t^2}{2} \text{ and } \frac{t(2+t)}{2(1+t)}.$$

20. Prove that, if $x>1$, $\log x<\frac{1}{2}\left(x-\frac{1}{x}\right)$.

21. Prove that, if $t>0$,

$$t-\frac{t^2}{2}+\frac{t^3}{3(1+t)}<\log(1+t)<t-\frac{t^2}{2}+\frac{t^3}{3}.$$

22. If $p>q$, prove that $p>\frac{pe^x+qe^{-x}}{e^x+e^{-x}}>q$.

23. Integrate x^2e^x, and prove that

$$\tfrac{8}{15}<\int_0^1 x^2e^{x^2}\,dx<e-2.$$

24. If $t>0$, prove that $\int_0^t x\log(1+x)\,dx$ lies between $\frac{1}{3}t^3$ and $\frac{1}{2}t(t-2)$.

25. If p and $t-1$ are positive, use the relation

$$\int_1^t \frac{1}{x}\,dx<\int_1^t \frac{1}{x^{1-p}}\,dx$$

to prove that $\log t<\frac{t^p}{p}$. Deduce that if m is positive, $\frac{\log t}{t^m}\to 0$ when $t\to\infty$.

What result is obtained by putting $t=e^y$? Deduce that $\frac{y^q}{e^{ry}}\to 0$ when $y\to\infty$, if $r>0$.

26. Prove that $\lim_{n\to\infty}\left(\frac{\sin x/n}{x/n}\right)^n=1$.

HARDER MISCELLANEOUS EXAMPLES

EXERCISE IV. g.

1. Prove that $1\cdot71 > \int_0^2 \frac{1}{\sqrt{(1+x^2)}}\,dx > 1\cdot15$.

2. If p is positive, show that $1^p + 2^p + 3^p + \ldots + n^p$ lies between

$$\frac{n^{p+1}}{p+1} \text{ and } \frac{(n+1)^{p+1}-1}{p+1}.$$

3. Prove that $\frac{1}{2} + \frac{1}{5} + \frac{1}{10} + \ldots + \frac{1}{n^2+1}$ lies between those values of $\tan^{-1} n$ and $\tan^{-1}\frac{n}{n+2}$ which are between 0 and $\frac{\pi}{2}$.

4. Prove that, if $x > 1$, $(x-1)(x-3) - 2x(x-2)\log x$ is negative.

5. Prove that $\int_0^t \frac{1}{\sqrt{(1+x^4)}}\,dx + \int_0^{\frac{1}{t}} \frac{1}{\sqrt{(1+x^4)}}\,dx$ is independent of t.

6. If $0 < a < b$ and $c < d < 0$, determine whether $\frac{a}{c}$ or $\frac{b}{d}$ is the greater.

7. If x and y are positive and less than unity, prove that

$$\frac{x(1-y)}{y} < \frac{\log(1-x)}{\log(1-y)} < \frac{x}{y(1-x)}.$$

8. If t is positive, prove that $\log(1+t) > \int_0^t \frac{1}{(1+\frac{1}{2}s)^2}\,ds = \frac{2t}{2+t}$.

9. If $x > 1$, prove that $\log x > \frac{2(x-1)}{x+1}$.

10. If $x > 1$, prove that $\log x < \int_1^x \frac{1}{2}\left(1 + \frac{1}{s^2}\right) ds = \frac{1}{2}\left(x - \frac{1}{x}\right)$.

11. If t is positive, prove that $\log(1+t) < \frac{t(2+t)}{2(1+t)}$.

12. If $x > 1$, prove that $\frac{4(\sqrt{x}-1)}{\sqrt{x}+1} < \log x < \frac{x-1}{\sqrt{x}}$.

13. Prove that $e^x < \frac{2+x}{2-x}$, if $0 < x < 2$.

14. Prove that $\log t < n(\sqrt[n]{t} - 1)$ where $t > 0$ and $t \neq 1$.

15. Prove that, as t decreases steadily down towards unity, $\frac{\log t}{t-1}$ increases steadily. Also state this result as a geometrical property of the hyperbola $xy = 1$.

16. By putting $1+\frac{1}{x}$ for t in the result of No. 15, show that $\left(1+\frac{1}{x}\right)^x$ steadily increases as x increases through positive values.

17. Prove that $\log\left\{\left(1+\frac{1}{x}\right)^x\right\}<1$, if $x>0$.

18. Prove that $\log(1+x)=\frac{x}{1+\theta x}$, if $x>-1$, for some value of θ between 0 and 1.

19. Prove that $(1+x)^{\frac{1}{x}}\to e$ when $x\to 0$ through positive values, and that $\left(1+\frac{1}{x}\right)^x\to e$ when $x\to+\infty$.

20. Prove that $\int_0^1 \frac{\log(1+x)}{1+x^2}dx$ lies between $\frac{1}{2}\log 2$ and $\frac{\pi}{8}-\frac{1}{4}\log 2$. [See also No. 28.]

21. Prove that $\int_0^\infty \frac{\log x}{1+x^2}dx=0$ by taking the range of integration in two parts, from 0 to 1 and 1 to ∞. What result is given by the substitution, $x=cy$?

22. If $\Gamma(n)\equiv\int_0^\infty x^{n-1}e^{-x}dx$, where n is positive and the existence of the integral is assumed, show by integration by parts that

$$\Gamma(n+1)=n\Gamma(n),$$

and deduce that if $(m-1)$ is a positive integer $\Gamma(m)=(m-1)!$

23. If $f(n)=\int_0^\infty x^n e^{-x^2}dx$, and assuming that this integral exists if $n>-1$, prove that $f(n+2)=\frac{1}{2}(n+1).f(n)$.

24. If $\mathrm{B}(m,n)=\int_0^1 x^{m-1}(1-x)^{n-1}dx$, and assuming that this integral exists if m and n are positive, prove that

(i) $\mathrm{B}(m,n)=\mathrm{B}(n,m)$; (ii) $\mathrm{B}(m+1,n)=\frac{m}{m+n}\mathrm{B}(m,n)$;

(iii) $\mathrm{B}(m,n)=2\int_0^{\frac{\pi}{2}}\sin^{2m-1}\theta\cos^{2n-1}\theta\,d\theta$.

Express $\mathrm{B}(m+1,n+1)$ in terms of $\mathrm{B}(m,n)$.

25. If $x>1$, prove that $\tan^{-1}x<\frac{\pi}{4}+\frac{1}{2}\log x$.

26. If $0<\theta<\frac{\pi}{2}$, prove that $\operatorname{cosec}\theta<\frac{2}{\theta}$. Also show that the integrals $\int_\epsilon^1 \log\frac{1}{\theta}d\theta$ and $\int_\epsilon^{\frac{\pi}{2}}\log\operatorname{cosec}\theta\,d\theta$ tend to limits when ϵ tends to zero through positive values.

27. Prove that $\int_0^{\frac{\pi}{2}} \log\sin\theta\,d\theta = \int_0^{\frac{\pi}{2}} \log\cos\theta\,d\theta$ and hence that each equals $\frac{1}{2}\int_0^{\frac{\pi}{2}} \log(\sin 2\theta)\,d\theta - \frac{\pi}{4}\log 2.$

Prove also that

$$\int_0^{\frac{\pi}{2}} \log\sin 2\theta\,d\theta = \frac{1}{2}\int_0^{\pi} \log\sin\psi\,d\psi = \int_0^{\frac{\pi}{2}} \log\sin\psi\,d\psi.$$

Deduce from these results that

$$\int_0^{\frac{\pi}{2}} \log\sin\theta\,d\theta = -\frac{\pi}{2}\log 2.$$

28. By using the two transformations, $x = \tan\theta$ and $x = \tan\left(\frac{\pi}{4} - \phi\right)$ and equating the results obtained, prove that

$$\int_0^1 \frac{\log(1+x)}{1+x^2}\,dx = \frac{\pi}{8}\log 2.$$

CHAPTER V

EXPANSIONS IN SERIES

Power Series. An expression of the form

$$a_0 + a_1x + a_2x^2 + \dots + a_rx^r + \dots$$

is called a **power series in x.**

Let $\phi_n(x)$ denote the sum of the first n terms,

then $$\phi_n(x) \equiv a_0 + a_1x + a_2x^2 + \dots + a_{n-1}x^{n-1}.$$

If, for some or all values of x, $\lim\limits_{n\to\infty} \phi_n(x)$ exists and is, say, $\phi(x)$, the series is called **convergent** for those values of x, and $\phi(x)$ is called the **sum to infinity.** Also, the series is called the *expansion* of $\phi(x)$ in powers of x, and we write

$$\phi(x) = a_0 + a_1x + a_2x^2 + \dots + a_rx^r + \dots. \quad \dots\dots\dots\dots\dots(1)$$

The most useful expansions are those which are "rapidly convergent," i.e. those in which $\phi_n(x)$ is a good approximation to $\phi(x)$ for reasonably small values of n.

It is most important to distinguish between the meanings of the following:

$$a_0 + a_1x + a_2x^2 + \dots + a_{n-1}x^{n-1},$$

and $$a_0 + a_1x + a_2x^2 + \dots + a_{n-1}x^{n-1} + \dots.$$

The first means the sum of n terms of the given series, *and is obtained by successive addition.*

The second means $\lim\limits_{n\to\infty} (a_0 + a_1x + a_2x^2 + \dots + a_{n-1}x^{n-1})$, *if this limit exists, and is undefined if this limit does not exist.* Sometimes, however, the second is written down when it is merely proposed to discuss the existence of the limit.

The Geometric Progression; $1 - x + x^2 - x^3 + \dots$.

$$\phi_n(x) \equiv 1 - x + x^2 - \dots + (-1)^{n-1}x^{n-1}$$

$$= \frac{1-(-x)^n}{1-(-x)} = \frac{1}{1+x} - \frac{(-1)^n \cdot x^n}{1+x}. \quad \dots\dots\dots\dots\dots(2)$$

But, if $-1 < x < 1$, $\lim\limits_{n\to\infty} x^n = 0$, see limit (i) below;

$$\therefore \phi(x) = \lim_{n\to\infty} \phi_n(x) = \frac{1}{1+x}.$$

Therefore, if $-1<x<1$, the power series $1-x+x^2-x^3+\ldots$ is convergent and has $\dfrac{1}{1+x}$ for sum to infinity.

Two Important Limits.

(i) If $-1<x<1$, $\lim\limits_{n\to\infty} x^n=0$. ....(3)

Consider first $0<x<1$; put $x=1-p$, so that $0<p<1$. Then

$$x^n=(1-p)^n<\frac{1}{(1+p)^n}, \text{ since } (1-p)(1+p)=1-p^2<1,$$

$$<\frac{1}{1+np}<\frac{1}{np}.$$

$\therefore$ x^n can be made less than any given positive number, ϵ, by taking n large enough, $\left(n>\dfrac{1}{\epsilon p}\right)$; but x^n is positive;

$$\therefore\ x^n\to 0 \quad\text{when}\quad n\to\infty.$$

Also $(-x)^n=(-1)^n\,.\,x^n$; therefore the result holds also for

$$-1<x<0.$$

(ii) For all values of x, $\lim\limits_{n\to\infty}\dfrac{x^n}{n!}=0$. ....(4)

Consider first $x>0$; take any fixed integer k greater than $2x$. Then, if $u_n\equiv\dfrac{x^n}{n!}$,

$$\frac{u_{k+1}}{u_k}=\frac{x}{k+1}<\tfrac{1}{2};\quad \frac{u_{k+2}}{u_{k+1}}=\frac{x}{k+2}<\tfrac{1}{2};\quad \frac{u_{k+3}}{u_{k+2}}<\tfrac{1}{2};\text{ etc.}$$

$\therefore$ by multiplication, $u_{k+r}<(\tfrac{1}{2})^r\,.\,u_k$; but u_n is positive;

$$\therefore\text{ by (3), }\lim_{r\to\infty}u_{k+r}=0;\quad\therefore\ \lim_{n\to\infty}u_n=0.$$

Also, as in (i), the result can be extended to all negative values of x.

The Symbol $|x|$. It is often convenient to use the symbol $|x|$ to denote the value of x if x is positive and the value of $-x$ if x is negative.

Thus, the condition $-1<x<1$ is written more shortly in the form $|x|<1$; the statement that x lies in the range of values

$a-\epsilon$ to $a+\epsilon$ is represented by $|x-a|<\epsilon$; the *positive* square root of a^2 may be written $|a|$; etc.

The statement in equation (3) above would therefore often be given in the form:

$$\text{if } |x|<1, \quad \lim_{n\to\infty} x^n = 0.$$

Expansions of sin x and cos x. We proceed to expand $\sin x$ and $\cos x$ in power series, and for the sake of completeness we include the fundamental results upon which the proof depends.

If $0<x<\frac{1}{2}\pi$, and the angles are measured in radians, we assume that

$$\cos x < \frac{\sin x}{x} < 1.$$

When $x \to 0$, $\cos x \to 1$, thus $\frac{\sin x}{x} \to 1$. Since the value of $\frac{\sin x}{x}$ is unaltered when x is changed to $-x$, it follows that $\lim \frac{\sin x}{x} = 1$ when $x \to 0$ in any manner. This result is required for the differentiation of $\sin x$ and $\cos x$.

By the definition of a differential coefficient,

$$\frac{d}{dx}(\sin x) = \lim_{h\to 0} \frac{\sin(x+h)-\sin x}{h} = \lim_{h\to 0} \frac{2\cos(x+\frac{1}{2}h)\sin\frac{1}{2}h}{h}$$

$$= \cos x \, . \lim_{h\to 0} \frac{\sin\frac{1}{2}h}{\frac{1}{2}h} = \cos x.$$

Similarly, $\frac{d}{dx}(\cos x) = -\sin x$, or this may be deduced from $\cos x = \sin\left(\frac{\pi}{2}-x\right)$. Also, results like

$$\frac{d}{dx}(\tan x) = \sec^2 x, \quad \frac{d}{dx}\tan^{-1} x = \frac{1}{1+x^2}$$

may be derived by the usual processes of the Calculus.

If $f(x)$ is a one-valued integrable function of x which is positive for $0<x<a$, then the function $f_1(x)$, defined by

$$f_1(x) \equiv \int_0^x f(t)\,dt,$$

where $0<x<a$, is also necessarily positive. Similarly, if $f_2(x)$ is defined by

$$f_2(x) \equiv \int_0^x f_1(t)\,dt,$$

this new function is also positive, and by continuing the process we get a series of functions all positive in the range $0 < x < a$.

Now take $f(x) \equiv x - \sin x$, and suppose that x is positive; then $f(x)$ is one-valued and positive, and therefore

$$f_1(x) \equiv \int_0^x (t - \sin t)\,dt = \frac{x^2}{2!} + \cos x - 1,$$

$$f_2(x) \equiv \int_0^x f_1(t)\,dt = \frac{x^3}{3!} + \sin x - x,$$

$$f_3(x) \equiv \int_0^x f_2(t)\,dt = \frac{x^4}{4!} - \cos x + 1 - \frac{x^2}{2!},$$

$$f_4(x) \equiv \int_0^x f_3(t)\,dt = \frac{x^5}{5!} - \sin x + x - \frac{x^3}{3!},$$

$$\vdots \qquad \vdots \qquad \vdots$$

are all positive. Thus, *if x is positive* and p is any positive integer,

$$\sin x > x - \frac{x^3}{3!} + \frac{x^5}{5!} - \dots - \frac{x^{4p-1}}{(4p-1)!} \equiv s_{2p}, \text{ say;}$$

and $$\sin x < x - \frac{x^3}{3!} + \frac{x^5}{5!} - \dots - \frac{x^{4p-1}}{(4p-1)!} + \frac{x^{4p+1}}{(4p+1)!} \equiv s_{2p+1}, \text{ say.}$$

These inequalities may be written: $s_{2p} < \sin x < s_{2p+1}$;

$$\therefore\ s_{2p+1} - \sin x < s_{2p+1} - s_{2p} = \frac{x^{4p+1}}{(4p+1)!}.$$

But by limit (ii), p. 78, $\dfrac{x^{4p+1}}{(4p+1)!} \to 0$, when $p \to \infty$.

Also $s_{2p+1} - \sin x$ is positive; $\therefore\ s_{2p+1} \to \sin x$ when $p \to \infty$.

Similarly, since

$$0 < \sin x - s_{2p} < s_{2p+1} - s_{2p}, \quad s_{2p} \to \sin x \text{ when } p \to \infty;$$

$\therefore$ $\sin x$ is the sum to infinity of the series, $\dfrac{x}{1!} - \dfrac{x^3}{3!} + \dfrac{x^5}{5!} - \dots$.

When x is changed into $-x$, every term of the series changes sign, and so does $\sin x$. Therefore the result holds also when x is negative; it is obviously true also if $x = 0$. We have therefore

$$\mathbf{\sin x = \frac{x}{1!} - \frac{x^3}{3!} + \frac{x^5}{5!} - \frac{x^7}{7!} + \dots} \quad \text{.....................(5)}$$

for all values of x.

Similarly, from the relations on p. 80,

$$\cos x > 1 - \frac{x^2}{2!} + \frac{x^4}{4!} \dots - \frac{x^{4p-2}}{(4p-2)!} \equiv c_{2p}, \text{ say,}$$

and $$\cos x < 1 - \frac{x^2}{2!} + \frac{x^4}{4!} \dots - \frac{x^{4p-2}}{(4p-2)!} + \frac{x^{4p}}{(4p)!} \equiv c_{2p+1}, \text{ say.}$$

Hence, $c_{2p} < \cos x < c_{2p+1}$, where $c_{2p+1} - c_{2p} = \dfrac{x^{4p}}{(4p)!}$.

Therefore, by the same argument as before, it follows that

$$c_{2p+1} \to \cos x \quad \text{and} \quad c_{2p} \to \cos x \quad \text{when} \quad p \to \infty.$$

We have therefore $$\mathbf{\cos x = 1 - \frac{x^2}{2!} + \frac{x^4}{4!} - \frac{x^6}{6!} + \dots\dots\dots\dots} \quad (6)$$

for all values of x.

Note. Attention should be called to a crucial point in the argument used in these proofs. The fact that $\lim_{p\to\infty} (s_{2p+1} - s_{2p}) = 0$ shows that *if* either s_{2p+1} or s_{2p} tends to a limit, the other must tend to the same limit; but it does not ensure that either of them actually tends to a limit. It is essential to prove that the limit exists. This is done by the inequality,

$$s_{2p} < \sin x < s_{2p+1},$$

which shows that $0 < (s_{2p+1} - \sin x) < (s_{2p+1} - s_{2p})$, and therefore

$$\lim_{p\to\infty} (s_{2p+1} - \sin x) = 0.$$

Hence $s_{2p+1} \to \sin x$, when $p \to \infty$. It *then* follows that $s_{2p} \to \sin x$, or this can be proved in the same way. *Both* these results are needed to show that the sum to n terms tends to sin x when $n \to \infty$.

Example 1. Calculate sin 36° to 4 significant figures.

$$\sin 36° = \sin\frac{\pi}{5} = \frac{\pi}{5} - \frac{\pi^3}{5^3 . 3!} + \frac{\pi^5}{5^5 . 5!} - \dots$$

$$= 0{\cdot}62832 - 0{\cdot}04134 + 0{\cdot}00082 - \dots;$$

Also, with the notation of p. 80,

$$s_3 - \sin x < s_3 - s_4 = \frac{\pi^7}{5^7 . 7!} < 10^{-5};$$

$$\therefore \sin 36° = 0{\cdot}5878 \text{ to 4 significant figures.}$$

Example 2. Find the first three terms of the expansion of $\tan x$ in powers of x.

Since $\tan(-x) = -\tan x$, $\tan x$ is an *odd* function of x. If then we *assume* that $\tan x$ can be expanded in powers of x, the expansion must be of the form $\tan x = Ax + Bx^3 + Cx^5 + \dots$.

$$\text{Then} \quad (Ax + Bx^3 + Cx^5 + \dots)\left(1 - \frac{x^2}{2!} + \frac{x^4}{4!} - \dots\right) = \tan x \,.\, \cos x$$

$$= \sin x = \frac{x}{1!} - \frac{x^3}{3!} + \frac{x^5}{5!} - \dots.$$

Equating coefficients: $A = 1$; $B - \frac{1}{2}A = -\frac{1}{6}$; $C - \frac{1}{2}B + \frac{1}{24}A = \frac{1}{120}$;

$$\therefore \; A = 1, \; B = \tfrac{1}{3}, \; C = \tfrac{2}{15}; \quad \therefore \; \tan x = x + \tfrac{1}{3}x^3 + \tfrac{2}{15}x^5 + \dots.$$

It should be noted that this process does not prove that $\tan x$ can be expanded as a convergent power series in x. This is, however, true, and, for small values of x, x, $x + \frac{1}{3}x^3$, $x + \frac{1}{3}x^3 + \frac{2}{15}x^5$, are successive approximations to $\tan x$.

Example 3. Show how to expand $\cos^2 x$ and $\sin^3 x$ in powers of x.

Use the formulae: $2\cos^2 x = 1 + \cos 2x$, $4\sin^3 x = 3\sin x - \sin 3x$.

Example 4. Solve $\cos\theta = \theta$, approximately.

Inspection of a rough graph shows that there is only one root and that its value is approximately $0{\cdot}7$. For a value of θ of this size, we have $\cos\theta \simeq 1 - \frac{1}{2}\theta^2$; $\therefore \; 1 - \frac{1}{2}\theta^2 \simeq \theta$;

$$\therefore \; \theta^2 + 2\theta \simeq 2; \quad (\theta + 1)^2 \simeq 3; \quad \theta \simeq \pm\sqrt{3} - 1; \quad \theta \simeq +0{\cdot}7.$$

Put $\theta = 0{\cdot}7 + \alpha$, then $\cos(0{\cdot}7 + \alpha) = 0{\cdot}7 + \alpha$;

$$\therefore \; \cos(0{\cdot}7)\cos\alpha - \sin(0{\cdot}7)\sin\alpha = 0{\cdot}7 + \alpha, \text{ where } \alpha \text{ is small};$$

$$\therefore \; \cos(0{\cdot}7) - \alpha \,.\, \sin(0{\cdot}7) = 0{\cdot}7 + \alpha, \text{ approximately};$$

$$\therefore \; \alpha \simeq \frac{0{\cdot}765 - 0{\cdot}7}{1 + 0{\cdot}644} \simeq 0{\cdot}04; \quad \therefore \; \theta \simeq 0{\cdot}74.$$

A closer approximation, $\theta \simeq 0{\cdot}739$, could be found by putting $\theta = 0{\cdot}74 + \beta$, and repeating the process just used.

EXERCISE V. a.

Find the sums to infinity of the series in Nos. **1-5.**

1. $\dfrac{1}{1!} - \dfrac{1}{3!} + \dfrac{1}{5!} - \dots$.

2. $1 - \dfrac{1}{2!} + \dfrac{1}{4!} - \dots$.

3. $1-\frac{2}{1!}+\frac{2^3}{3!}-\frac{2^5}{5!}+\ldots.$ **4.** $\frac{2}{3!}-\frac{4}{5!}+\frac{6}{7!}-\ldots.$

5. $\frac{\pi^2}{2.4}-\frac{\pi^4}{2.4.6.8}+\frac{\pi^6}{2.4.6.8.10.12}-\ldots.$

6. Show that the positive square root of the sum of

$$1+\frac{2}{1!}-\frac{2^3}{3!}+\frac{2^5}{5!}-\ldots \text{ is the sum of } 2-\frac{4}{3!}+\frac{6}{5!}-\frac{8}{7!}+\ldots.$$

7. Calculate from the series the cosine of 1 radian, correct to 3 significant figures.

8. Calculate from the series the sine of 3°, correct to 3 significant figures.

9. Prove that $\tan x-\sin x \simeq \frac{1}{2}x^3$, if x is small.

10. Prove that $\sin^2\theta \simeq \theta^2\left(1-\frac{\theta^2}{3}+\frac{2\theta^4}{45}\right)$, if θ is small.

11. Express $x \operatorname{cosec} x$ in powers of x, neglecting x^6 and higher powers.

12. Express $\sin\left(\frac{\pi}{4}+x\right)\cos x$ as a power series in x and give the general term. Also express it as a power series in $\frac{\pi}{4}+x$.

13. Find the general term in the expansion of $\cos^3 x$ in powers of x.

14. Show that $\frac{3\sin\theta}{2+\cos\theta}$ differs from θ by about $\frac{\theta^5}{180}$ when θ is small.

15. Find whether $\tan x-24\tan\frac{x}{2}$ or $4\sin x-15x$ is the greater when x is small and positive.

16. Prove that $\lim_{x\to 0}\left(\frac{1}{x^2}-\cot^2 x\right)=\frac{2}{3}$.

17. If α is small, prove that $\frac{\sin\theta+\sin\alpha}{\sin(\theta+\alpha)} \simeq 1+\alpha\tan\frac{\theta}{2}$.

18. Find an approximate solution of $\cos\theta=2\theta$.

19. Find an approximate solution near to $\frac{3\pi}{2}$ of $\tan\theta=\theta$.

20. If $\tan(\theta-\phi)=(1+\lambda)\tan\phi$ and λ is small, prove that one value of $\tan\phi$ is approximately $(1-\frac{1}{2}\lambda)\tan\frac{\theta}{2}$.

21. Prove that, for $0<x<\pi$,

(i) $2(1-\cos x)>x\sin x$; (ii) $x(2+\cos x)>3\sin x$.

22. Prove that $\dfrac{\tan x}{x} > \dfrac{x}{\sin x}$, when $0 < x < \dfrac{\pi}{2}$.

23. From $\sin\theta = \theta - \dfrac{\theta^3}{3!} + \dfrac{\theta^5}{5!} - \ldots$, obtain the successive approximations, $\theta \simeq \sin\theta$, $\theta \simeq \sin\theta + \frac{1}{6}\sin^3\theta$, $\theta \simeq \sin\theta + \frac{1}{6}\sin^3\theta + \frac{3}{40}\sin^5\theta$, θ being small.

24. If $nt = \phi - \epsilon\sin\phi$, and ϵ^3 is negligible, prove that

$$\phi = nt + \epsilon\sin nt + \tfrac{1}{2}\epsilon^2 \sin 2nt.$$

25. By the method of p. 80, show that, if x is positive, $e^x - 1$, $e^x - 1 - x$, $e^x - 1 - x - \dfrac{x^2}{2!}$, ..., $e^x - 1 - x - \dfrac{x^2}{2!} - \ldots - \dfrac{x^n}{n!}$ are all positive.

26. By the method of p. 80, show that, if x is positive,

$$e^{-x} = 1 - \frac{x}{1!} + \frac{x^2}{2!} - \frac{x^3}{3!} + \ldots.$$

The Logarithmic Series. In equation (1), p. 77, a_0 is the value of $\phi(x)$ for $x = 0$, and $a_0 + a_1 x$ is its approximate value for a small, positive or negative, value of x. Thus the fact that $\log x$ is meaningless when $x \leqslant 0$ suggests that it cannot be expanded as a power series in x. But the function $\log(1+x)$ is capable of expansion for a certain range of values of x.

Using the sum of a G.P. given in equation (2) we have

$$\frac{1}{1+x} = 1 - x + x^2 - \ldots + (-1)^{n-1}x^{n-1} + \frac{(-x)^n}{1+x}.$$

We shall suppose that y is a positive number; then

$$\log(1+y) \equiv \int_1^{1+y} \frac{1}{t}\,dt = \int_0^y \frac{1}{1+x}\,dx, \text{ by putting } t = 1+x,$$

$$= \int_0^y \left\{1 - x + x^2 - \ldots + (-1)^{n-1}x^{n-1} + \frac{(-x)^n}{1+x}\right\} dx$$

$$= y - \frac{y^2}{2} + \frac{y^3}{3} - \ldots \text{ to } n \text{ terms} + (-1)^n \int_0^y \frac{x^n}{1+x}\,dx. \quad \ldots(7)$$

Also $$\mathsf{K} \equiv \int_0^y \frac{x^n}{1+x}\,dx < \int_0^y x^n\,dx = \frac{y^{n+1}}{n+1};$$

$$\therefore \text{ if } y \leqslant 1, \quad \mathsf{K} < \frac{1}{n+1} \text{ and } \to 0 \text{ when } n \to \infty.$$

$\therefore$ from (7),

$$\log(1+y) = \lim_{n\to\infty}\left\{y - \frac{y^2}{2} + \frac{y^3}{3} - \ldots \text{ to } n \text{ terms}\right\}, \text{ if } 0 < y \leqslant 1. \quad (8)$$

Again, if $0<y<1$,

$$-\log(1-y) = -\int_1^{1-y}\frac{1}{t}\,dt = \int_0^y \frac{1}{1-x}\,dx, \text{ by putting } t=1-x,$$

$$= \int_0^y \left\{1+x+x^2+\ldots+x^{n-1}+\frac{x^n}{1-x}\right\}dx$$

$$= y+\frac{y^2}{2}+\frac{y^3}{3}+\ldots+\frac{y^n}{n}+\int_0^y \frac{x^n}{1-x}\,dx. \quad \ldots\ldots\ldots\ldots\ldots\ldots(9)$$

Also
$$\mathsf{H} \equiv \int_0^y \frac{x^n}{1-x}\,dx < \int_0^y \frac{x^n}{1-y}\,dx;$$

$\therefore$ since $0<y<1$, $\mathsf{H} < \dfrac{1}{1-y}\cdot\dfrac{y^{n+1}}{n+1} < \dfrac{1}{(1-y)(n+1)}$

and $\mathsf{H}\to 0$ when $n\to\infty$;

$\therefore$ from (9),

$$-\log(1-y) = \lim_{n\to\infty}\left\{y+\frac{y^2}{2}+\frac{y^3}{3}+\ldots \text{ to } n \text{ terms}\right\}, \text{ if } 0<y<1. \quad (10)$$

The results of (8) and (10) may be combined into the single statement that $\log(1+x)$ is the sum to infinity of the series $x-\dfrac{x^2}{2}+\dfrac{x^3}{3}-\ldots$ provided that $0<x\leqslant 1$ or $-1<x<0$. Also the result is true for $x=0$. We therefore write

$$\log(1+\mathbf{x}) = \mathbf{x}-\frac{\mathbf{x}^2}{2}+\frac{\mathbf{x}^3}{3}-\frac{\mathbf{x}^4}{4}+\ldots \text{ if } -1<\mathbf{x}\leqslant 1. \quad \ldots\ldots\ldots(11)$$

Note. Care must be taken about the insertion in (11) of such a value as $-1+\dfrac{1}{n}$ for x. This gives a true result if n is positive. If, however, it were now proposed to make $n\to\infty$, it could not be assumed that either side had a limit, or that if the limits existed they must be equal. Actually in this case the limits do not exist.

The proof above that $\mathsf{H}\to 0$ definitely requires $y<1$, not merely $y\leqslant 1$.

From
$$\log(1+x) = x-\frac{x^2}{2}+\frac{x^3}{3}-\frac{x^4}{4}+\ldots \qquad (-1<x\leqslant 1)$$

and
$$\log(1-x) = -x-\frac{x^2}{2}-\frac{x^3}{3}-\frac{x^4}{4}-\ldots \qquad (-1\leqslant x<1)$$

by subtracting and dividing by 2, we have

$$\tfrac{1}{2}\log\frac{1+\mathbf{x}}{1-\mathbf{x}} = \mathbf{x}+\frac{\mathbf{x}^3}{3}+\frac{\mathbf{x}^5}{5}+\frac{\mathbf{x}^7}{7}+\ldots \quad (-1<\mathbf{x}<1). \quad \ldots\ldots(12)$$

An alternative form of this result is obtained by putting

$$\frac{1+x}{1-x}=y, \quad \text{then} \quad x=\frac{y-1}{y+1};$$

$$\therefore \ \frac{1}{2}\log y=\left(\frac{y-1}{y+1}\right)+\frac{1}{3}\left(\frac{y-1}{y+1}\right)^3+\frac{1}{5}\left(\frac{y-1}{y+1}\right)^5+\dots \ (y>0). \ \dots(13)$$

Equations (12) and (13) may be used for the numerical computation of logarithms; convenient methods of proceeding are indicated in Ex. V. b, No. 4 and Ex. V. e, Nos. 17, 18.

Example 5. Find the sum to infinity of

$$\frac{2}{1.2.3}+\frac{3}{3.4.5}+\frac{4}{5.6.7}+\dots.$$

The nth term is $\dfrac{n+1}{(2n-1)2n(2n+1)}$, and may be expressed in Partial Fractions (see p. 231) in the form

$$\frac{\frac{3}{4}}{2n-1}-\frac{1}{2n}+\frac{\frac{1}{4}}{2n+1};$$

$\therefore$ the sum to n terms is

$$\frac{3}{4}\left(1+\frac{1}{3}+\frac{1}{5}+\dots+\frac{1}{2n-1}\right)-\left(\frac{1}{2}+\frac{1}{4}+\dots+\frac{1}{2n}\right)+\frac{1}{4}\left(\frac{1}{3}+\frac{1}{5}+\dots+\frac{1}{2n+1}\right)$$

$$=\left(1-\frac{1}{2}+\frac{1}{3}-\frac{1}{4}+\dots-\frac{1}{2n}\right)-\frac{1}{4}\left(1-\frac{1}{2n+1}\right).$$

When $n\to\infty$ the limits of the two brackets are $\log 2$ and 1;

$\therefore$ the sum to infinity is $\log 2-\frac{1}{4}$.

EXERCISE V. b.

1. Write down the sums to infinity of the series

(i) $1-\frac{1}{2}+\frac{1}{3}-\frac{1}{4}+\dots$; (ii) $\frac{1}{2}-\frac{1}{2^2.2}+\frac{1}{2^3.3}-\frac{1}{2^4.4}+\dots$;

(iii) $\frac{1}{3}+\frac{1}{3^3.3}+\frac{1}{3^5.5}+\frac{1}{3^7.7}+\dots$.

2. Prove the following results, finding the conditions under which they hold:

(i) $\log(x+a)=\log a+\dfrac{x}{a}-\dfrac{x^2}{2a^2}+\dfrac{x^3}{3a^3}-\dots$;

(ii) $\log x-\log y=\dfrac{x-y}{x}+\dfrac{(x-y)^2}{2x^2}+\dfrac{(x-y)^3}{3x^3}+\dots$;

(iii) $\dfrac{1}{x^2}+\dfrac{1}{2x^4}+\dfrac{1}{3x^6}+\dots=\dfrac{2}{2x^2-1}+\dfrac{2}{3(2x^2-1)^3}+\dots$.

3. Expand the following functions as power series in x, giving the coefficients of x^n and the conditions of validity :

(i) $\log\left(1-\frac{x}{2}\right)$; (ii) $\log\{(1-x)(1+3x)\}$;

(iii) $\log(1+5x+6x^2)$; (iv) $\log(x^2+2x+1)$;

(v) $\log(x+2)$; (vi) $\log(x^2+3x+2)$;

(vii) $\log\frac{1-2x^2}{1-x}$; (viii) $\log(1+x+x^2)$.

4. (i) Use series (13) to calculate log 2 to 4 places of decimals;

(ii) Use series (13) to calculate $\log\frac{3}{2}$, $\log\frac{5}{4}$, and $\log\frac{7}{6}$, each to 4 places of decimals ;

(iii) Use the results of (i) and (ii) to obtain the logarithms of 3, 4, 5, 6, 7, 8, 9 and 10 ;

(iv) Prove that $\log_{10} \mathrm{N} = \log_e \mathrm{N} \div \log_e 10$ and use the results of (iii) to deduce the corresponding logarithms to base 10.

5. What is the coefficient of x^n in the expansions of the following functions as power series in x and for what values of x are the expansions valid ?

(i) $(1-2x)\log(1-2x)$; (ii) $(1+3x)^2\log(1+3x)$;

(iii) $(1-x)\log\left(1+\frac{x^2}{4}\right)$.

6. Given that $|x| < 1$, find the sums to infinity of the series whose nth terms are

(i) $\frac{nx^{n+1}}{n+1}$; (ii) $\frac{x^{2n}}{2n(2n-1)}$; (iii) $\frac{(-x)^n}{n(n+1)}$.

7. Express $\log\frac{x+1}{x-1}$ in powers of $\frac{1}{x}$ when $|x| > 1$.

8. Express $2\log n - \log(n+1) - \log(n-1)$ as a power series in $\frac{1}{n}$ when $n > 1$.

9. Express $\log(x+2) - 2\log(x+1) + 2\log(x-1) - \log(x-2)$ as a series of powers of $\frac{2}{x^3-3x}$, and find for what values of x the expansion is valid.

Sum to infinity the following series :

10. $\frac{1}{1.2}+\frac{1}{2.3}+\frac{1}{3.4}+\ldots$. **11.** $\frac{1}{1.2}-\frac{1}{2.3}+\frac{1}{3.4}-\ldots$.

12. $\frac{1}{1.2.3}-\frac{1}{2.3.4}+\frac{1}{3.4.5}-\ldots$.

13. $\frac{1}{1.2.3}+\frac{1}{3.4.5}+\frac{1}{5.6.7}+\dots$.   14. $\frac{1}{1.3}+\frac{1}{2.5}+\frac{1}{3.7}+\dots$

15. $\frac{1}{1.2.3}+\frac{1}{2.3.5}+\frac{1}{3.4.7}+\dots$.

16. Prove that $\log|\cot\frac{1}{2}A|=\cos A+\frac{1}{3}\cos^3 A+\frac{1}{5}\cos^5 A+\dots$ unless $A\neq n\pi$; deduce that the sum to infinity of $1+\frac{1}{3.2^2}+\frac{1}{5.2^4}+\dots$ is $\log 3$.

17. Evaluate $\lim_{x\to 0}\frac{(2+x)\log(1+x)+(2-x)\log(1-x)}{x^4}$.

18. Evaluate $\lim_{x\to 1}\frac{1-x+\log x}{1-\sqrt{(2x-x^2)}}$.

19. If $x\log x+x-1=\epsilon$, which is small, prove that $x\simeq 1+\frac{1}{2}\epsilon-\frac{1}{16}\epsilon^2$.

20. Find approximate solutions of the equation $5\log x=2{\cdot}7-\frac{2}{x}$.

Gregory's Expansion of $\tan^{-1}x$. From the sum of a G.P., equation (2), p. 77, we have

$$\frac{1}{1+x^2}=1-x^2+x^4-\dots+(-1)^{n-1}x^{2n-2}+(-1)^n\frac{x^{2n}}{1+x^2};$$

$$\therefore \int_0^y\frac{1}{1+x^2}dx=y-\frac{y^3}{3}+\frac{y^5}{5}-\dots+(-1)^{n-1}\frac{y^{2n-1}}{2n-1}+(-1)^n.K,$$

where $K\equiv\int_0^y\frac{x^{2n}}{1+x^2}dx.$

If $-1\leqslant y\leqslant +1$, the *numerical* value of $K<\int_0^1 x^{2n}dx=\frac{1}{2n+1}$;

$$\therefore K\to 0,\quad\text{when}\quad n\to\infty;$$

$\therefore$ the sum to infinity of $y-\frac{y^3}{3}+\frac{y^5}{5}-\dots$ is $\int_0^y\frac{1}{1+x^2}dx$.

But $\int_0^y\frac{1}{1+x^2}dx$ is the value of $\tan^{-1}y$ between $-\frac{\pi}{2}$ and $+\frac{\pi}{2}$;

$$\therefore \tan^{-1}y=y-\frac{y^3}{3}+\frac{y^5}{5}-\dots,\ \dots\dots(14)$$

provided that $-1\leqslant y\leqslant 1$, and that the value taken for $\tan^{-1}y$ lies in the range from $-\frac{\pi}{4}$ to $+\frac{\pi}{4}$, inclusive.

Evaluation of π. By using (14), we can obtain π as the sum to infinity of a series. Putting $y=1$, we have

$$\tfrac{1}{4}\pi = 1 - \tfrac{1}{3} + \tfrac{1}{5} - \tfrac{1}{7} + \dots \quad \text{.......................(15)}$$

This series converges so slowly that for practical calculation it is necessary to employ alternative series (Ex. V. c, Nos. 3, 4).

The reader should verify the following results:

(i) *Machin's formula,* $4\tan^{-1}\dfrac{1}{5} - \tan^{-1}\dfrac{1}{239} = \dfrac{\pi}{4}$.

(ii) *Rutherford's formula,* $4\tan^{-1}\dfrac{1}{5} - \tan^{-1}\dfrac{1}{70} + \tan^{-1}\dfrac{1}{99} = \dfrac{\pi}{4}$.

These give π as the sum to infinity of rapidly convergent series.

EXERCISE V. c.

1. Find the sums to infinity of

$$\text{(i) } 1 - \frac{1}{3\,.\,2^2} + \frac{1}{5\,.\,2^4} - \dots;$$

$$\text{(ii) } (1 - 3^{-\frac{1}{2}}) - \tfrac{1}{3}(1 - 3^{-\frac{3}{2}}) + \tfrac{1}{5}(1 - 3^{-\frac{5}{2}}) - \dots.$$

2. Give the sum to infinity of $\tan x - \frac{1}{3}\tan^3 x + \frac{1}{5}\tan^5 x - \dots$ when

(i) $\dfrac{3\pi}{4} < x < \dfrac{5\pi}{4}$; (ii) $\dfrac{7\pi}{4} < x < \dfrac{9\pi}{4}$; (iii) $n\pi - \dfrac{\pi}{4} < x < n\pi + \dfrac{\pi}{4}$.

3. Calculate π to five places of decimals by Machin's or Rutherford's formula.

4. Calculate π to four places of decimals by the formula,

$$\frac{\pi}{4} = 2\tan^{-1}\tfrac{1}{3} + \tan^{-1}\tfrac{1}{7}.$$

5. Simplify $\tan^{-1}\frac{1}{2} + \tan^{-1}\frac{1}{3}$ and use the result to express π as the sum to infinity of a series.

6. Find the sum to infinity of

$$\left(\frac{2}{3} + \frac{1}{7}\right) - \frac{1}{3}\left(\frac{2}{3^3} + \frac{1}{7^3}\right) + \frac{1}{5}\left(\frac{2}{3^5} + \frac{1}{7^5}\right) - \dots.$$

7. Find, when possible, the sum to infinity of

$$x + \tfrac{1}{5}x^5 + \tfrac{1}{9}x^9 + \tfrac{1}{13}x^{13} + \dots.$$

8. Expand, when possible, $\tan^{-1}\left(\dfrac{\cos\theta + \sin\theta}{\cos\theta - \sin\theta}\right)$ as a power series in $\tan\theta$.

9. If $y = x - \frac{x^3}{3} + \frac{x^5}{5}$, and x is so small that x^7 is negligible, obtain the successive approximations $x \simeq y$, $x \simeq y + \frac{y^3}{3}$, $x \simeq y + \frac{y^3}{3} + \frac{2y^5}{15}$. Interpret this with $x = \tan\theta$.

10. If ϵ is small, prove that one root of $\tan^{-1}x = \epsilon - x$ is given by $x \simeq \frac{1}{2}\epsilon + \frac{1}{48}\epsilon^3$, and find the next approximation.

The Exponential Series.

If we assume that the function exp (x) or e^x can be expanded in the form

$$\exp(x) = e^x = a_0 + a_1 x + a_2 x^2 + \ldots + a_r x^r + \ldots$$

and if we also assume that

$$\frac{d}{dx}(e^x) = \frac{d}{dx}(a_0) + \frac{d}{dx}(a_1 x) + \ldots + \frac{d}{dx}(a_r x^r) + \ldots$$

and that we may continue to differentiate in this way, it is easy to find the values of $a_0, a_1, a_2, \ldots$.

Putting $x = 0$ in the first equation, we have $1 = a_0$.

The second equation is

$$e^x = a_1 + 2a_2 x + 3a_3 x^2 + 4a_4 x^3 + \ldots.$$

The equations obtained by continuing the process are

$$e^x = 1 \,.\, 2a_2 + 2 \,.\, 3a_3 x + 3 \,.\, 4a_4 x^2 + \ldots,$$

$$e^x = 1 \,.\, 2 \,.\, 3a_3 + 2 \,.\, 3 \,.\, 4a_4 x + \ldots$$

$$\ldots\ldots\ldots\ldots\ldots\ldots\ldots\ldots\ldots\ldots\ldots\ldots$$

Putting $x = 0$ in these, we have

$$1 = a_1, \quad 1 = 1 \,.\, 2a_2, \quad 1 = 1 \,.\, 2 \,.\, 3a_3, \ldots;$$

$$\therefore\ a_0 = 1, \quad a_1 = 1, \quad a_2 = \frac{1}{2!}, \quad a_3 = \frac{1}{3!}, \text{ etc.}$$

Therefore the expansion is

$$\exp(\mathbf{x}) = \mathbf{e^x} = \mathbf{1} + \frac{\mathbf{x}}{1!} + \frac{\mathbf{x}^2}{2!} + \frac{\mathbf{x}^3}{3!} + \ldots + \frac{\mathbf{x}^r}{r!} + \ldots. \qquad \ldots\ldots\ldots\ldots(16)$$

But the assumptions stated above are not easy to justify. A valid process which sometimes replaces this method is based on Maclaurin's Theorem. We shall now, however, proceed to obtain the result by a different method, based on integration by parts.

To prove that the series $1 + \frac{x}{1!} + \frac{x^2}{2!} + \ldots + \frac{x^r}{r!} + \ldots$ **is convergent for all values of x** *and that its sum to infinity is* e^x.

Put $u_n \equiv \int_0^x e^{-t} t^n \, dt$, n being a positive integer.

If $n > 1$, $u_n = \Big[-e^{-t} . t^n \Big]_0^x + \int_0^x e^{-t} . n t^{n-1} dt = -e^{-x} . x^n + n . u_{n-1}$;

$$\therefore \frac{x^n}{n\,!} = e^x \left\{ \frac{u_{n-1}}{(n-1)!} - \frac{u_n}{n!} \right\}.$$

If $n = 1$, $u_1 = \Big[-e^{-t} . t \Big]_0^x + \int_0^x e^{-t} dt = -e^{-x} . x + 1 - e^{-x}$;

$$\therefore 1 + x = e^x \left\{ 1 - \frac{u_1}{1!} \right\};$$

$\therefore$ by adding the results for $n = 1, 2, 3, \ldots, m$,

$$1 + \frac{x}{1!} + \frac{x^2}{2!} + \ldots + \frac{x^m}{m!} = e^x \left\{ 1 - \frac{u_m}{m!} \right\}. \quad \ldots\ldots\ldots\ldots\ldots(17)$$

We shall now prove that $\lim_{m \to \infty} \frac{u_m}{m!} = 0$.

Consider first $x > 0$; then for $0 < t < x$, $e^{-t} = \frac{1}{e^t} < 1$;

$$\therefore 0 < \frac{u_m}{m!} \equiv \int_0^x e^{-t} \frac{t^m}{m\,!} dt < \int_0^x \frac{t^m}{m!} dt = \frac{x^{m+1}}{(m+1)!};$$

$\therefore$ by limit (ii) on p. 78, $\frac{u_m}{m!} \to 0$ when $m \to \infty$.

Next suppose $x < 0$ and put $x = -y$ so that $y > 0$.

$$\frac{u_m}{m!} = \int_0^{-y} e^{-t} \frac{t^m}{m!} dt = (-1)^{m+1} \int_0^y e^s \frac{s^m}{m!} ds, \text{ putting } s = -t.$$

But for $0 < s < y$, $e^s < e^y$;

$$\therefore 0 < \int_0^y e^s . \frac{s^m}{m!} ds < e^y \int_0^y \frac{s^m}{m!} ds = e^y \frac{y^{m+1}}{(m+1)!};$$

$\therefore$ by limit (ii) on p. 78, $\int_0^y e^s . \frac{s^m}{m!} ds \to 0$ when $m \to \infty$;

$$\therefore \frac{u_m}{m!} \to 0 \text{ when } m \to \infty.$$

The required result therefore follows from equation (17).

Calculation of e. Putting $x = 1$ in (16) we have

$$e = 1 + \frac{1}{1!} + \frac{1}{2!} + \frac{1}{3!} + \ldots;$$

$\therefore$ e is greater than the sum, s_n, to n terms of this series; but

$$s_{n+p}-s_n=\frac{1}{n!}+\frac{1}{(n+1)!}+\ldots+\frac{1}{(n+p-1)!}$$

$$<\frac{1}{n!}\left(1+\frac{1}{n}+\frac{1}{n^2}+\ldots+\frac{1}{n^{p-1}}\right)<\frac{1}{n!}\,\frac{1}{1-\frac{1}{n}}=\frac{1}{(n-1)\,.\,(n-1)!}$$

and as this is true for all positive values of p, it follows that

$$e-s_n\leqslant\frac{1}{(n-1)\,.\,(n-1)!}.$$

For example, taking $n=10$, we get

$$1+\frac{1}{1!}+\frac{1}{2!}+\ldots+\frac{1}{9!}<e\leqslant\left(1+\frac{1}{1!}+\frac{1}{2!}+\ldots+\frac{1}{9!}\right)+\frac{1}{9\,.\,9!},$$

and this is found to give the value of e to 6 places of decimals.

$(e=2{\cdot}7182818\ldots.)$

Note. If a function $f(p)$, which $\rightarrow$ a limit l when $p\rightarrow\infty$, satisfies the inequality $f(p)<\mathsf{K}$ for all values of p, K being independent of p, the correct conclusion is not $l<\mathsf{K}$ but $l\leqslant\mathsf{K}$. Thus in the above work the conclusion $e-s_n<\dfrac{1}{(n-1)\,.\,(n-1)!}$ would not be justified; it can however be proved thus:

$$s_{n+p}-s_n<\frac{1}{n!}+\frac{1}{(n+1)!}+\frac{1}{n!}\left(\frac{1}{n^2}+\frac{1}{n^3}+\ldots+\frac{1}{n^{p-1}}\right)$$

$$=\frac{1}{n!}\left(1+\frac{1}{n}+\frac{1}{n^2}+\ldots+\frac{1}{n^{p-1}}\right)-\frac{1}{n\,.\,n!}+\frac{1}{(n+1)!}$$

$$<\frac{1}{(n-1)\,.\,(n-1)!}-\frac{1}{n\,.\,n!}+\frac{1}{(n+1)!},\text{ as above,}$$

$$\therefore\ e-s_n\leqslant\frac{1}{(n-1)\,.\,(n-1)!}-\frac{1}{n\,.\,(n+1)!}<\frac{1}{(n-1)\,.\,(n-1)!}.$$

Nature of e. It is easy to see that e is *not rational.* For if $e=\dfrac{p}{q}$, where p, q are integers, $s_{q+1}<e=\dfrac{p}{q}<s_{q+1}+\dfrac{1}{q\cdot q!}$ and multiplication by $q!$ gives $\mathsf{K}<p\,.\,(q-1)!<\mathsf{K}+\dfrac{1}{q}$, where K is an integer, but $p\,.\,(q-1)!$ is also an integer, so the inequalites cannot be true.

The Compound Interest Law. It was proved on p. 64 that $\frac{de^x}{dx}=e^x$. We shall now show that every function y which has the property $\frac{dy}{dx}=y$ is of the form $\mathsf{A}e^x$ where A is constant.

If $$\frac{dy}{dx}=y, \quad \frac{dx}{dy}=\frac{1}{y}; \quad \therefore\ x=\int\frac{1}{y}\,dy=\log y+\mathsf{C},$$

and, if we put $\mathsf{C}=-\log\mathsf{A}$, we have

$$x=\log y-\log\mathsf{A}=\log\frac{y}{\mathsf{A}}; \quad \therefore\ y=\mathsf{A}e^x.$$

The equation $\frac{dy}{dx}=y$ means that y is a function whose rate of increase with respect to x is y. This is the rate that occurs if money is lent at Compound Interest at 100 per cent. per unit time, *the interest being added continuously*. Thus if £A is lent under these conditions, and the unit of time is a year, the amount after x years is £$(\mathsf{A}e^x)$.

If the interest is compounded at intervals of $\frac{1}{k}$th of a year, the amount after kx periods for each of which the interest is $\frac{100}{k}$ per cent. would be £$\left\{\mathsf{A}\left(1+\frac{1}{k}\right)^{kx}\right\}$. For continuous addition of interest we make $k\to\infty$; we may therefore expect that

$$\lim_{k\to\infty}\left(1+\frac{1}{k}\right)^{kx}=e^x.$$

Writing $\frac{n}{y}$ for k, and successively x, $-x$, for y, we have

$$\lim_{n\to\infty}\left(1+\frac{x}{n}\right)^{n}=e^{x} \qquad \text{.........................(18)}$$

and $$\lim_{n\to\infty}\left(1-\frac{x}{n}\right)^{n}=e^{-x}. \qquad \text{.........................(19)}$$

Formal proofs of these limits will be given in the companion volume on Analysis; another method of proof is indicated in Ex. IV, g, Nos. 15, 16, 19.

Example 6. Find (in terms of e) the sum to infinity of the series

$$\frac{1}{1!}+\frac{4}{2!}+\frac{7}{3!}+\dots.$$

$$r\text{th term}=\frac{3r-2}{r!}=\frac{3r}{r!}-\frac{2}{r!}, \quad \dots\dots\dots\dots\text{(i)}$$

which, if $r>1$,

$$=\frac{3}{(r-1)!}-\frac{2}{r!}. \quad \dots\dots\dots\dots\text{(ii)}$$

Thus, by (i), $\quad 1\text{st term}=3-\frac{2}{1!},$

and (ii) gives $\quad 2\text{nd term}=\frac{3}{1!}-\frac{2}{2!},$

$$\vdots \qquad \vdots$$

$$n\text{th term}=\frac{3}{(n-1)!}-\frac{2}{n!};$$

∴ the sum to n terms is

$$3\left\{1+\frac{1}{1!}+\frac{1}{2!}+\dots+\frac{1}{(n-1)!}\right\}-2\left\{\frac{1}{1!}+\frac{1}{2!}+\frac{1}{3!}+\dots+\frac{1}{n!}\right\}.$$

When $n\to\infty$ the sums within the brackets tend to e and to $(e-1)$ respectively, thus the sum to infinity $=3e-2(e-1)=e+2$.

Example 7. Find the sum to infinity of

$$5+\frac{2\,.\,6}{1!}+\frac{3\,.\,7}{2!}+\frac{4\,.\,8}{3!}+\dots.$$

If $r>1$, the rth term is

$$\frac{r(r+4)}{(r-1)!}=\frac{(r-1)(r-2)+7(r-1)+5}{(r-1)!}, \quad \dots\dots\dots\dots\text{(i)}$$

and this, if $r>2$,

$$=\frac{r-2}{(r-2)!}+\frac{7}{(r-2)!}+\frac{5}{(r-1)!}, \quad \dots\dots\dots\dots\text{(ii)}$$

and this, if $r>3$,

$$=\frac{1}{(r-3)!}+\frac{7}{(r-2)!}+\frac{5}{(r-1)!} \quad \dots\dots\dots\dots\text{(iii)}$$

The 1st term is $\quad 5,$

the 2nd term $=0+7+\frac{5}{1!}$ by (i),

the 3rd term $=1+\frac{7}{1!}+\frac{5}{2!}$ by (ii),

the 4th term $=\frac{1}{1!}+\frac{7}{2!}+\frac{5}{3!}$ by (iii),

and (iii) gives all the later terms.

Thus the sum to infinity is found, by the method of Example 6, to be $e+7e+5e=13e$.

Note. In the above example, the exceptional terms at the beginning of the series could be found by (iii) if conventions were made to the effect that $0!=1$ and $\frac{1}{x!}=0$ when x is a negative integer. The reason that they can be found in this way is shown by (i) and (ii).

Example 8. Find, by successive approximations, x in terms of a, when $x+e^x=1+a$, and a is small.

$$x=1+a-e^x=a-x-\frac{x^2}{2!}-\frac{x^3}{3!}-\ldots;$$

$$\therefore\ 2x=a-\frac{x^2}{2!}-\frac{x^3}{3!}-\ldots.$$

For first approximation, neglect a^2; then

$$2x=a;\quad \therefore\ x=\tfrac{1}{2}a.$$

For second approximation, neglect a^3; then

$$2x=a-\frac{1}{2!}\left(\frac{a}{2}\right)^2=a-\frac{a^2}{8};\quad \therefore\ x=\frac{a}{2}-\frac{a^2}{16}.$$

For third approximation, neglect a^4; then

$$2x=a-\frac{1}{2!}\left(\frac{a}{2}-\frac{a^2}{16}\right)^2-\frac{1}{3\,!}\left(\frac{a}{2}\right)^3=a-\frac{a^2}{8}+\frac{a^3}{96};$$

$$\therefore\ x=\frac{a}{2}-\frac{a^2}{16}+\frac{a^3}{192}.$$

For fourth approximation, neglect a^5; then

$$2x=a-\frac{1}{2!}\left(\frac{a}{2}-\frac{a^2}{16}+\frac{a^3}{192}\right)^2-\frac{1}{3!}\left(\frac{a}{2}-\frac{a^2}{16}\right)^3-\frac{1}{4!}\left(\frac{a}{2}\right)^4$$

$$=a-\frac{a^2}{8}+\frac{a^3}{96}+\frac{a^4}{2}\left(-\frac{1}{256}-\frac{1}{192}+\frac{1}{64}-\frac{1}{192}\right)=a-\frac{a^2}{8}+\frac{a^3}{96}+\frac{a^4}{1536};$$

$$\therefore\ x=\frac{a}{2}-\frac{a^2}{16}+\frac{a^3}{192}+\frac{a^4}{3072},$$

and so on.

EXERCISE V. d.

Find (in terms of e) the sums to infinity of the series in Nos. 1-16.

1. $1-\frac{1}{1!}+\frac{1}{2!}-\frac{1}{3!}+\ldots.$

2. $\frac{1}{1!}+\frac{1}{3!}+\frac{1}{5!}+\ldots.$

3. $1+\frac{2}{1!}+\frac{3}{2!}+\frac{4}{3!}+\ldots.$

4. $\frac{1}{2!}+\frac{2}{3!}+\frac{3}{4!}+\ldots.$

5. $\frac{1^2}{2!}+\frac{2^2}{3!}+\frac{3^2}{4!}+\ldots.$

6. $1+\frac{2^2}{1!}+\frac{3^2}{2!}+\ldots.$

7. $\frac{2}{3!}+\frac{4}{5!}+\frac{6}{7!}+\ldots.$

8. $1+\frac{2}{2!}+\frac{3}{4!}+\frac{4}{6!}+\ldots.$

9. $\frac{1}{1!}+\frac{1+2}{2!}+\frac{1+2+3}{3!}+\ldots.$

10. $\frac{1}{1!}+\frac{1+2}{2!}+\frac{1+2+2^2}{3!}+\ldots.$

11. $\frac{1^3}{2!}+\frac{2^3}{3!}+\frac{3^3}{4!}+\ldots.$

12. $\frac{1^2}{3!}+\frac{2^2}{5!}+\frac{3^2}{7!}+\ldots.$

13. $\frac{3}{2!}+\frac{8}{4!}+\frac{13}{6!}+\frac{18}{8!}+\ldots.$

14. $\frac{1.3}{2!}+\frac{2.4}{3!}+\frac{3.5}{4!}+\ldots.$

15. $\frac{1^3}{1!}-\frac{2^3}{2!}+\frac{3^3}{3!}-\ldots.$

16. $\frac{2^3}{1!}-\frac{3^3}{2!}+\frac{4^3}{3!}-\ldots.$

Find the values of the following:

17. $\left(1+\frac{1}{2!}+\frac{1}{4!}+\ldots\right)^2-\left(1+\frac{1}{3!}+\frac{1}{5!}+\ldots\right)^2.$

18. $\left(\frac{1}{2!}+\frac{1}{4!}+\frac{1}{6!}+\ldots\right)\div\left(1+\frac{1}{3!}+\frac{1}{5!}+\frac{1}{7!}+\ldots\right).$

Find the sums to infinity of the series whose rth terms are

19. $\frac{r^4}{(r+1)!}.$

20. $\frac{1^2+2^2+\ldots+r^2}{r!}.$

21. $\frac{1^3+2^3+\ldots+r^3}{(r+1)!}.$

22. $\frac{x^r}{(r+1)!}.$

23. $\frac{x^r}{(r+2).r!}.$

24. $\frac{x^r}{(2r+1)!}.$

25. $\frac{x^r}{(r+3).r!}.$

Expand the following in power series, giving the coefficient of x^n in each:

26. $e^x(2+3x).$

27. $\frac{1+2x+3x^2}{e^x}.$

28. $1+\frac{(x+1)}{1!}+\frac{(x+1)^2}{2!}+\ldots.$

29. $\frac{e^{5x}+e^x}{e^{3x}}.$

30. $e^{\frac{1}{2}x}(1+2x-4x^2).$

31. Sum the series $(x^2-y^2)-\frac{1}{2!}(x^4-y^4)+\frac{1}{3!}(x^6-y^6)-\ldots.$

32. Sum the series $\frac{\log 2}{1!}-\frac{(\log 2)^2}{2!}+\frac{(\log 2)^3}{3!}-\ldots.$

33. Find the coefficient of x^4 in $(e^x+e^{-x})^n$.

34. Evaluate $\lim_{x\to 0}\{e^x+\log(1+x)-1-2x\}\div x^3.$

35. Evaluate $\lim\limits_{x\to 1}\dfrac{\log x - x^x + 1}{\log x - x + 1}$.

36. If x is small, prove that $(1+x)^{\frac{1}{1-x}} \simeq 1 + x + x^2 + \dfrac{3x^3}{2}$.

37. Show that (i) $e^x < \dfrac{1}{1-x}$, if $0 < x < 1$;

(ii) $e^x < \dfrac{2+x}{2-x}$, if $0 < x < 2$;

and examine the results when x is negative.

EASY MISCELLANEOUS EXAMPLES.

EXERCISE V. e.

1. Give the sums to infinity of :

(i) $1 - \dfrac{\pi^2}{3!} + \dfrac{\pi^4}{5!} - \dots$; (ii) $\frac{1}{2} - \frac{1}{4} + \frac{1}{6} - \frac{1}{8} + \dots$;

(iii) $1 + \dfrac{1}{2.3} + \dfrac{1}{2^2.5} + \dfrac{1}{2^3.7} + \dots$; (iv) $1 - \dfrac{1}{3.3} + \dfrac{1}{5.3^2} - \dfrac{1}{7.3^3} + \dots$;

(v) $\dfrac{1}{2!} + \dfrac{1}{4!} + \dfrac{1}{6!} + \dots$; (vi) $\dfrac{1}{1!} + \dfrac{1}{5!} + \dfrac{1}{9!} + \dots$.

2. Show that the sum to infinity of $\dfrac{2}{2!} - \dfrac{2^3}{4!} + \dfrac{2^5}{6!} - \dots$ is the square of that of $1 - \dfrac{1}{3!} + \dfrac{1}{5!} - \dots$.

3. Prove that $\theta \cot\theta \simeq 1 - \frac{1}{3}\theta^2 - \frac{1}{45}\theta^4$, if θ is small.

4. Show that the error involved in replacing θ by $\dfrac{1}{3}\left(8\sin\dfrac{\theta}{2} - \sin\theta\right)$ is about $\dfrac{\theta^5}{480}$ if θ is small. Hence solve $\sin\theta = \frac{5}{6}\theta$ approximately.

5. Find an approximation to $\dfrac{\sin\theta - \sin(\theta+2a) + \sin a}{\cos\theta - \cos(\theta+2a) + \cos a}$ when a is so small that a^3 is negligible.

6. Prove that $\dfrac{1+3\cos\theta}{3+\cos\theta} = \sqrt{(\cos\theta)}$, if θ^6 is negligible.

7. If $\cos(\alpha+\theta) = \cos\alpha\cos\phi - \cos\beta\sin\alpha\sin\phi$, where ϕ is small, prove that one value of θ is nearly $\phi\cos\beta + \frac{1}{2}\phi^2\cot\alpha\sin^2\beta$.

8. If ϵ is small and positive, prove that $x - \epsilon = \dfrac{\pi}{2}\sin^2 x$ has three roots.

9. Show that $\theta = \frac{\pi}{12}$ is an approximate solution of

$$16 \sin \theta = 12\theta + 1,$$

and find a better approximation.

10. If x is small, and $e^x \tan \frac{x}{2} = a$, prove that

$$x \simeq 2a - 4a^2 + \frac{34a^3}{3} - \frac{112a^4}{3}.$$

11. If θ is small, and $\theta \cot \theta = 1 - \epsilon$, prove that $\theta \simeq \sqrt{(3\epsilon)} \cdot \left(1 - \frac{\epsilon}{10}\right)$.

12. Prove that the sum to infinity of

$$1 + \left(\frac{1}{2} + \frac{1}{3}\right)\frac{1}{4} + \left(\frac{1}{4} + \frac{1}{5}\right)\frac{1}{4^2} + \left(\frac{1}{6} + \frac{1}{7}\right)\frac{1}{4^3} + \dots \text{ is } \tfrac{1}{2}\log 12.$$

13. If n is positive, prove that the sum to infinity of

$$\frac{1}{1 \cdot 2\,(n+1)} + \frac{1}{2 \cdot 3\,(n+1)^2} + \frac{1}{3 \cdot 4\,(n+1)^3} + \dots$$

is $1 - \log\left(1 + \frac{1}{n}\right)^n$.

14. Expand as power series in x, giving the general terms:

(i) $\log\left(1 + \frac{1}{x-3}\right)$; (ii) $\log(1 - x^2 + x^4)$.

15. Express $\log(x+y) - \log(x-y)$ as a series of powers of $\frac{y}{x}$, stating when this is possible.

16. Express $2\log(x+h) - \log x - \log(x+2h)$ as a series of powers of $\frac{h}{x+h}$ and state when this is possible.

17. Prove that $\log 10 = 3\log 2 + 2\left(\frac{1}{9} + \frac{1}{3 \cdot 9^3} + \frac{1}{5 \cdot 9^5} + \dots\right)$ and hence evaluate $\log 10$, given that $\log 2 = \cdot 693147$. Deduce the value of $\log_{10} 2$.

18. If $a = \log \frac{6}{5}$, $b = \log \frac{10}{9}$, and $c = \log \frac{25}{24}$, prove that

$$\log 2 = 3a + b + c,$$

and hence calculate $\log 2$ to 3 places of decimals.

19. If $0 < \theta < \frac{\pi}{2}$, prove that $\sum_{1}^{\infty} \frac{\cos^{2r-1} 2\theta}{2r-1} = \log \cot \theta$. What happens if $\frac{\pi}{2} < \theta < \pi$?

20. Find the sum to infinity of $\dfrac{x}{1+x^2}+\dfrac{x^3}{3(1+x^2)^3}+\dfrac{x^5}{5(1+x^2)^5}+\dots$, and the values of x for which it converges. Expand the sum in another way and find the coefficient of x^{3n} in the new expansion.

21. Find the sums to infinity of :

(i) $\dfrac{1}{1.2}+\dfrac{1}{3.4}+\dfrac{1}{5.6}+\dots$; (ii) $\dfrac{1}{2.3}+\dfrac{1}{4.5}+\dfrac{1}{6.7}+\dots$;

(iii) $\dfrac{1}{1.2.3}+\dfrac{1}{5.6.7}+\dfrac{1}{9.10.11}+\dots$;

(iv) $\dfrac{5}{1.2.3}+\dfrac{8}{2.3.5}+\dfrac{11}{3.4.7}+\dots$;

and prove that in (iv) the sum to n terms differs from the sum to infinity by less than $\dfrac{3}{2n}$.

22. Prove that the sum to n terms of the series

$$\frac{1}{1.2.3.4}+\frac{1}{3.4.5.6}+\frac{1}{5.6.7.8}+\dots$$

is $\dfrac{2}{3}s_n-\dfrac{5}{12}+\dfrac{4n+5}{12(n+1)(2n+1)}$, where s_n is the sum to $2n$ terms of $1-\frac{1}{2}+\frac{1}{3}-\frac{1}{4}+\dots$. Hence find the sum to infinity.

23. If x is small, prove that

(i) $\log\log(1+x)^{\frac{1}{x}} \simeq -\dfrac{x}{2}+\dfrac{5x^2}{24}-\dfrac{x^3}{8}$;

(ii) $\frac{1}{2}\log\dfrac{1+x}{1-x}-\sin x\sqrt{(1+x^2)} \simeq \dfrac{2x^5}{5}$.

24. Prove that $\log\sin\theta=\log\theta-\dfrac{\theta^2}{6}-\dfrac{\theta^4}{180}-\dots$ if $0<\theta<\pi$.

25. Prove that $\lim\limits_{x\to 1}\left(\dfrac{x}{x-1}-\dfrac{1}{\log x}\right)=\frac{1}{2}$.

26. Find the sums to infinity of

(i) $\dfrac{1}{1.3}+\dfrac{1}{5.7}+\dfrac{1}{9.11}+\dots$; (ii) $\dfrac{x^3}{3}+\dfrac{x^7}{7}+\dfrac{x^{11}}{11}+\dots$,

where $|x|<1$.

27. Evaluate $\lim\limits_{x\to 0}\dfrac{\tan^{-1}x-\sin x}{x^2-2x+2\log(1+x)}$.

28. Neglecting x^8, choose numerical values for a and b, so that

$$ax\sin x+b\sin^2\frac{x}{2}-x\tan^{-1}x=-\tfrac{13}{72}x^6.$$

29. Expand when possible $\frac{1}{2}\cos^{-1}\left(\frac{\cos a+\cos\theta}{1+\cos a\cos\theta}\right)$ as a power series in $\tan\frac{\theta}{2}$, and state the conditions of validity.

Find the sums to infinity of the series whose rth terms are

30. $\frac{r}{(2r-1)!}$.

31. $\frac{1+2+3+\ldots+r}{(r+1)!}$.

32. $\frac{2r^2+r-1}{r!}$.

33. $\frac{r+1}{(r+2)\,.\,r!}$.

34. $\frac{5r+1}{(2r+1)!}$.

35. $\frac{(2r+1)^3}{(r+1)!}$.

36. Express $1+\frac{(a+bx)}{1!}+\frac{(a+bx)^2}{2!}+\ldots$ as a power series in x, proving that the coefficient of x^n is $\frac{e^a b^n}{n!}$.

37. Prove that $e^{3x}-4e^x+6e^{-x}-4e^{-3x}+e^{-5x}\simeq 16(x^4-x^5)$ if x is small.

38. Prove that $(1+x)^{1+x}\simeq 1+x+x^2+\frac{1}{2}x^3$ if x is small.

39. If x is large, prove that $\left(1+\frac{1}{x}\right)^x\simeq e\left(1-\frac{1}{2x}\right)$.

40. If x is large, prove that $\left(1+\frac{1}{x}\right)^{x+\frac{1}{2}}\simeq e\left(1+\frac{1}{12x^2}\right)$.

41. If $x^{n+p}=a^n$, p is small compared to n, and $a>0$, prove that $x\simeq a\left(1-\frac{p}{n}\log a\right)$.

42. Expand $1-a\cos u$ in ascending powers of a as far as a^3, if a is small and $u=k+a\sin u$.

HARDER MISCELLANEOUS EXAMPLES.

EXERCISE V. f.

1. Find an approximation for $2\theta-\frac{28\sin\theta+\sin 2\theta}{9+6\cos\theta}$, when θ is small.

2. If $a>2b>0$ and $0<x<\pi$, prove that $x>\frac{(a+b)\sin x}{a+b\cos x}$.

3. Prove the inequalities (due in effect to Archimedes)

$$\tfrac{1}{3}(2\sin\theta+\tan\theta)>\theta>\frac{3}{2\operatorname{cosec}\theta+\cot\theta},\ \text{if}\ 0<\theta<\frac{\pi}{2}.$$

4. If x and y are the lengths of the sides of regular polygons of n sides inscribed in a circle and circumscribed about it, prove that the circumference of the circle is approximately $\frac{n}{3}(2x+y)$.

5. If A and B are the areas of the polygons in No. 4, prove that the area of the circle is approximately $\frac{1}{3}(2B+A)$.

6. Two regular polygons of m and n sides have equal perimeters l. Prove that if m and n are large the areas of the polygons differ by about $\frac{\pi l^2}{12}\left(\frac{1}{m^2}\sim\frac{1}{n^2}\right)$.

7. Show that $\log 7$ differs from $2\log 3+2\log 5-5\log 2-\frac{2}{449}$ by less than $\frac{2}{3\,.\,448\,.\,449\,.\,450}$.

8. If x is small, prove that $\log(\sec x)=2\tan^2\frac{x}{2}$, neglecting x^6.

9. If $\log\frac{1+x+x^2}{1-x+x^2}=\sum(a_n x^n)$ where $x^2<1$, prove that if n is even $a_n=0$, and that if n is odd and a multiple of 3, $a_n=-\frac{4}{n}$, and that if n is prime to 6, $a_n=\frac{2}{n}$.

10. Find the coefficient of x^n in the expansion of

$$\log(1+2x+2x^2+x^3),\ \text{ when } x^2<1.$$

Consider separately the cases when $n=0$, 1, 2, 3, 4, 5, (mod 6).

11. If a, b, c are consecutive positive integers, prove that

$$\log b-\tfrac{1}{2}(\log a+\log c)=\sum_1^\infty\frac{x^{2n-1}}{2n-1},\ \text{ where } x=\frac{1}{2ac+1}.$$

12. Assuming that the coefficients of x^{3n} may be equated when the two sides of the identity $\log(1+x^3)\equiv\log(1+x)+\log(1-x+x^2)$ are expanded in powers of x, find the sum of the series

$$1-\frac{3n-3}{2!}+\frac{(3n-4)(3n-5)}{3!}-\frac{(3n-5)(3n-6)(3n-7)}{4!}+\dots.$$

13. From the identity $\log(1-\alpha x)(1-\beta x)\equiv\log(1-sx+px^2)$, where $s=\alpha+\beta$, $p=\alpha\beta$, by expanding and equating coefficients of various powers of x, show that

(i) $\alpha^5+\beta^5=s(s^4-5s^2p+5p^2)$;

(ii) $\alpha^{13}+\beta^{13}=s(s^{12}-13s^{10}p+65s^8p^2-156s^6p^3+182s^4p^4$
$-91s^2p^5+13p^6)$.

14. If $\alpha+\beta+\gamma=0$, $\beta\gamma+\gamma\alpha+\alpha\beta=-s$, $\alpha\beta\gamma=p$, prove by a method similar to that of No. 13, that

(i) $\alpha^3+\beta^3+\gamma^3=3p$;

(ii) $\alpha^5+\beta^5+\gamma^5=5ps$;

(iii) $\alpha^7+\beta^7+\gamma^7=7s^2p$.

15. Deduce by expansion of $\log(1-\alpha x)(1-\beta x)(1-\gamma x)(1-\delta x)$ that

$$\Sigma\alpha^3 - 3\Sigma\alpha\beta\gamma = \Sigma\alpha\{\Sigma\alpha^2 - \Sigma\alpha\beta\}.$$

16. Express $\alpha^n + \beta^n$ in terms of p and q, where α, β are the roots of $x^2 - px + q = 0$.

17. Use the identity $(1 + y\sqrt{2} + y^2)(1 - y\sqrt{2} + y^2) \equiv 1 + y^4$ and the expansion of $\log(1+x)$ to prove that

$$\frac{2^n}{2n} - \frac{2^{n-1}}{2n-1}\cdot\frac{2n-1}{1!} + \frac{2^{n-2}}{2n-2}\cdot\frac{(2n-2)(2n-3)}{2!} - \frac{2^{n-3}}{2n-3}\cdot\frac{(2n-3)(2n-4)(2n-5)}{3!} + \ldots = 0,$$

where n is an odd positive integer.

18. If $|x| < 1$, prove that $(1+x)^{1+x}(1-x)^{1-x} \geqslant 1$.

Deduce that $a^a b^b > \left(\dfrac{a+b}{2}\right)^{a+b}$ if a, b are positive and unequal.

19. If $x > y > z > 0$, prove that $\left(\dfrac{x+z}{x-z}\right)^x < \left(\dfrac{y+z}{y-z}\right)^y$.

20. If x is small, show that the following functions can be arranged in ascending order of magnitude by expanding in powers of x, as far as x^3 only, and arrange them :

(i) $\sin(\tan^{-1} x)$; (ii) $\tan(\sin^{-1} x)$; (iii) $\tan^{-1}(\tan^{-1} x)$;
(iv) $\tan(\tan x)$; (v) $\sin(\sin x)$; (vi) $\sin^{-1}(\sin^{-1} x)$.

21. Evaluate $\displaystyle\sum_1^\infty \frac{2r^2 + 3r - 1}{(r+2)!}$.

22. If $x > 0$, prove that $(x-3)e^x + \frac{1}{2}x^2 + 2x + 3 > 0$.

23. If x is large prove that $\left(1 + \dfrac{1}{x}\right)^x \simeq e\left(1 - \dfrac{1}{2x} + \dfrac{11}{24x^2} - \dfrac{7}{16x^3}\right)$.

24. If p is small, prove that successive approximations to a root of $x^{2+p} = a^2$ are a, $a - \frac{1}{2}ap\log a$, and $a\{1 - \frac{1}{2}p\log a + \frac{1}{8}p^2(2 + \log a)\log a\}$, where $a > 0$.

25. Show that the coefficient of x^n in $e^{(e^x)}$ is

$$\frac{1}{n!}\left\{\frac{1^n}{1!} + \frac{2^n}{2!} + \ldots + \frac{r^n}{r!} + \ldots\right\},$$

and hence find the sum of the series in the bracket for $n = 4$.

26. Prove that $\left\{\dfrac{1}{2!} - \dfrac{1}{3!} + \dfrac{1}{4!} - \ldots + \dfrac{(-1)^n}{n!}\right\} . n!$ differs from $\dfrac{n!}{e}$ by less than $\dfrac{1}{n+1}$.

27. Apply the inequalities $\frac{1}{n+1} < \log\left(1+\frac{1}{n}\right) < \frac{1}{n}$, where n is positive, to show that $\left(1+\frac{1}{n^2}\right)\left(1+\frac{2}{n^2}\right)\dots\left(1+\frac{n}{n^2}\right)$ is less than $\sqrt{e}\,.\sqrt[2n]{e}$, but greater than $\sqrt{e}$.

28. If n is a positive integer, by expanding $(e^x-1)^n$ in two ways and comparing the coefficients of various powers of x, prove that

$$n^{n-1}-\frac{n}{1}(n-1)^{n-1}+\frac{n(n-1)}{1.2}(n-2)^{n-1}-\dots=0,$$

and find the values of

(i) $n^{n-1}-(n-1)^n+\frac{n-1}{1.2}(n-2)^n-\frac{(n-1)(n-2)}{1.2.3}(n-3)^n+\dots$;

(ii) $n^n-(n-1)^{n+1}+\frac{n-1}{1.2}(n-2)^{n+1}-\frac{(n-1)(n-2)}{1.2.3}(n-3)^{n+1}+\dots$;

(iii) $n^{n+1}-(n-1)^{n+2}+\frac{n-1}{1.2}(n-2)^{n+2}-\frac{(n-1)(n-2)}{1.2.3}(n-3)^{n+2}+\dots$.

29. If n is a positive integer, prove that

$$n^n-\frac{n}{1}(n-2)^n+\frac{n(n-1)}{1.2}(n-4)^n-\dots=2^n.n!.$$

30. If n is a positive integer, prove that

$$1^n-n.2^n+\frac{n(n-1)}{1.2}3^n-\dots \text{ to } \overline{n+1} \text{ terms} =(-1)^n.n!.$$

31. By expanding $(e^x+1)^n-(e^x-1)^n$ in two ways, prove that $c_1(n-1)^3+c_3(n-3)^3+\dots\dots=n^2(n+3).2^{n-4}$,

where $c_0+c_1x+c_2x^2+\dots\equiv(1+x)^n$, and $n>3$.

32. If n is a fixed positive integer and x is positive, prove by differentiation that $\frac{(n+1+x)^{n+1}}{(n+x)^n}$ increases with x. Deduce that

$$\left(1+\frac{x}{n}\right)^n < \left(1+\frac{x}{n+1}\right)^{n+1}.$$

33. If n is a positive integer and $0<x<n$, use the method of No. 32 to find whether $\left(1-\frac{x}{n}\right)^n$ or $\left(1-\frac{x}{n+1}\right)^{n+1}$ is the greater.

CHAPTER VI

THE SPECIAL HYPERBOLIC FUNCTIONS

From the expansions in Ch. V, we have

$$e^x = 1 + \frac{x}{1!} + \frac{x^2}{2!} + \frac{x^3}{3!} + \dots$$

and

$$e^{-x} = 1 - \frac{x}{1!} + \frac{x^2}{2!} - \frac{x^3}{3!} + \dots.$$

Therefore, by addition and subtraction,

$$\tfrac{1}{2}(e^x + e^{-x}) = 1 + \frac{x^2}{2!} + \frac{x^4}{4!} + \dots \qquad \text{.......................(1)}$$

and

$$\tfrac{1}{2}(e^x - e^{-x}) = x + \frac{x^3}{3!} + \frac{x^5}{5!} + \dots. \qquad \text{.......................(2)}$$

These results should be compared with the expansions of $\cos x$ and $\sin x$ in Ch. V, pp. 80, 81. The precise connection will be explained after complex numbers and functions of a complex variable have been defined. But equations (1) and (2) suggest that the functions $\frac{1}{2}(e^x + e^{-x})$ and $\frac{1}{2}(e^x - e^{-x})$ possess properties analogous to those of $\cos x$ and $\sin x$. We therefore define these functions as the "hyperbolic cosine" and the "hyperbolic sine" of x and we write

$$\operatorname{ch} x \equiv \tfrac{1}{2}(e^x + e^{-x}); \quad \operatorname{sh} x \equiv \tfrac{1}{2}(e^x - e^{-x}), \qquad \text{..............(3)}$$

and we speak of these functions as "cosh x" and "shine x" (or else "sinsh x"): they are sometimes written "cosh x" and "sinh x."

We therefore have

$$\operatorname{ch} x = 1 + \frac{x^2}{2!} + \frac{x^4}{4!} + \dots, \qquad \text{.......................(4)}$$

$$\operatorname{sh} x = x + \frac{x^3}{3!} + \frac{x^5}{5!} + \dots. \qquad \text{.......................(5)}$$

We also define the hyperbolic tangent, hyperbolic secant, hyperbolic cosecant, hyperbolic cotangent, which are written $\operatorname{th} x$, $\operatorname{sech} x$, $\operatorname{cosech} x$, $\coth x$, by the relations

$$\operatorname{th} x = \frac{\operatorname{sh} x}{\operatorname{ch} x}; \quad \operatorname{sech} x = \frac{1}{\operatorname{ch} x}; \quad \operatorname{cosech} x = \frac{1}{\operatorname{sh} x}; \quad \coth x = \frac{\operatorname{ch} x}{\operatorname{sh} x}.$$

Note. $\operatorname{th} x$ is pronounced "than x" or "tansh x," and is sometimes written "tanh x."

Formulae for the Hyperbolic Functions. Putting $e^x = t$, we have

$$\operatorname{ch} x = \frac{1}{2}\left(t + \frac{1}{t}\right) \quad \text{and} \quad \operatorname{sh} x = \frac{1}{2}\left(t - \frac{1}{t}\right);$$

$$\therefore \operatorname{ch}^2 x - \operatorname{sh}^2 x = \frac{1}{4}\left(t + \frac{1}{t}\right)^2 - \frac{1}{4}\left(t - \frac{1}{t}\right)^2 = \frac{1}{4} \cdot 2t \cdot \frac{2}{t} = 1;$$

$$\therefore \mathbf{ch^2 x - sh^2 x = 1.} \qquad \text{.........................(6)}$$

Similarly, $\operatorname{sh} x_1 \operatorname{ch} x_2 + \operatorname{ch} x_1 \operatorname{sh} x_2$

$$= \tfrac{1}{4}\left\{\left(t_1 - \frac{1}{t_1}\right)\left(t_2 + \frac{1}{t_2}\right) + \left(t_1 + \frac{1}{t_1}\right)\left(t_2 - \frac{1}{t_2}\right)\right\}$$

$$= \tfrac{1}{4}\left\{2t_1 t_2 - \frac{2}{t_1 t_2}\right\} = \tfrac{1}{2}\{e^{x_1 + x_2} - e^{-x_1 - x_2}\}$$

$$= \operatorname{sh}(x_1 + x_2);$$

$$\therefore \mathbf{sh(x_1 + x_2) = sh\, x_1\, ch\, x_2 + ch\, x_1\, sh\, x_2.} \qquad \text{...............(7)}$$

Also $\operatorname{ch} x_1 \operatorname{ch} x_2 + \operatorname{sh} x_1 \operatorname{sh} x_2$

$$= \tfrac{1}{4}\left\{\left(t_1 + \frac{1}{t_1}\right)\left(t_2 + \frac{1}{t_2}\right) + \left(t_1 - \frac{1}{t_1}\right)\left(t_2 - \frac{1}{t_2}\right)\right\}$$

$$= \tfrac{1}{4}\left\{2t_1 t_2 + \frac{2}{t_1 t_2}\right\} = \tfrac{1}{2}\{e^{x_1 + x_2} + e^{-x_1 - x_2}\}$$

$$= \operatorname{ch}(x_1 + x_2);$$

$$\therefore \mathbf{ch(x_1 + x_2) = ch\, x_1\, ch\, x_2 + sh\, x_1\, sh\, x_2.} \qquad \text{...............(8)}$$

We have also from the definitions the general relations

$$\mathbf{ch(-x) = ch\, x; \quad sh(-x) = -sh\, x,} \qquad \text{...............(9)}$$

and the special values

$$\mathbf{ch\, 0 = 1; \quad sh\, 0 = 0.} \qquad \text{........................(10)}$$

By comparing formulae (6)-(9) with the corresponding trigonometrical formulae, the reader will see that to every (general) trigonometrical formula there corresponds an analogous formula for the hyperbolic functions which may be written down by *Osborn's* rule: *In any formula connecting the circular functions of* general *angles, replace each circular function by the corresponding hyperbolic function and change the sign of every product* (*or implied product*) *of two sines.*

Thus, from $\tan(A + B) = \dfrac{\tan A + \tan B}{1 - \tan A \tan B}$, we may infer that $\operatorname{th}(A + B) = \dfrac{\operatorname{th} A + \operatorname{th} B}{1 + \operatorname{th} A \operatorname{th} B}$, since $\tan A \tan B \equiv \dfrac{\sin A \sin B}{\cos A \cos B}$ implies a product of two sines.

The rule does not apply to properties depending on the periodicity of the circular functions or the values of the ratios of special angles; e.g. the rule must not be used in connection with

$$\sin(2\pi - x) = \sin x \quad \text{or} \quad \cos\left(x + \frac{\pi}{4}\right) = \frac{1}{\sqrt{2}}(\cos x - \sin x).$$

For the present, this rule should be regarded merely as a mnemonic. Its justification is best left till circular functions of a complex variable have been defined, see Chapter X.

EXERCISE VI. a.

Prove some of the following formulae in Nos. 1–10, and check the others by the rule on p. 105.

1. $\text{ch}(-x) = \text{ch}\, x$; $\text{sh}(-x) = -\text{sh}\, x$; $\text{th}(-x) = -\text{th}\, x$.
2. $\text{sh}\, 2\theta = 2\, \text{sh}\, \theta\, \text{ch}\, \theta$.
3. $\text{ch}\, 2\theta = \text{ch}^2\theta + \text{sh}^2\theta = 2\,\text{ch}^2\theta - 1 = 1 + 2\,\text{sh}^2\theta$.
4. $1 + \text{ch}\, \alpha = 2\, \text{ch}^2 \frac{\alpha}{2}$; $1 - \text{ch}\, \alpha = -2\, \text{sh}^2 \frac{\alpha}{2}$.
5. $\text{th}\, 2x = \dfrac{2\, \text{th}\, x}{1 + \text{th}^2 x}$.
6. (i) $\text{ch}(\alpha - \beta) = \text{ch}\, \alpha\, \text{ch}\, \beta - \text{sh}\, \alpha\, \text{sh}\, \beta$;
   (ii) $\text{sh}(\alpha - \beta) = \text{sh}\, \alpha\, \text{ch}\, \beta - \text{ch}\, \alpha\, \text{sh}\, \beta$.
7. (i) $\text{sh}\, \theta - \text{sh}\, \phi = 2\, \text{ch}\, \dfrac{\theta + \phi}{2}\, \text{sh}\, \dfrac{\theta - \phi}{2}$;
   (ii) $\text{ch}\, \theta - \text{ch}\, \phi = 2\, \text{sh}\, \dfrac{\theta + \phi}{2}\, \text{sh}\, \dfrac{\theta - \phi}{2}$.

   Write down the corresponding formulae for $\text{sh}\, \theta + \text{sh}\, \phi$ and $\text{ch}\, \theta + \text{ch}\, \phi$.
8. $\text{th}(\theta - \phi) = \dfrac{\text{th}\, \theta - \text{th}\, \phi}{1 - \text{th}\, \theta\, \text{th}\, \phi}$.
9. (i) $\text{sh}\, 3\theta = 3\, \text{sh}\, \theta + 4\, \text{sh}^3\theta$; (ii) $\text{ch}\, 3\theta = 4\, \text{ch}^3\theta - 3\, \text{ch}\, \theta$.

   Write down the formula for $\text{th}\, 3\theta$ in terms of $\text{th}\, \theta$.
10. $\text{sech}^2 x = 1 - \text{th}^2 x$. What is the corresponding formula for $\text{cosech}^2 x$?

Write down alternative expressions for the following:

11. $1 - \text{coth}^2 x$.
12. $\text{sh}^3 x$.
13. $\text{sh}^2 x - \text{sh}^2 y$.
14. $\text{sh}\, \theta\, \text{sh}\, \phi$.
15. $\text{sh}\, \theta\, \text{ch}\, \phi$.
16. $\text{ch}\, \theta\, \text{ch}\, \phi$.
17. $(\text{ch}\, x - \text{sh}\, x)^{-1}$.
18. $(\text{ch}\, x + \text{sh}\, x)^n$.
19. $(\text{ch}\, x - \text{sh}\, x)^n$.

20. Expand th $(x+y+z)$.

21. Prove that $\text{ch}(x+y)\,\text{ch}(x-y)=\text{ch}^2 x+\text{sh}^2 y$.

22. Prove that $\dfrac{1+\text{th}\,\theta}{1-\text{th}\,\theta}=\text{ch}\,2\theta+\text{sh}\,2\theta$.

23. Express $\text{ch}\,\theta$ and $\text{th}\,\theta$ in terms of $\text{sh}\,\theta$.

24. Express $\text{ch}\,\theta$ and $\text{sh}\,\theta$ in terms of $\text{th}\,\theta$.

25. Express $\text{sh}\,\theta$ and $\text{th}\,\theta$ in terms of k, where $k=\text{ch}\,2\theta$.

26. Express $\text{sh}\,\theta$ and $\text{ch}\,\theta$ in terms of t, where $t=\text{th}\,\frac{1}{2}\theta$.

27. If $x=\sin u\,\text{ch}\,v$ and $y=\cos u\,\text{sh}\,v$, find a relation between (i) x, y, u; (ii) x, y, v.

28. Prove that $\coth\dfrac{\theta}{2}-\coth\theta=\text{cosech}\,\theta$.

29. If $\tan\theta=\tan\alpha\,\text{th}\,\beta$ and $\tan\phi=\cot\alpha\,\text{th}\,\beta$, prove that
$$\tan(\theta+\phi)=\frac{\text{sh}\,2\beta}{\sin 2\alpha}.$$

30. Prove that $\text{ch}^2(\theta+\phi)-\text{ch}^2(\theta-\phi)=\text{sh}\,2\theta\,\text{sh}\,2\phi$.

31. Prove that $\sin^2\theta\,\text{ch}^2\phi+\cos^2\theta\,\text{sh}^2\phi=\frac{1}{2}(\text{ch}\,2\phi-\cos 2\theta)$.

32. Simplify $\dfrac{1+\text{sh}\,\phi+\text{ch}\,\phi}{1-\text{sh}\,\phi-\text{ch}\,\phi}$.

33. Express $\text{sh}\,2x+\text{sh}\,2y+\text{sh}\,2z-\text{sh}(2x+2y+2z)$ in factors.

34. If $\sin x\,\text{ch}\,y=\cos\alpha$ and $\cos x\,\text{sh}\,y=\sin\alpha$, prove that
$$\text{sh}^2 y=\cos^2 x=\pm\sin\alpha.$$

35. Simplify $\text{sh}(\log x)$ and $\text{ch}(\log x)$.

36. Prove that $\text{ch}\,x+\text{ch}\,2x+\text{ch}\,3x+\ldots+\text{ch}\,nx$ equals
$$\tfrac{1}{2}\,\text{sh}\,(n+\tfrac{1}{2})x\;\text{cosech}\,\tfrac{1}{2}x-\tfrac{1}{2}.$$

Differential Coefficients and Integrals. Using the definitions, we have

$$\frac{\mathbf{d}}{\mathbf{dx}}(\mathbf{sh\,x})=\frac{d}{dx}\left(\frac{e^x-e^{-x}}{2}\right)=\frac{e^x+e^{-x}}{2}=\mathbf{ch\,x} \qquad \text{............(11)}$$

$$\frac{\mathbf{d}}{\mathbf{dx}}(\mathbf{ch\,x})=\frac{d}{dx}\left(\frac{e^x+e^{-x}}{2}\right)=\frac{e^x-e^{-x}}{2}=\mathbf{sh\,x} \qquad \text{............(12)}$$

$$\therefore \int \mathbf{sh\,x\,.\,dx}=\mathbf{ch\,x};\quad \int \mathbf{ch\,x\,.\,dx}=\mathbf{sh\,x}.$$

Further,
$$\frac{d}{dx}(\text{th}\,x)=\frac{d}{dx}\left(\frac{\text{sh}\,x}{\text{ch}\,x}\right)=\frac{\text{ch}^2 x-\text{sh}^2 x}{\text{ch}^2 x};$$

$$\therefore \frac{\mathbf{d}}{\mathbf{dx}}(\mathbf{th\,x})=\mathbf{sech^2 x}. \qquad \text{........................(13)}$$

In general, expressions involving the hyperbolic functions are integrated by methods similar to those used for the circular functions.

It should also be noted that the general solution of the equation $\frac{d^2y}{dx^2}=y$ may be written in the form $y=\mathsf{A}\,\text{sh}\,x+\mathsf{B}\,\text{ch}\,x$ where A, B are arbitrary constants, just as that of $\frac{d^2y}{dx^2}=-y$ may be written in the form $y=\mathsf{A}\sin x+\mathsf{B}\cos x$. (See Ex. VI. b, No. 27.)

Example 1. Find $\frac{d}{dx}\left[\tan^{-1}\left(\text{th}\,\frac{x}{2}\right)\right]$.

$$\frac{d}{dx}\left[\tan^{-1}\left(\text{th}\,\frac{x}{2}\right)\right]=\frac{1}{1+\text{th}^2\frac{x}{2}}\cdot\tfrac{1}{2}\,\text{sech}^2\frac{x}{2}$$

$$=\frac{1}{2\left(\text{ch}^2\frac{x}{2}+\text{sh}^2\frac{x}{2}\right)}=\frac{1}{2\,\text{ch}\,x}$$

$$=\tfrac{1}{2}\,\text{sech}\,x.$$

Example 2. Evaluate $\int \text{sh}^2x\,dx$.

$$\int \text{sh}^2x\,dx=\int \tfrac{1}{2}\{\text{ch}\,2x-1\}\,dx$$

$$=\tfrac{1}{4}\,\text{sh}\,2x-\tfrac{1}{2}x+c.$$

EXERCISE VI. b.

Differentiate with respect to x:

1. $\text{sh}\,x+\text{ch}\,x$.
2. $\text{ch}^2\,x$.
3. $\text{sh}^2\,x$.
4. $\text{sh}\,x\,\text{ch}\,x$.
5. $\text{cosech}\,x$.
6. $\text{sech}\,x$.
7. $\text{coth}\,x$.
8. $\log(\text{sh}\,x)$.
9. $\log(\text{ch}\,x)$.
10. $\log\left(\text{th}\,\frac{x}{2}\right)$.
11. $\tan^{-1}(\text{coth}\,x)$.
12. $\log(\text{sh}\,x+\text{ch}\,x)$.

Integrate with respect to x:

13. $\text{ch}\,2x$.
14. $\text{sh}\,3x$.
15. $\text{th}\,x$.
16. $\text{coth}\,x$.
17. $\text{sh}^2\,x$.
18. $\text{cosech}^2\,x$.
19. $\text{th}^2\,x$.
20. $\text{coth}^2\,x$.
21. $\text{sech}\,x$.
22. $\text{cosech}\,x$.
23. $\text{sh}\,x\,\text{sh}\,2x$.
24. ch^3x.

25. What is $\frac{d}{dx}(\text{ch}\,x\cos x+\text{sh}\,x\sin x)$?

26. Find the value of $\int \text{ch}\,x\sin x\,dx$.

27. If $y=a\,\text{sh}\,nx+b\,\text{ch}\,nx$ where a, b, n are constants, prove that $\frac{d^2y}{dx^2}=n^2y$.

28. (*Behaviour of sh x and ch x*).

(i) Prove that $\text{ch}\,x$ is always positive and that $\text{sh}\,x$ has the same sign as x.

(ii) Deduce from (i) that $\text{sh}\,x$ steadily increases as x increases, that $\text{ch}\,x$ steadily decreases if x is negative and steadily increases if x is positive, as x increases.

(iii) What is the minimum value of $\text{ch}\,x$?

(iv) How does $\text{ch}\,x$ behave when $x\to\infty$ and when $x\to-\infty$?

(v) How does $\text{sh}\,x$ behave when $x\to\infty$ and when $x\to-\infty$?

(vi) Find the limit of $\frac{\text{ch}\,x}{e^x}$ when $x\to+\infty$ and of $\frac{\text{ch}\,x}{e^{-x}}$ when $x\to-\infty$.

(vii) Find the limit o $\frac{\text{sh}\,x}{e^x}$ when $x\to+\infty$ and of $\frac{\text{sh}\,x}{e^{-x}}$ when $x\to-\infty$.

(viii) Draw in the margin the graphs of $\text{sh}\,x$ and $\text{ch}\,x$. Compare each with the graphs of e^x and e^{-x}. [The graph of $\text{ch}\,x$ is called a *Catenary*, because it is the curve in which a uniform flexible chain with fixed ends hangs.]

29. (*Behaviour of th x and coth x*).

(i) Prove that $\text{th}\,x$ and $\coth x$ are both odd functions of x.

(ii) Prove that $\text{th}\,x$ steadily increases as x increases. What conclusion can be drawn from the fact that

$$\frac{d}{dx}\coth x=-\text{cosech}^2\,x\ ?$$

(iii) Find the limits of $\text{th}\,x$ when $x\to+\infty$ and when $x\to-\infty$. What are the limits of $\coth x$ in these cases?

(iv) Discuss the behaviour of $\text{th}\,x$ when $x\to 0$, (*a*) through positive values, (*b*) through negative values.

(v) What is the slope of $y=\text{th}\,x$ at $x=0$?

(vi) Prove that $|\text{th}\,x|<1$ and $|\coth x|>1$ for all values of x.

(vii) Draw in the margin the graphs of $\text{th}\,x$ and $\coth x$.

30. Draw the graphs of $\text{sech}\,x$ and $\text{cosech}\,x$.

Inverse Hyperbolic Functions. If $y = \operatorname{sh} x$, then $e^x - e^{-x} = 2y$;

$$\therefore \quad e^{2x} - 2ye^x + y^2 = 1 + y^2;$$

$$\therefore \quad (e^x - y)^2 = 1 + y^2;$$

$$\therefore \quad e^x = y \pm \sqrt{(1 + y^2)}.$$

But $e^x > 0$; $\therefore$ $y - \sqrt{(1+y^2)}$ is not a possible value of e^x;

$$\therefore \quad e^x = y + \sqrt{(1 + y^2)};$$

$$\therefore \quad x = \log [y + \sqrt{(1 + y^2)}].$$

Since $\operatorname{sh} x$ increases steadily as x increases from $-\infty$ to $+\infty$, it is clear that for any value of $\operatorname{sh} x$ there is only one value of x. If $y = \operatorname{sh} x$, we write $x = \operatorname{sh}^{-1} y$. The function $\operatorname{sh}^{-1} y$ is therefore a one-valued function of y given by the relation

$$\operatorname{sh}^{-1} \mathrm{y} = \log [\mathrm{y} + \sqrt{(1 + \mathrm{y}^2)}]. \quad \text{......................(14)}$$

This inverse function, $\operatorname{sh}^{-1} y$, is therefore not really a new function, but nevertheless the notation is useful.

The reader has seen (Ex. VI. b, No. 28) that, if $y = \operatorname{ch} x$, y has no value less than 1, and that to any value of y greater than 1 there correspond two values of x, numerically equal but of opposite sign.

The function $x = \operatorname{ch}^{-1} y$ is therefore only defined for values of $y \geqslant 1$ and is a two-valued function.

The reader should prove, by the same method as that used above for $\operatorname{sh}^{-1} y$, that

$$\operatorname{ch}^{-1} \mathrm{y} = \log [\mathrm{y} \pm \sqrt{(\mathrm{y}^2 - 1)}] = \pm \log [\mathrm{y} + \sqrt{(\mathrm{y}^2 - 1)}]. \quad \text{.......(15)}$$

Similarly, the reader will see from the results of Ex. VI. b, No. 29, that, if $y = \operatorname{th} x$, $-1 < y < 1$, and that to any value of y in this range there corresponds one value of x.

The function $x = \operatorname{th}^{-1} y$ is therefore only defined for the range of values $-1 < y < 1$ and is a one-valued function. By the same method as before, it may be shown that

$$\operatorname{th}^{-1} \mathrm{y} = \tfrac{1}{2} \log \left(\frac{1 + \mathrm{y}}{1 - \mathrm{y}} \right). \quad \text{........................(16)}$$

Applications to Geometry and Integration. The equation

$$\operatorname{ch}^2 \theta - \operatorname{sh}^2 \theta = 1$$

shows that the coordinates of any point P on the hyperbola

$$\frac{x^2}{a^2} - \frac{y^2}{b^2} = 1$$

may be written $(a \operatorname{ch} \theta, b \operatorname{sh} \theta)$; this is analogous to the use of the eccentric angle for the ellipse, (cf. Ex. VI. c, No. 13). Further, if O is the centre and if A is the vertex $(a, 0)$, it may be shown that the area of the sector AOP is $\frac{1}{2}ab\theta$ (Ex. VI. c, No. 14).

Another important application occurs in integration. Just as integrals involving $\sqrt{(1-x^2)}$ or $\sqrt{(a^2-x^2)}$ can often be evaluated by the substitution $x=\sin\theta$ or $x=a\sin\theta$, so integrals involving $\sqrt{(1+x^2)}$ or $\sqrt{(a^2+x^2)}$ can often be evaluated by putting $x=\operatorname{sh}\theta$ or $x=a\operatorname{sh}\theta$, and those involving $\sqrt{(x^2-1)}$ or $\sqrt{(x^2-a^2)}$ by putting $x=\operatorname{ch}\theta$ or $x=a\operatorname{ch}\theta$.

Example 3. Evaluate $\int \frac{1}{\sqrt{(a^2+x^2)}}dx$.

Put $\quad x=|a|.\operatorname{sh}\theta;\quad \therefore\ dx=|a|.\operatorname{ch}\theta.d\theta;$

$$\therefore \text{ the integral } =\int\frac{|a|.\operatorname{ch}\theta}{\sqrt{(a^2+a^2\operatorname{sh}^2\theta)}}.d\theta=\int 1.d\theta=\theta+c$$

$$=\operatorname{sh}^{-1}\left(\frac{x}{|a|}\right)+c$$

$$=\log[x+\sqrt{(a^2+x^2)}]+c,$$

where c is a constant.

Example 4. Evaluate $\int\sqrt{(x^2-4)}dx$, where $x<-2$.

Here it is not possible to put $x=2\operatorname{ch}\theta$, because x is negative, while $2\operatorname{ch}\theta$ is positive. But we can put $x=-2\operatorname{ch}\theta$, and we can take θ as positive.

Then $\sqrt{(x^2-4)}=+2\operatorname{sh}\theta$; also $dx=-2\operatorname{sh}\theta.d\theta$;

$$\therefore \text{ the integral } =\int(2\operatorname{sh}\theta).(-2\operatorname{sh}\theta)\,d\theta=2\int(1-\operatorname{ch}2\theta)\,d\theta$$

$$=2\theta-\operatorname{sh}2\theta+c=2\theta-2\operatorname{sh}\theta\operatorname{ch}\theta+c$$

$$=2\operatorname{ch}^{-1}\left(-\frac{x}{2}\right)+\frac{1}{2}x\sqrt{(x^2-4)}+c,$$

where c is a constant.

The difficulty of sign illustrated in this Example does not arise in numerical work because, if an integral such as $\int_{-4}^{-3}\sqrt{(x^2-4)}dx$ occurs, it is natural to begin by substituting $x=-\xi$, and this reduces the integral to $\int_3^4\sqrt{(\xi^2-4)}d\xi$, so that ξ is positive throughout the given range of values.

EXERCISE VI. c

1. Prove that, if $y \geqslant 1$, $\text{ch}^{-1}y = \pm\log\{y+\sqrt{(y^2-1)}\}$.

2. Prove that, if $|y|<1$, $\text{th}^{-1}y = \frac{1}{2}\log\frac{1+y}{1-y}$.

3. Draw the graphs of $\text{ch}^{-1}x$ and $\text{sh}^{-1}x$.

4. Draw the graphs of $\text{th}^{-1}x$ and $\text{coth}^{-1}x$.

5. Prove that, if $0<y\leqslant 1$, $\text{sech}^{-1}y = \pm\log\frac{1+\sqrt{(1-y^2)}}{y}$.

6. Express $\text{cosech}^{-1}y$ in logarithmic form, (i) if $y>0$, (ii) if $y<0$.

7. Prove that $\frac{d}{dx}(\text{sh}^{-1}x) = +\frac{1}{\sqrt{(1+x^2)}}$.

8. Prove that $\frac{d}{dx}(\text{ch}^{-1}x) = \pm\frac{1}{\sqrt{(x^2-1)}}$, and explain the ambiguous sign, showing how to distinguish between the two cases.

9. Prove that, if $|x|<1$, $\frac{d}{dx}(\text{th}^{-1}x) = \frac{1}{1-x^2}$.

10. Prove that, if $|x|>1$, $\frac{d}{dx}(\text{coth}^{-1}x) = -\frac{1}{x^2-1}$.

11. Eliminate u from the equations:

$$\text{(i) } x = a\,\text{ch}\,u,\quad y = b\,\text{sh}\,u;$$

$$\text{(ii) } x = a\,\text{ch}\,(u+\alpha),\quad y = b\,\text{sh}\,(u+\beta).$$

12. Prove that $x = a\,\text{ch}\,(u+\alpha)$, $y = b\,\text{ch}\,(u+\beta)$ are parametric equations of a hyperbola.

13. Prove that the chord of the hyperbola $x^2-y^2=a^2$ joining the points $(a\,\text{ch}\,\theta, a\,\text{sh}\,\theta)(a\,\text{ch}\,\phi, a\,\text{sh}\,\phi)$ is

$$x\,\text{ch}\frac{\theta+\phi}{2} - y\,\text{sh}\frac{\theta+\phi}{2} = a\,\text{ch}\frac{\theta-\phi}{2}.$$

14. Prove that the area between the hyperbola $\frac{x^2}{a^2}-\frac{y^2}{b^2}=1$, the x-axis, and the ordinate from P, $(a\,\text{ch}\,\theta, b\,\text{sh}\,\theta)$, is $\frac{1}{4}ab\,(\text{sh}\,2\theta-2\theta)$ and that the area of the sector bounded by the curve, the x-axis and OP is $\frac{1}{2}ab\theta$.

15. Evaluate $\int\frac{1}{a^2-x^2}dx$ by putting $x = a\,\text{th}\,\theta$, and compare the result with that obtained by expressing the integrand in partial fractions.

16. Evaluate $\int\frac{dx}{\sqrt{(x^2-a^2)}}$ where $x^2>a^2$, (i) for $x>0$, (ii) for $x<0$.

17. Evaluate $\int \sqrt{(x^2+a^2)}\,dx$.

Evaluate the following integrals:

18. $\int_2^{2\frac{1}{2}} \sqrt{(x^2-4)}\,dx$.

19. $\int_2^{2\frac{1}{2}} \frac{1}{\sqrt{(x^2-4)}}\,dx$.

20. $\int_{-5}^{-4} \frac{1}{\sqrt{(x^2+9)}}\,dx$.

21. $\int \sqrt{(x^2+9)}\,dx$.

22. $\int \sqrt{(x^2-a^2)}\,dx$, if $x > |a|$.

23. $\int \frac{x^2}{\sqrt{(x^2+1)}}\,dx$.

24. Prove that $\int \sec x\,dx = 2\,\text{th}^{-1}\left(\tan\frac{x}{2}\right)$ if $\tan^2\frac{x}{2} < 1$, and find the corresponding result when $\tan^2\frac{x}{2} > 1$ (cf. Nos. 9, 10).

EASY MISCELLANEOUS EXAMPLES.

EXERCISE VI. d.

1. Prove that $(\text{ch}\,x+\text{sh}\,x)(\text{ch}\,y+\text{sh}\,y) = \text{ch}\,(x+y)+\text{sh}\,(x+y)$.

2. Prove that $\text{ch}^6 x - \text{sh}^6 x = 1 + \frac{3}{4}\,\text{sh}^2 2x$ and express it in terms of $\text{ch}\,4x$.

3. Simplify
$\{\text{sh}\,(x-y)+\text{sh}\,x+\text{sh}\,(x+y)\} \div \{\text{ch}\,(x-y)+\text{ch}\,x+\text{ch}\,(x+y)\}$.

4. Prove that $\left(\frac{1+\text{th}\,\theta}{1-\text{th}\,\theta}\right)^3 = \text{ch}\,6\theta + \text{sh}\,6\theta$.

Differentiate with respect to x:

5. $x\sqrt{1+x^2} + \text{sh}^{-1}x$.

6. $x\sqrt{x^2-a^2} - a^2\,\text{ch}^{-1}\frac{x}{a}$.

7. $\text{sech}^{-1}x$.

8. $\text{cosech}^{-1}\frac{x}{a}$.

9. $x^{\text{sh}\,x}$.

10. $e^{ax}\,\text{sh}\,bx$.

Integrate with respect to x:

11. $e^x(\text{th}\,x + \text{sech}^2 x)$.

12. $\text{sh}\,x\,\text{sh}\,2x\,\text{sh}\,3x$.

13. $e^{ax}\,\text{sh}\,bx$.

14. $\frac{\text{sh}\,x}{(1+\text{ch}\,x)(2+\text{ch}\,x)}$.

15. Find the parabola which most closely approximates to $y = \text{ch}\,x$ near the point (0, 1), and deduce the radius of curvature of the catenary at that point.

16. Find the angle of intersection of the curves $y = 1 + \text{ch}\,x$ and $y = e^x$.

17. If $\theta > 0$, prove that $\operatorname{ch}\theta > \operatorname{sh}\theta > \theta > \operatorname{th}\theta$.

18. Evaluate $\lim\limits_{x\to 0}\dfrac{\operatorname{sh}x}{x}$ and $\lim\limits_{x\to 0}\dfrac{x}{\operatorname{th}x}$.

19. Evaluate $\lim\limits_{x\to 0}\left(\dfrac{1}{x^2}-\coth^2 x\right)$.

20. Prove that $\lim\limits_{x\to 0}\dfrac{\operatorname{sh}x-\sin x}{x^3}=\frac{1}{3}$.

21. Show that $\operatorname{sh}x-\operatorname{th}x \simeq \frac{1}{2}x^3$, if x is small.

22. Express $x\operatorname{cosech}x$ in terms of powers of x when x is so small that x^6 is negligible.

23. Prove that $\operatorname{ch}x < \frac{5}{3}$, if $|x| < 1$.

24. Prove that $\operatorname{sh}x < \dfrac{6x}{5}$, if $0 < x < 1$.

25. Prove that $2(\operatorname{ch}x-1) < x\operatorname{sh}x$.

26. Show that $x = 1{\cdot}9$ is an approximate solution of $x = 2\operatorname{th}x$, and find a closer approximation.

27. If $\tan x = \operatorname{th}y$, prove that $2\tan^{-1}(\sin 2x) = \tan^{-1}(\operatorname{sh}4y)$.

28. Prove that $\operatorname{sh}^{-1}(\cot\theta) = \log(\cot\theta + |\operatorname{cosec}\theta|)$.

29. (i) Express $\operatorname{th}^{-1}x+\operatorname{th}^{-1}y$ in the form $\operatorname{th}^{-1}p$;

(ii) Prove that if x, y are the coordinates of a point P and $\operatorname{th}^{-1}x+\operatorname{th}^{-1}y=c$, a constant, then P lies on a hyperbola with asymptotes parallel to the axes.

30. If P, Q are the points $(a\operatorname{ch}\theta,\ b\operatorname{sh}\theta)$, $(a\operatorname{ch}\phi,\ b\operatorname{sh}\phi)$ on the hyperbola $\dfrac{x^2}{a^2}-\dfrac{y^2}{b^2}=1$, prove that

(i) the area of the segment cut off by PQ is $\frac{1}{2}ab\,\{\operatorname{sh}(\theta-\phi)-\theta+\phi\}$;

(ii) the tangent at $\left(a\operatorname{ch}\dfrac{\theta+\phi}{2},\ b\operatorname{sh}\dfrac{\theta+\phi}{2}\right)$ is parallel to PQ.

(iii) the pole of PQ is $\left(a\operatorname{ch}\dfrac{\theta+\phi}{2}\operatorname{sech}\dfrac{\theta-\phi}{2},\ b\operatorname{sh}\dfrac{\theta+\phi}{2}\operatorname{sech}\dfrac{\theta-\phi}{2}\right)$.

31. By expressing $\operatorname{ch}\theta$, $\operatorname{sh}\theta$ in terms of $\operatorname{th}\dfrac{\theta}{2}\,(=t)$, find from $x=a\operatorname{ch}\theta$, $y=b\operatorname{sh}\theta$, rational algebraic parametric equations to the hyperbola, $\dfrac{x^2}{a^2}-\dfrac{y^2}{b^2}=1$. Show that two points of the curve on opposite branches cannot be represented by one set of parametric equations in terms of θ, but can be represented by one set of parametric equations in terms of an arbitrary parameter t.

32. Use the formula, $s=\int\sqrt{\left\{1+\left(\frac{dy}{dx}\right)^2\right\}}\,dx$ to show that $y^2-s^2=1$ for the catenary, $y=\operatorname{ch}x$.

33. What curve is represented by the parametric equations, $x=a\operatorname{sh}^2\theta, y=2a\operatorname{sh}\theta$? Apply the formula, $s=\int\sqrt{\left\{\left(\frac{dx}{d\theta}\right)^2+\left(\frac{dy}{d\theta}\right)^2\right\}}\,d\theta$, to show that the length of the arc of this curve, measured from the origin, is $a(\theta+\operatorname{sh}\theta\operatorname{ch}\theta)$.

HARDER MISCELLANEOUS EXAMPLES.

EXERCISE VI. e.

1. If $\frac{\operatorname{sh}2x}{a}=\frac{\sin 2y}{b}=\operatorname{ch}2x+\cos 2y$, find $\operatorname{th}2x$ and $\tan 2y$ in terms of a and b.

2. If $\tan x=\tan\lambda\operatorname{th}\mu$ and $\tan y=\cot\lambda\operatorname{th}\mu$, prove that

$$\tan(x+y)=\operatorname{sh}2\mu\operatorname{cosec}2\lambda.$$

3. Prove that $\tan\{2\tan^{-1}(\tan\alpha\operatorname{th}\beta)\}=\frac{\sin 2\alpha\operatorname{sh}2\beta}{1+\cos 2\alpha\operatorname{ch}2\beta}$.

4. If $\operatorname{ch}u=\sec\theta$, where $-\pi<\theta<\pi$, and if $u\theta$ is positive, prove that $\operatorname{sh}u=\tan\theta$, $u=\log(\sec\theta+\tan\theta)$, and $\operatorname{th}\frac{u}{2}=\tan\frac{\theta}{2}$. How are the results affected if $u\theta$ is negative ?

5. Evaluate $\frac{d}{dx}\operatorname{th}^{-1}\frac{axx_1+b(x+x_1)+c}{yy_1}$, where

$$y=\sqrt{(ax^2+2bx+c)} \quad\text{and}\quad y_1=\sqrt{(ax_1{}^2+2bx_1+c)}.$$

6. Evaluate $\int\frac{1}{(x-x_1)\sqrt{(ax^2+2bx+c)}}\,dx$ by means of No. 5.

7. If $n>1$, prove that $\int_0^\infty\{\sqrt{(x^2+1)}-x\}^n\,dx=\frac{n}{n^2-1}$.

8. Prove that $x(2+\operatorname{ch}x)>3\operatorname{sh}x$, for $x>0$.

9. Find whether $\frac{\operatorname{th}x}{x}$ or $\frac{x}{\operatorname{sh}x}$ is the greater when x is small.

10. If θ is small, prove that $\frac{\operatorname{cosec}\theta+\operatorname{cosech}\theta}{\theta^3}-\frac{2}{\theta^4}\simeq\frac{7}{180}$.

11. Solve the equation $\operatorname{ch}(\log x)=\operatorname{sh}(\log\frac{1}{2}x)+\frac{7}{4}$.

12. Show that $x=\log 2+\frac{1}{8}a$ is an approximate solution of

$$2e^x\operatorname{sh}x=3+a$$

where a is small, and find a closer approximation.

13. Prove that $\sin x = \text{th}\, mx$ where m is positive, has an infinity of roots, and that the large positive ones occur in pairs near $(2n+\frac{1}{2})\pi$, and that closer approximations are

$$(2n+\tfrac{1}{2})\,\pi \pm a - ma^2 \quad \text{where} \quad a = \text{sech}\,(2n+\tfrac{1}{2})\,m\pi.$$

14. Prove that $(\text{ch}\,\theta + \text{sh}\,\theta)^n = \text{ch}\, n\theta + \text{sh}\, n\theta$.

15. Prove that $2\,\text{ch}\, na = (\text{ch}\, a + \text{sh}\, a)^n + (\text{ch}\, a - \text{sh}\, a)^n$.

16. Express $\text{ch}\, 5x$ in terms of $\text{ch}\, x$.

17. Express $\text{sh}\, 5x$ in terms of $\text{sh}\, x$.

18. Express $\dfrac{\text{sh}\, 6x}{\text{ch}\, x}$ in terms of $\text{sh}\, x$.

19. Prove that $64\,\text{ch}^7 x = \text{ch}\, 7x + 7\,\text{ch}\, 5x + 21\,\text{ch}\, 3x + 35\,\text{ch}\, x$, and express $\text{sh}^7 x$ in terms of hyperbolic sines of multiples of x.

20. Express $\text{sh}^6 x$ in terms of hyperbolic cosines of multiples of x.

21. Prove that

$$(1 + \text{ch}\,\theta + \text{sh}\,\theta)^n = 2^n\,\text{ch}^n \frac{\theta}{2}\left(\text{ch}\frac{n\theta}{2} + \text{sh}\frac{n\theta}{2}\right).$$

22. Find an expression for $(\text{ch}\,\theta + \text{sh}\,\theta - 1)^n$ similar to that in No. 21.

23. Prove that

$$\tan^{-1}\left(\frac{\tan 2\theta + \text{th}\, 2\phi}{\tan 2\theta - \text{th}\, 2\phi}\right) + \tan^{-1}\left(\frac{\tan\theta - \text{th}\,\phi}{\tan\theta + \text{th}\,\phi}\right) = \tan^{-1}(\cot\theta\,\coth\phi).$$

24. Sum the series $\text{sh}\, a + \text{sh}\, 2a + \text{sh}\, 3a + \ldots$ to n terms.

25. Sum the series $\text{ch}\, a + \text{ch}\,(a+\beta) + \text{ch}\,(a+2\beta) + \ldots$ to n terms.

26. Sum the series $\text{ch}\,\theta + 2\,\text{ch}\, 2\theta + 3\,\text{ch}\, 3\theta + \ldots$ to n terms.

27. If n is a positive integer, prove that

$$\text{ch}\, x + n\,\text{ch}\, 2x + \frac{n(n-1)}{1.2}\text{ch}\, 3x + \ldots \text{ to } (n+1) \text{ terms}$$
$$= \left(2\,\text{ch}\frac{x}{2}\right)^n \text{ch}\,\frac{(n+2)\,x}{2}.$$

28. Prove that the sum to infinity of

$$1 + \text{ch}\,\theta + \frac{1}{2!}\text{ch}\, 2\theta + \frac{1}{3!}\text{ch}\, 3\theta + \ldots \text{ is } e^{\text{ch}\,\theta}\,\text{ch}\,(\text{sh}\,\theta).$$

29. Find the sum to infinity of

$$\text{sh}\,\theta + \frac{1}{2!}\text{sh}\, 2\theta + \frac{1}{3!}\text{sh}\, 3\theta + \ldots.$$

30. If $0 < a < \beta$, show that the sum to infinity of

$$\frac{\text{sh}\, a}{e^{\beta}} - \frac{\text{sh}\, 2a}{2e^{2\beta}} + \frac{\text{sh}\, 3a}{3e^{3\beta}} - \ldots \text{ is } \tfrac{1}{2}\log\frac{e^{a} + e^{\beta}}{e^{-a} + e^{\beta}}.$$

31. If $-\beta < \alpha < \beta$, show that the sum to infinity of

$$\frac{\operatorname{sh}\alpha}{e^{\beta}} - \frac{\operatorname{sh}3\alpha}{3e^{3\beta}} + \frac{\operatorname{sh}5\alpha}{5e^{5\beta}} - \ldots \text{ is } \tfrac{1}{2}\tan^{-1}\left(\frac{\operatorname{sh}\alpha}{\operatorname{ch}\beta}\right).$$

32. Find the sum to infinity of

$$\operatorname{ch}\theta + \frac{\sin\theta}{1!}\operatorname{ch}2\theta + \frac{\sin^2\theta}{2!}\operatorname{ch}3\theta + \ldots.$$

33. If $|x| < e^{-\alpha}$, prove that if $\alpha > 0$

$$\sum_0^{\infty}[x^n \operatorname{sh}(n+1)\alpha] = \frac{\operatorname{sh}\alpha}{1-2x\operatorname{ch}\alpha + x^2}.$$

CHAPTER VII

PROJECTION AND FINITE SERIES

Projection. The *projection of a point* on a straight line is the foot of the perpendicular from the point to the line.

If, in Fig. 45, N_1, N_2 are the projections of A_1, A_2 on Ox, then N_1N_2 is called the *projection of* A_1A_2 *on* Ox.

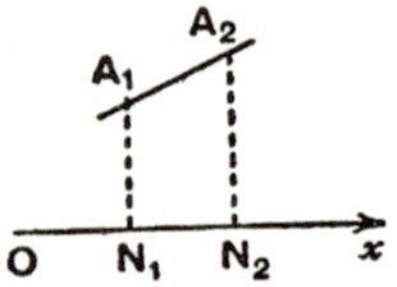

FIG. 45.

If O is the origin, Ox the x-axis and (x_1, y_1), (x_2, y_2) the coordinates of A_1, A_2, then

Projection of A_1A_2 on Ox $=N_1N_2 = x_2 - x_1$. ......(1)

This relation is true for all positions of A_1 and A_2, provided that the usual sign-conventions of coordinate geometry are observed.

The coordinates of a point are directed (i.e. positive or negative) numbers; N_1N_2 represents the displacement from N_1 to N_2 and *is measured by* a positive number if $N_1 \rightarrow N_2$ and $O \rightarrow x$ have the same sense, and by a negative number if they have opposite senses. (See also *M.G.*, p. 37.)

If A_1, A_2, A_3 *are any three points in a plane and if all the projections are taken on the same line* Ox, *then*

projection of A_1A_3 = projection of A_1A_2 + projection of A_2A_3. .....(2)

If N_1, N_2, N_3 are the projections, and (x_1, y_1), (x_2, y_2), (x_3, y_3) are the coordinates, of A_1, A_2, A_3, then

$$\begin{aligned}\text{projection of } A_1A_3 &= N_1N_3 = x_3 - x_1 = (x_2 - x_1) + (x_3 - x_2)\\ &= N_1N_2 + N_2N_3\\ &= \text{sum of projections of } A_1A_2 \text{ and } A_2A_3.\end{aligned}$$

A similar argument shows that, for any number of points in a plane, the projection of AC is the sum of the projections of AB_1, B_1B_2, B_2B_3, ... , $B_{n-1}B_n$, B_nC. (See *M.G.* Ch. V.)

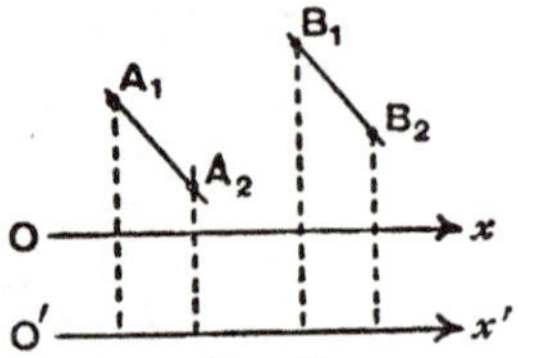

FIG. 46.

The following results are evident from Fig. 46.

(i) If $O'x'$ is parallel to and in the same sense as Ox, the projections of A_1A_2 on Ox and $O'x'$ are equal.

(ii) If two lines A_1A_2, B_1B_2 are equal and parallel and in the same sense, their projections on any line Ox are equal.

(iii) The projection of A_1A_2 on any line Ox is *equal in magnitude and opposite in sign* to that of A_2A_1 on Ox. For, if A_1 is (x_1, y_1) and A_2 is (x_2, y_2), the projections are $x_2 - x_1$ and $x_1 - x_2$ respectively.

Measurement of Angles. An angle from a directed line Ox to a directed line A_1A_2 is *defined* as follows, (see Fig. 47). Through A_1 draw A_1x_1 parallel to Ox and in the same sense; then if a rotation through an angle θ in the anti-clockwise direction will bring A_1x_1 into the position A_1A_2, the angle θ is called an angle from Ox to A_1A_2. This is sometimes written

$$\angle(A_1A_2, Ox) = \theta. \quad \text{..................(3)}$$

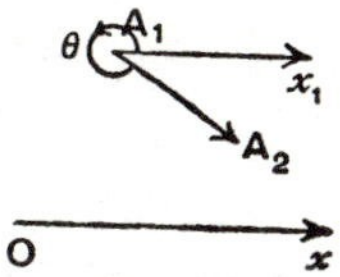

FIG. 47.

The following facts deserve notice:

(i) If θ is an angle from Ox to A_1A_2, then $\theta + 2n\pi$ is also an angle from Ox to A_1A_2, n being any positive or negative integer.

(ii) The angles from Ox to A_2A_1 are not the same as those from Ox to A_1A_2. If one value of $\angle(A_1A_2, Ox)$ is θ, then one value of $\angle(A_2A_1, Ox)$ is $\theta + \pi$.

(iii) With the notation of (ii), one value of $\angle(Ox, A_1A_2)$ is $-\theta$, and the general value is $2n\pi - \theta$.

The equality sign in equation (3) is in fact the sign of congruence (mod 2π); and the order of the elements in the symbol, $\angle(A_1A_2, Ox)$, is relevant.

Evaluation of Projections. If the length of A_1P is l units, and if $\angle(A_1P, Ox) = \theta$, the projection of A_1P on $Ox = l\cos\theta$. ....(4)

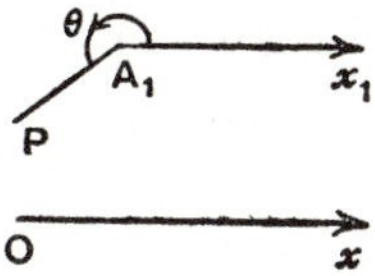

FIG. 48.

This is a direct consequence of the definition of the cosine of the general angle (*E.T.*, Ch. VII, p. 99). For, if A_1x_1 is parallel to and in the same sense as Ox, $\angle(A_1P, A_1x_1) = \angle(A_1P, Ox) = \theta$;

$$\therefore \text{ projection of } A_1P \text{ on } Ox$$
$$= \text{projection of } A_1P \text{ on } A_1x_1$$
$$= l\cos\theta.$$

Here, l is a signless number and the statement is true for all values of θ; $l\cos\theta$ may, of course, be either positive or negative.

Position of a point on a directed line. On a directed line A_1A_2, there are two positions of P such that the length of A_1P is l units, see Fig. 49 (a) and Fig. 49 (b).

This ambiguity can be removed by a natural use of a sign-convention.

If the sense of A_1P is the same as that of A_1A_2, we write $A_1P = +l$, and if it is opposite to the sense of A_1A_2 we write $A_1P = -l$. By means of this convention, the position of a point P on a *directed* line is fixed

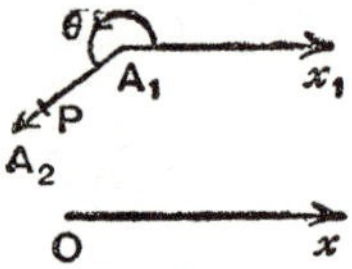

FIG. 49 (*a*).

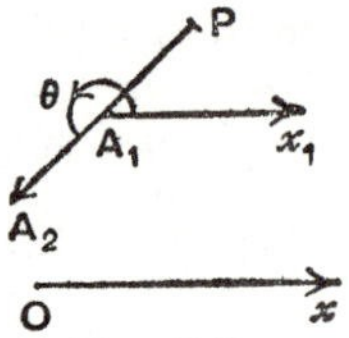

FIG. 49 (*b*).

uniquely by the *directed* (*positive or negative*) *number* which is the measure of A_1P. It is often convenient to represent this directed number by A_1P.

We can now replace equation (4) by the following :

If the directed line A_1A_2 *makes an angle* θ *with* Ox, *and if the position of* P *on* A_1A_2 *is given by the directed number* A_1P, *then*

$$\textbf{the projection of } A_1P \textbf{ on } Ox = A_1P \,.\, \cos\theta. \quad \text{............(5)}$$

For, in Fig. 49 (*a*), $\angle(A_1P, Ox) = \angle(A_1A_2, Ox) = \theta$ and $A_1P = +l$, where l units is the length of A_1P ;

$\therefore$ by (4), projection of A_1P on $Ox = l\cos\theta = A_1P \,.\, \cos\theta$.

In Figs. 49 (*a*), 49 (*b*), the projections of A_1P on Ox are equal in magnitude and opposite in sign ;

$\therefore$ in Fig. 49 (*b*), the projection of A_1P on $Ox = -l\cos\theta$;

but in Fig. 49 (*b*), $A_1P = -l$;

$\therefore$ the projection is $A_1P\cos\theta$, as before.

Projection on the axis of y. The axis of y is the directed line through O, which makes $+\frac{\pi}{2}$ with Ox, and the y-coordinate of any point P is the projection of OP on Oy.

It follows, by the same argument as before, from the definition of the sine of the general angle, that if $\angle(A_1A_2, Ox) = \theta$, and if P is any point on the directed line A_1A_2, given by the directed number A_1P, then

$$\textbf{the projection of } A_1P \textbf{ on } Oy = A_1P \,.\, \sin\theta. \quad \text{............(6)}$$

Example 1. Find, with the data of Fig. 50, the projections of OB on Ox and Oy.

$\angle(OA, Ox) = \alpha$; but the directed line AB is $\frac{3\pi}{2}$ ahead of OA ;

$$\therefore \angle(AB, Ox) = \alpha + \frac{3\pi}{2};$$

$\therefore$ the projections of OA, AB on Ox are $l \cos a$, $m \cos\left(a+\frac{3\pi}{2}\right)$;

$$\therefore \text{ projection of OB on O}x = l \cos a + m \cos\left(a+\frac{3\pi}{2}\right)$$

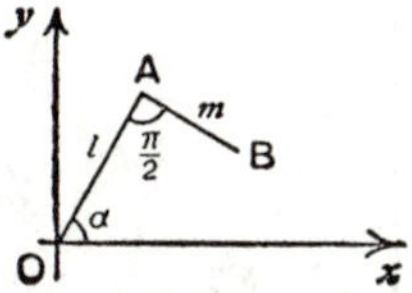

FIG. 50.

$$= l \cos a + m \sin a.$$

Similarly, projection of OB on O$y = l \sin a + m \sin\left(a+\frac{3\pi}{2}\right)$

$$= l \sin a - m \cos a.$$

Note. It saves time in working examples to adjust the signs of the terms by inspection of the figure ; a glance at Fig. 50 shows that the projection of OB on OY is $l \sin a - m \cos a$, not $l \sin a + m \cos a$.

EXERCISE VII. a.

1. In Fig. 51, ABC is equilateral and $\angle$(AB, Ox) $=\theta$; find expressions for (i) $\angle$(BC, Ox); (ii) $\angle$(CA, Ox); (iii) $\angle$(BA, Ox); (iv) $\angle$(Ox, AC).

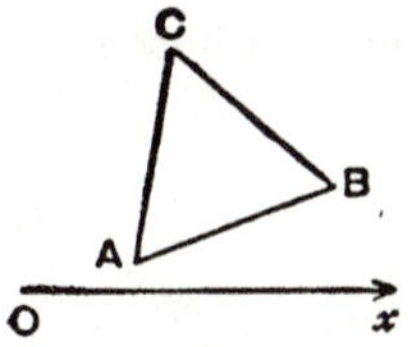

FIG. 51.

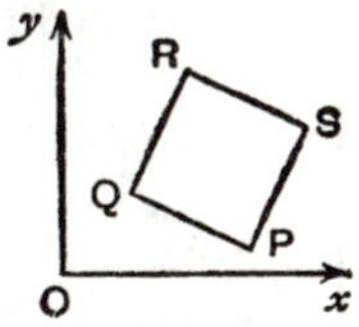

FIG. 52.

2. In Fig. 52, PQRS is a square and $\angle$(PQ, Ox) $=\phi$; find expressions for (i) $\angle$(PS, Ox); (ii) $\angle$(RS, Ox); (iii) $\angle$(SQ, Ox); (iv) $\angle$(RP, Ox).

3. With the data of No. 1, name a directed line such that the angle from Ox to it equals (i) $\frac{\pi}{3}+\theta$; (ii) $\theta-\frac{2\pi}{3}$.

4. With the data of No. 2, name a directed line such that the angle from Ox to it equals (i) $\phi - \frac{3\pi}{2}$; (ii) $\phi + \frac{3\pi}{2}$.

5. With the data of Fig. 53, find the projections of the directed lines AB, CD, EF, (i) on Ox, (ii) on Oy.

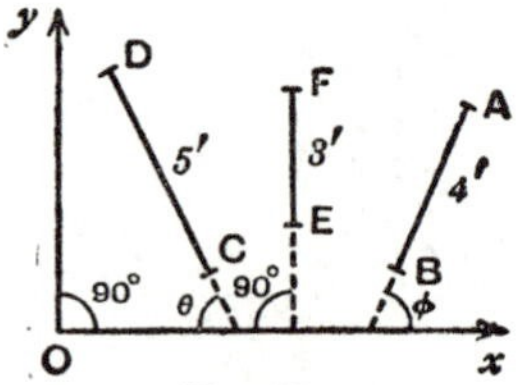

FIG. 53.

6. With the data of No. 2, if the length of PQ is c units, find the projections of the directed lines RQ, QP, QS, (i) on Ox, (ii) on Oy.

7. In Fig. 54, ABCDEF is a regular hexagon; the length of AB is a units; find the projections on AK of the directed lines AB, BC, AC, AD, CF.

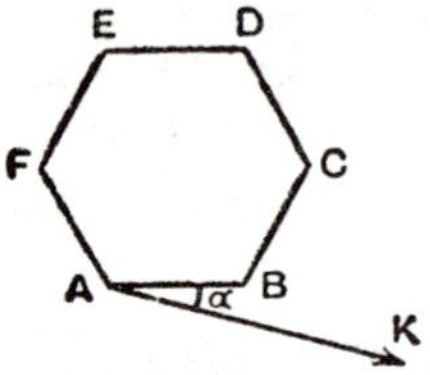

FIG. 54.

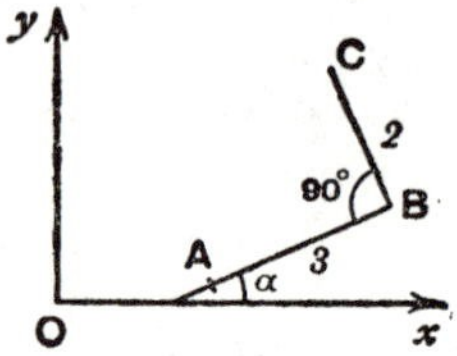

FIG. 55.

8. With the data of Fig. 55, find the projections of AC, (i) on Ox, (ii) on Oy.

9. With the data of Fig. 56, find the coordinates of C and D.

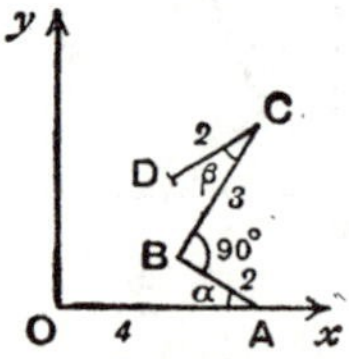

FIG. 56.

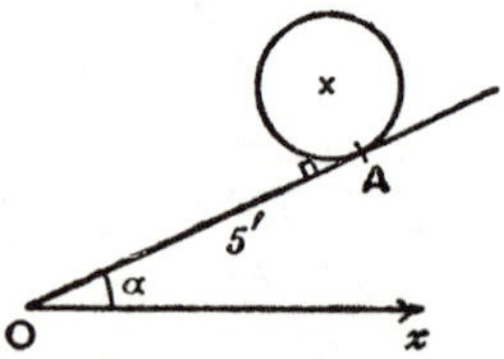

FIG. 57.

10. Fig. 57 represents a wheel of radius 1 ft. on an inclined plane; OA = 5 ft. Find the height of its centre above the horizontal Ox.

11. If from the point (h, k) a line of length r is drawn in a direction making an angle θ with Ox, what are the coordinates of its other extremity?

12. If the directed line AB is of length s, and if $\angle(\text{AB}, \text{O}x)=\phi$, and if B is the point (h, k), what are the coordinates of A ?

13. If AB, BC are of lengths r_1, r_2 and make angles θ_1, θ_2 with Ox, what is the length of AC ?

14. In Fig. 58, OD and PR are perpendiculars to AB, the length of OD is p, and the coordinates of P are (h, k); find the length of RP by taking the projections of RD, DO, ON, NP on OD.

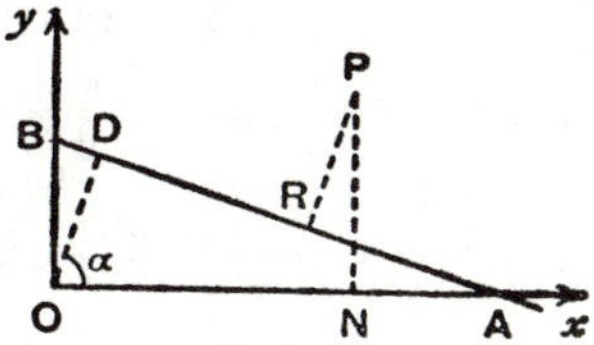

FIG. 58.

Addition Theorems. To prove that

$$\text{(i) } \cos(A+B)=\cos A \cos B-\sin A \sin B;$$
$$\text{(ii) } \sin(A+B)=\sin A \cos B+\cos A \sin B;$$

for angles of any magnitude.

Let the directed lines Oξ, OP, Oη make angles A, A + B, A $+\frac{\pi}{2}$ with Ox; and let the projections of P on Oξ, Oη be N, M. Suppose that OP contains l units of length.

The positions of N, M on the directed lines Oξ, Oη are given by the directed numbers which measure ON, OM, and these are, by the definitions of the cosine and sine of the general angle, $l\cos$ B, $l \sin$ B.

∴ by equation (5), p. 120,

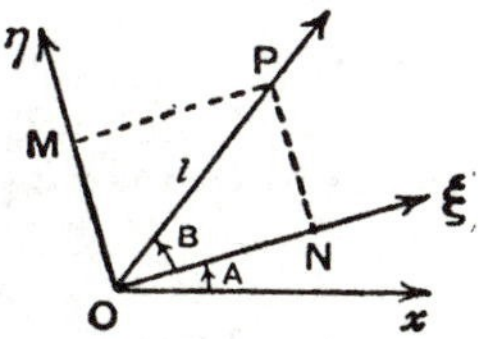

FIG. 59.

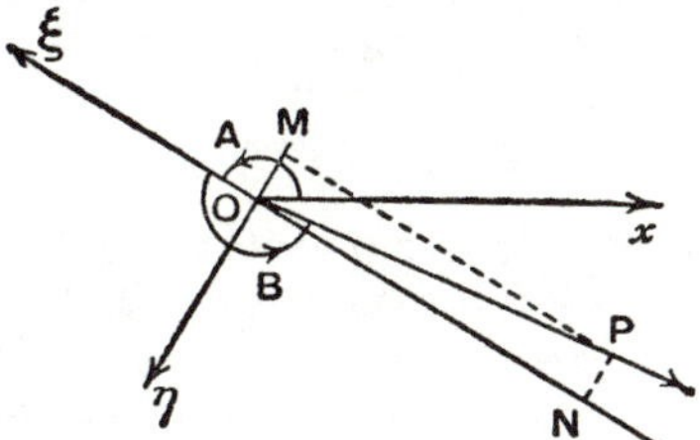

FIG. 60.

$$\text{Projection of ON on O}x = l\cos B \,.\, \cos A.$$

$$\text{Projection of OM on O}x = l\sin B \,.\, \cos\left(A+\frac{\pi}{2}\right).$$

Also the projection of OP on Ox $=l\cos(A+B)$.

But the projection of OP on Ox is equal to the sum of the projections of ON, NP, i.e. to the sum of the projections of ON, OM, on Ox.

$$\therefore\ l \cos(A+B) = l \cos B \cos A + l \sin B \cos\left(A+\frac{\pi}{2}\right).$$

But $\cos\left(A+\frac{\pi}{2}\right) = -\sin A$, see *E.T.*, pp. 199, 200;

$$\therefore\ \cos(A+B) = \cos A \cos B - \sin A \sin B. \quad \ldots\ldots\ldots\ldots(7)$$

Further, if the directed line Oy makes $+\frac{\pi}{2}$ with Ox, the projections of ON, OM, OP on Oy are

$$l \cos B \sin A,\ l \sin B \sin\left(A+\frac{\pi}{2}\right),\ l \sin(A+B).$$

$\therefore$ as before,

$$l \sin(A+B) = l \cos B \sin A + l \sin B \sin\left(A+\frac{\pi}{2}\right).$$

But $\sin\left(A+\frac{\pi}{2}\right) = \cos A$, see *E.T.*, pp. 199, 200;

$$\therefore\ \sin(A+B) = \sin A \cos B + \cos A \sin B. \quad \ldots\ldots\ldots\ldots(8)$$

This proof holds good for values of A and B of any magnitude, positive or negative. Figs. 59, 60 show two possible cases; the

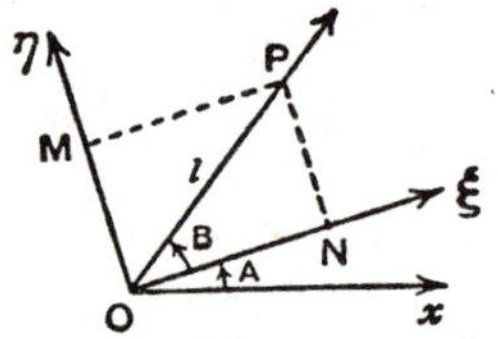

FIG. 59.

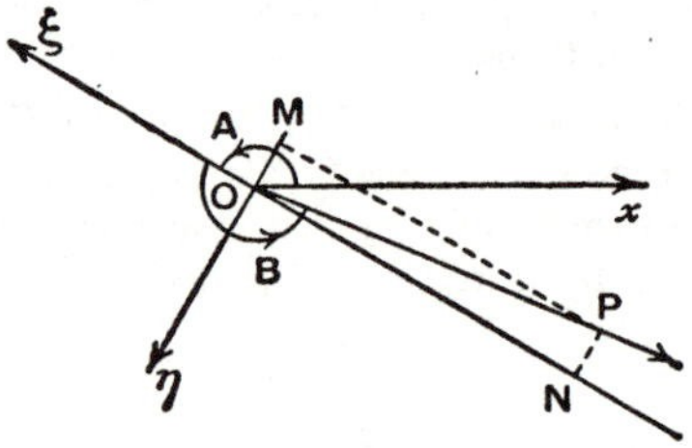

FIG. 60.

reader should draw other figures (e.g. $A=100°$, $B=50°$ or $A=220°$, $B=160°$) and satisfy himself that the proof applies to them, *without any modification*.

Since the results of this chapter and their proofs hold for negative angles (see *E.T.*, Ch. XIV, p. 198), we may write $-B$ for B in (7) and (8). This gives

$$\cos(A-B) = \cos A \cos(-B) - \sin A \sin(-B);$$

$$\therefore\ \cos(A-B) = \cos A \cos B + \sin A \sin B \quad \ldots\ldots\ldots\ldots(9)$$

and $$\sin(A-B) = \sin A \cos(-B) + \cos A \sin(-B);$$

$$\therefore\ \sin(A-B) = \sin A \cos B - \cos A \sin B. \quad \ldots\ldots\ldots\ldots(10)$$

Application of Projection to the Summation of Certain Series.

Sum to n terms the series

(i) $\cos\alpha+\cos(\alpha+\beta)+\cos(\alpha+2\beta)+\ldots$;

(ii) $\sin\alpha+\sin(\alpha+\beta)+\sin(\alpha+2\beta)+\ldots$.

In Fig. 61, OA_1, A_1A_2, ... , $A_{n-1}A_n$ are equal chords of a circle of radius R, forming an open polygon with exterior angles β.

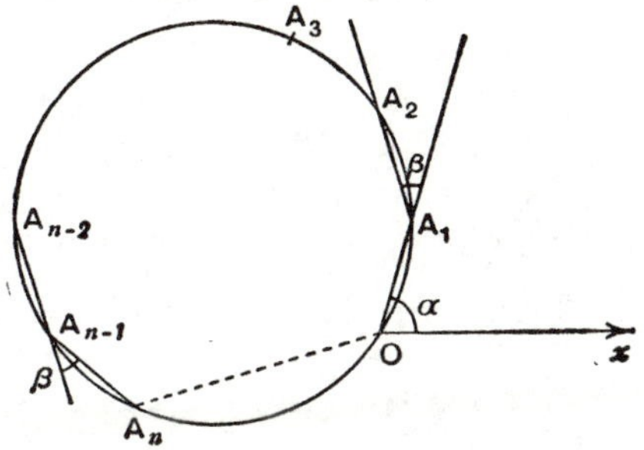

FIG. 61.

Each chord subtends an angle $\frac{1}{2}\beta$ at the circumference and $\therefore$ from the formula $R=\frac{a}{2\sin A}$, its length is $2R\sin\frac{\beta}{2}$; also OA_n subtends an angle $\frac{n\beta}{2}$ at the circumference, therefore $OA_n=2R\sin\frac{n\beta}{2}$.

Draw Ox so that $\angle(OA_1, Ox)=\alpha$.

Then $\angle(A_1A_2, Ox)=\alpha+\beta$; $\angle(A_2A_3, Ox)=\alpha+2\beta$; ...

$\angle(A_{n-1}A_n, Ox)=\alpha+(n-1)\beta$. Also $\angle(OA_n, Ox)=\alpha+\frac{1}{2}(n-1)\beta$.

Now the projection of OA_n on Ox is the sum of the projections on Ox of OA_1, A_1A_2, ... , $A_{n-1}A_n$;

$$\therefore\ 2R\sin\frac{n\beta}{2}\cos\left(\alpha+\frac{n-1}{2}\beta\right)$$

$$=2R\sin\frac{\beta}{2}\{\cos\alpha+\cos(\alpha+\beta)+\ldots+\cos(\alpha+\overline{n-1}\,\beta)\};$$

$$\therefore\ \cos\alpha+\cos(\alpha+\beta)+\cos(\alpha+2\beta)+\ldots+\cos(\alpha+\overline{n-1}\,\beta)$$

$$=\frac{\cos\left(\alpha+\frac{n-1}{2}\beta\right).\sin\frac{n\beta}{2}}{\sin\frac{\beta}{2}}\ \ldots\ldots\ldots\ldots\ldots\ldots\ldots(11)$$

Similarly, taking the projections on Oy,

$$\sin \alpha + \sin(\alpha+\beta) + \sin(\alpha+2\beta) + \ldots + \sin(\alpha+\overline{n-1}\,\beta)$$

$$= \frac{\sin\left(\alpha + \frac{n-1}{2}\beta\right) . \sin\frac{n\beta}{2}}{\sin\frac{\beta}{2}}. \quad \ldots\ldots\ldots\ldots\ldots\ldots(12)$$

Relation (12) may be deduced from relation (11) by writing $\alpha - \frac{\pi}{2}$ for α.

EXERCISE VII. b.

1. Examine the proof on pp. 123, 124 for the expansion of $\cos(A+B)$, drawing appropriate figures, in the following cases:

(i) $\pi < A < \frac{3\pi}{2}$, $\frac{\pi}{2} < B < \pi$; (ii) $\frac{3\pi}{2} < A < 2\pi$, $\pi < B < \frac{3\pi}{2}$;

(iii) $0 < A < \frac{\pi}{2}$, $-\frac{\pi}{2} < B < 0$; (iv) $\frac{\pi}{2} < A < \pi$, $-\frac{3\pi}{2} < B < -\pi$.

2. If in Fig. 59 the coordinates of P referred to Ox and the line Oy which makes $+\frac{\pi}{2}$ with Ox as axes, are x and y, what are the coordinates of P referred to $O\xi$ and $O\eta$?

3. Answer the same question as in No. 2 for Fig. 60.

4. Write out in full the proof by the method of pp. 123, 124, that

$$\cos(A-B) = \cos A \cos B + \sin A \sin B.$$

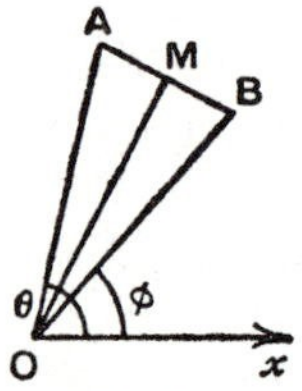

FIG. 62.

5. In Fig. 62, $OA = OB$, $AM = MB$, $\angle xOA = \theta$, $\angle xOB = \phi$; express the projections of OA, OB in terms of those of OM, MA, MB, and, by adding, prove that

(i) $\cos\theta + \cos\phi = 2\cos\frac{\theta+\phi}{2}\cos\frac{\theta-\phi}{2}$;

(ii) $\sin\theta + \sin\phi = 2\sin\frac{\theta+\phi}{2}\cos\frac{\theta-\phi}{2}$.

6. With the data of No. 5, by subtracting, prove the corresponding formulae for $\cos\theta - \cos\phi$ and $\sin\theta - \sin\phi$.

7. By projecting the sides of a regular pentagon on suitable lines, prove that

(i) $\cos 5^\circ + \cos 77^\circ + \cos 149^\circ + \cos 221^\circ + \cos 293^\circ = 0$;
(ii) $\sin 5^\circ + \sin 77^\circ + \sin 149^\circ + \sin 221^\circ + \sin 293^\circ = 0$.

8. Prove the results of No. 7 by formulae (11) and (12).

9. Prove by projection that

$$\cos\theta + \cos\left(\theta + \frac{2\pi}{n}\right) + \cos\left(\theta + \frac{4\pi}{n}\right) + \dots \text{ to } n \text{ terms} = 0.$$

Is a similar result true for sines?

10. Use formulae (11) and (12) to verify the results of No. 9.

11. By means of the identity

$$2\sin\theta\cos k\theta = \sin(k+1)\theta - \sin(k-1)\theta,$$

prove that $2\sin\theta\,\{\cos\theta + \cos 3\theta + \dots + \cos(2n-1)\theta\} = \sin 2n\theta$.
Prove this also by formula (11)

12. By means of the identity

$$2\sin\theta\sin k\theta = \cos(k-1)\theta - \cos(k+1)\theta,$$

find the sum of the series $\sin\theta + \sin 3\theta + \sin 5\theta + \dots$ to n terms. Check your result by formula (12).

Series. The formulae (11), (12) give the sums of series of sines or cosines of angles which are in A.P. Their utility justifies the addition of an analytical proof, which also illustrates an important method of summation.

Sum to n terms the series

$$\cos\alpha + \cos(\alpha+\beta) + \cos(\alpha+2\beta) + \dots .$$

Multiply each term by $2\sin\frac{1}{2}\beta$.

$$2\cos\alpha \,.\, \sin\tfrac{1}{2}\beta = \sin(\alpha+\tfrac{1}{2}\beta) - \sin(\alpha-\tfrac{1}{2}\beta),$$
$$2\cos(\alpha+\beta) \,.\, \sin\tfrac{1}{2}\beta = \sin(\alpha+\tfrac{3}{2}\beta) - \sin(\alpha+\tfrac{1}{2}\beta),$$
$$\dots\dots\dots\dots$$
$$2\cos(\alpha+\overline{n-1}\,\beta) \,.\, \sin\tfrac{1}{2}\beta = \sin(\alpha+\overline{n-\tfrac{1}{2}}\beta) - \sin(\alpha+\overline{n-\tfrac{3}{2}}\beta).$$

By addition,

$$2\sin\tfrac{1}{2}\beta \,.\, (\text{sum of series}) = \sin(\alpha+\overline{n-\tfrac{1}{2}}\beta) - \sin(\alpha-\tfrac{1}{2}\beta)$$
$$= 2\cos(\alpha+\tfrac{1}{2}\overline{n-1}\beta)\sin\tfrac{1}{2}n\beta.$$

This gives for the sum of the series the same expression as was obtained in relation (11), p. 125.

The sum can be expressed in words as follows :

$$\frac{\cos(\text{average angle}) \,.\, \sin(\text{n times semi-difference})}{\sin(\text{semi-difference})}, \quad \ldots\ldots(13)$$

and it is best to remember it in this form.

The reader should show that the series

$$\sin\alpha + \sin(\alpha+\beta) + \sin(\alpha+2\beta) + \ldots + \sin(\alpha+\overline{n-1}\,\beta)$$

can be summed by multiplying each term by the *same factor* as before, $2\sin\frac{\beta}{2}$, and that the sum may be written

$$\frac{\sin(\text{average angle}) \,.\, \sin(\text{n times semi-difference})}{\sin(\text{semi-difference})}. \quad \ldots\ldots(14)$$

The fact that the second series can be deduced from the first by writing $\alpha - \frac{1}{2}\pi$ for α shows that the multiplier $2\sin\frac{\beta}{2}$ required for the first must equally suit the second series.

If $\pi + \beta$ is written for β in the two series, we obtain

$$\text{(i)}\ \cos\alpha - \cos(\alpha+\beta) + \cos(\alpha+2\beta) - \ldots .$$

$$\text{(ii)}\ \sin\alpha - \sin(\alpha+\beta) + \sin(\alpha+2\beta) - \ldots .$$

These could be summed directly by using the multiplier,

$$2\sin\frac{\pi+\beta}{2} = 2\cos\frac{\beta}{2}.$$

The sums of the sine and cosine series are deducible from one another by differentiation with respect to α.

The application of differentiation or integration to deduce the sum of one series from that of another is frequently useful.

It is justified by the identities :

$$\frac{du}{dx} + \frac{dv}{dx} = \frac{d(u+v)}{dx}\,; \quad \int_a^b u\,dx + \int_a^b v\,dx = \int_a^b (u+v)\,dx.$$

But this argument *does not apply to an* infinite series, because the sum of an infinite series is not the sum of its terms, and term-by-term differentiation or integration of an infinite series need not in fact give the differential coefficient or integral of the sum to infinity, unless special conditions are satisfied.

Example 2. Sum to n terms

$$\cos^2\theta+\cos^2 2\theta+\cos^2 3\theta+\ldots;$$

The series $=\frac{1}{2}(1+\cos 2\theta)+\frac{1}{2}(1+\cos 4\theta)+\ldots+\frac{1}{2}(1+\cos 2n\theta)$

$$=\tfrac{1}{2}n+\tfrac{1}{2}(\cos 2\theta+\cos 4\theta+\ldots+\cos 2n\theta)$$

$$=\tfrac{1}{2}n+\frac{\cos(n+1)\theta\sin n\theta}{2\sin\theta}$$

$$=\frac{n}{2}+\frac{1}{4\sin\theta}(\sin\overline{2n+1}\theta-\sin\theta)$$

$$=\tfrac{1}{4}(2n-1)+\frac{\sin(2n+1)\theta}{4\sin\theta}.$$

Example 3. Sum to n terms:

$$\sin^3\alpha+\sin^3(\alpha+\beta)+\sin^3(\alpha+2\beta)+\ldots.$$

$\sin 3\theta=3\sin\theta-4\sin^3\theta$; $\therefore\ \sin^3\theta=\frac{1}{4}(3\sin\theta-\sin 3\theta)$;

$\therefore$ the series $=\frac{3}{4}\{\sin\alpha+\sin(\alpha+\beta)+\ldots+\sin(\alpha+\overline{n-1}\,\beta)\}$

$$-\tfrac{1}{4}\{\sin 3\alpha+\sin 3(\alpha+\beta)+\ldots+\sin 3(\alpha+\overline{n-1}\,\beta)\}$$

$$=\frac{3\sin\left(\alpha+\frac{n-1}{2}\beta\right)\sin\frac{n\beta}{2}}{4\sin\frac{\beta}{2}}-\frac{\sin\left(3\alpha+\frac{3n-3}{2}\beta\right)\sin\frac{3n\beta}{2}}{4\sin\frac{3\beta}{2}}.$$

EXERCISE VII. c.

1. Sum to n terms: $\cos\frac{\theta}{2}+\cos\theta+\cos\frac{3\theta}{2}+\ldots;$

2. Prove that $\cos\frac{2\pi}{7}+\cos\frac{4\pi}{7}+\cos\frac{6\pi}{7}=-\frac{1}{2}$.

3. Prove that $\cos\frac{\pi}{11}+\cos\frac{3\pi}{11}+\cos\frac{5\pi}{11}+\cos\frac{7\pi}{11}+\cos\frac{9\pi}{11}=\frac{1}{2}$.

4. If $n-1$ is a positive integer, prove that

   $\Sigma\sin\frac{2r\pi}{n}$ and $\Sigma\cos\frac{2r\pi}{n}$, for $r=1$ to n, are both zero.

5. Prove that $\Sigma\cos\frac{2r\pi}{2n+1}$, for $r=1$ to n, is $-\frac{1}{2}$.

6. Prove that $\dfrac{\sin\theta+\sin 2\theta+\ldots+\sin(n-1)\theta}{\cos\theta+\cos 2\theta+\ldots+\cos(n-1)\theta}=\tan\frac{n\theta}{2}$.

7. Sum: $\cos\alpha - \cos(\alpha+\beta) + \cos(\alpha+2\beta) - \cos(\alpha+3\beta) + \dots$

(i) to $2n$ terms; (ii) to $(2n+1)$ terms; (iii) to m terms.

Sum to n terms the series in Nos. 8-17.

8. $\sin\alpha - \sin(\alpha+\beta) + \sin(\alpha+2\beta) - \dots$.

9. $\cos\theta - \cos 2\theta + \cos 3\theta - \dots$.

10. $\sin^2\theta + \sin^2 2\theta + \sin^2 3\theta + \dots$.

11. $\cos(2n-1)\theta + \cos(2n-3)\theta + \cos(2n-5)\theta + \dots$.

12. $\cos\theta\sin 2\theta + \cos 2\theta\sin 3\theta + \cos 3\theta\sin 4\theta + \dots$.

13. $\cos\theta - \sin 2\theta - \cos 3\theta + \sin 4\theta + \cos 5\theta - \sin 6\theta - \dots$.

14. $\cos^2\theta + \cos^2(\theta+\phi) + \cos^2(\theta+2\phi) + \dots$.

15. $\sin^2\theta\sin 2\theta + \sin^2 2\theta\sin 3\theta + \sin^2 3\theta\sin 4\theta + \dots$.

16. $\cos^3\theta + \cos^3 2\theta + \cos^3 3\theta + \dots$.

17. $\cos^4\theta + \cos^4 2\theta + \cos^4 3\theta + \dots$.

18. Find the sum to n terms of $\sin\theta + \sin 2\theta + \sin 3\theta + \dots$ and deduce the sum to n terms of $\cos\theta + 2\cos 2\theta + 3\cos 3\theta + \dots$.

19. Prove that $\sin\theta + 3\sin 3\theta + 5\sin 5\theta + \dots$ to n terms

$$= \frac{\cos\theta\sin 2n\theta - 2n\sin\theta\cos 2n\theta}{2\sin^2\theta}.$$

20. If $C = \Sigma\, r\cos(\alpha + \overline{r-1}\beta)$, for $r=1$ to n, prove that

$$2(1-\cos\beta)C = (n+1)\cos(\alpha+\overline{n-1}\beta) - \cos(\alpha-\beta) - n\cos(\alpha+n\beta).$$

21. If $-\frac{\pi}{4} < \theta < \frac{\pi}{4}$, prove that (i) the sum of any number of terms of the series $\cos\theta - \cos 3\theta + \cos 5\theta - \cos 7\theta + \dots$ is positive or zero and less than $\sqrt{2}$; and that (ii) the sum of n terms of the series

$$\sin\theta - \tfrac{1}{3}\sin 3\theta + \tfrac{1}{5}\sin 5\theta - \dots \text{ is } \int_0^\theta \tfrac{1}{2}\sec x\{1 + (-1)^{n-1}\cos 2nx\}\,dx.$$

The Difference Method. The series on p. 127 were summed by expressing each term as a difference. No rule can be given which shows exactly when or how to apply the method. Considerable experience and ingenuity are sometimes required.

The essence of the method consists in expressing the general (rth) term, u_r, in the form $f(r+1) - f(r)$.

Then $u_1 + u_2 + u_3 + \dots + u_n = \Sigma\{f(r+1) - f(r)\}$, for $r=1$ to n,

$$= f(n+1) - f(1).$$

The difficulty disappears if the reader is asked to prove that the sum to any number of terms, say r terms, is $\phi(r)$, that is to say, if he knows what the answer is to be.

For $$u=(u_1+{}_ru_2+\dots+u_r)-(u_1+u_2+\dots+u_{r-1})$$
$$=\phi(r)-\phi(r-1),$$

and the known form of the answer therefore supplies the form of the difference which he must obtain. In such cases, the work is substantially equivalent to the method of induction.

Example 4. Prove that the sum to n terms of

$\tan\theta+2\tan 2\theta+4\tan 4\theta+8\tan 8\theta+\dots$ is $\cot\theta-2^n\cot(2^n\theta)$.

If this form for the sum is correct, the 1st term, $\tan\theta$, must equal $\cot\theta-2\cot 2\theta$; we therefore start by proving that this is so.

Now $$\cot\theta-\tan\theta=\frac{\cos\theta}{\sin\theta}-\frac{\sin\theta}{\cos\theta}=\frac{\cos^2\theta-\sin^2\theta}{\sin\theta\cos\theta}$$
$$=\frac{\cos 2\theta}{\frac{1}{2}\sin 2\theta}=2\cot 2\theta;$$
$$\therefore\ \tan\theta=\cot\theta-2\cot 2\theta;$$

$\therefore$ writing 2θ for θ and multiplying by 2,

$$2\tan 2\theta=2(\cot 2\theta-2\cot 4\theta)=2\cot 2\theta-2^2\cot 4\theta.$$

Similarly, $$2^2\tan 4\theta=2^2\cot 4\theta-2^3\cot 8\theta.$$

. . . .

$$2^{n-1}\tan(2^{n-1}\theta)=2^{n-1}\cot(2^{n-1}\theta)-2^n\cot(2^n\theta);$$

$\therefore$ by addition, the sum to n terms of the given series is

$$\cot\theta-2^n\cot(2^n\theta).$$

Example 5. Sum to n terms:

(i) $\operatorname{cosec}2\theta+\operatorname{cosec}4\theta+\operatorname{cosec}8\theta+\dots$;

(ii) $2\operatorname{cosec}2\theta\cot 2\theta+4\operatorname{cosec}4\theta\cot 4\theta+8\operatorname{cosec}8\theta\cot 8\theta+\dots$.

(i) We have $$\operatorname{cosec}2\theta=\frac{1}{\sin 2\theta}=\frac{2\cos^2\theta-\cos 2\theta}{\sin 2\theta};$$

$$\therefore\ \operatorname{cosec}2\theta=\frac{2\cos^2\theta}{2\sin\theta\cos\theta}-\frac{\cos 2\theta}{\sin 2\theta}=\cot\theta-\cot 2\theta.$$

Similarly, writing 2θ for θ,

$$\operatorname{cosec}4\theta=\cot 2\theta-\cot 4\theta,$$
$$\operatorname{cosec}8\theta=\cot 4\theta-\cot 8\theta,$$

. . . .

$$\operatorname{cosec}(2^n\theta)=\cot(2^{n-1}\theta)-\cot(2^n\theta);$$

$\therefore$ by addition, the sum to n terms of the given series is

$$\cot\theta - \cot(2^n\theta).$$

(ii) Since $\dfrac{d}{d\theta}(\operatorname{cosec} n\theta) = -n \operatorname{cosec} n\theta \cot n\theta$, it follows at once that the sum of the second series is

$$-\frac{d}{d\theta}[\cot\theta - \cot(2^n\theta)] = \operatorname{cosec}^2\theta - 2^n \operatorname{cosec}^2(2^n\theta).$$

If a series such as (ii) occurred apart from the series (i) which has here served as a guide, the method would become apparent by using integration. Thus

$$\int 2 \operatorname{cosec} 2\theta \cot 2\theta \, d\theta = \int \frac{2\cos 2\theta \, d\theta}{\sin^2 2\theta} = -\operatorname{cosec} 2\theta.$$

EXERCISE VII. d.

1. Prove that $\tan\theta \sec 2\theta = \tan 2\theta - \tan\theta$; hence find the sum to n terms of $\tan\theta \sec 2\theta + \tan 2\theta \sec 4\theta + \tan 4\theta \sec 8\theta + \ldots$.

2. Prove that $\tan\theta = \cot\theta - 2\cot 2\theta$; hence find the sum to n terms of $\tan\theta + \frac{1}{2}\tan\dfrac{\theta}{2} + \frac{1}{4}\tan\dfrac{\theta}{4} + \frac{1}{8}\tan\dfrac{\theta}{8} + \ldots$.

3. Prove that $\tan(\alpha + r\beta)\tan(\alpha + \overline{r-1}\,\beta)$

$$= \cot\beta\{\tan(\alpha + r\beta) - \tan(\alpha + \overline{r-1}\,\beta)\} - 1;$$

use this result to sum to n terms a certain series.

4. Prove that $\tan 2\theta - 2\tan\theta = \tan^2\theta \tan 2\theta$; use this result to sum to n terms a certain series.

5. Prove that $\sin^2\theta - 2\sin^2\dfrac{\theta}{2} = 2\cos\theta \sin^2\dfrac{\theta}{2}$; use this result to sum to n terms a certain series.

6. (i) Prove that $\cot r\theta - \cot(r+1)\theta = \dfrac{\sin\theta}{\sin r\theta \sin(r+1)\theta}$.

(ii) Sum to n terms, $\operatorname{cosec}\theta \operatorname{cosec} 2\theta + \operatorname{cosec} 2\theta \operatorname{cosec} 3\theta + \ldots$.

7. Prove that the sum to n terms of

$$\tan\frac{\theta}{2}\sec\theta + \tan\frac{\theta}{4}\sec\frac{\theta}{2} + \tan\frac{\theta}{8}\sec\frac{\theta}{4} + \ldots$$

equals $\tan\theta - \tan\dfrac{\theta}{2^n}$.

8. Prove that the sum to n terms of

$$\frac{\sin^3\theta}{\cos 3\theta} + \frac{\sin^3 3\theta}{3\cos 9\theta} + \frac{\sin^3 9\theta}{9\cos 27\theta} + \ldots$$

equals $\frac{3}{8}\{3^{-n}\tan(3^n\theta) - \tan\theta\}$.

Sum to n terms the series in Nos. 9-15.

9 $\sec\theta\sec 2\theta+\sec 2\theta\sec 3\theta+\sec 3\theta\sec 4\theta+\dots$.

10. $\tan\theta\tan 2\theta+\tan 2\theta\tan 3\theta+\tan 3\theta\tan 4\theta+\dots$.

11. $\cot\theta\cot 2\theta+\cot 2\theta\cot 3\theta+\cot 3\theta\cot 4\theta+\dots$.

12. $\dfrac{1}{\cos\theta+\cos 3\theta}+\dfrac{1}{\cos\theta+\cos 5\theta}+\dfrac{1}{\cos\theta+\cos 7\theta}+\dots$.

13. $\sin\theta\sec 3\theta+\sin 3\theta\sec 9\theta+\sin 9\theta\sec 27\theta+\dots$.

14. $\sec^2\theta+\dfrac{1}{2^2}\sec^2\dfrac{\theta}{2}+\dfrac{1}{4^2}\sec^2\dfrac{\theta}{4}+\dots$.

15. $\tan^2\theta+\dfrac{1}{2^2}\tan^2\dfrac{\theta}{2}+\dfrac{1}{4^2}\tan^2\dfrac{\theta}{4}+\dots$.

16. Prove that $\tan^{-1}(n+1)-\tan^{-1}n=\cot^{-1}(1+n+n^2)$. Hence sum the series, $\cot^{-1}3+\cot^{-1}7+\cot^{-1}13+\dots+\cot^{-1}(1+n+n^2)$.

Sum to n terms the following series :

17. $\tan^{-1}\left(\dfrac{2}{1+1.3}\right)+\tan^{-1}\left(\dfrac{2}{1+3.5}\right)+\tan^{-1}\left(\dfrac{2}{1+5.7}\right)+\dots$.

18. $\tan^{-1}\left(\dfrac{2}{1^2}\right)+\tan^{-1}\left(\dfrac{2}{2^2}\right)+\tan^{-1}\left(\dfrac{2}{3^2}\right)+\dots$.

19. $\cot^{-1}\left(\dfrac{1^2}{8}\right)+\cot^{-1}\left(\dfrac{2^2}{8}\right)+\cot^{-1}\left(\dfrac{3^2}{8}\right)+\dots$.

20. $\cot^{-1}(2.1^2)+\cot^{-1}(2.2^2)+\cot^{-1}(2.3^2)+\dots$.

EASY MISCELLANEOUS EXAMPLES.

EXERCISE VII. e.

1. In Fig. 63, AB, BC are of unit length ; prove by projection that

$$\cos\alpha-\sin\alpha=\sqrt{2}.\cos\left(\alpha+\frac{\pi}{4}\right).$$

What is the maximum value of $\cos\alpha-\sin\alpha$?

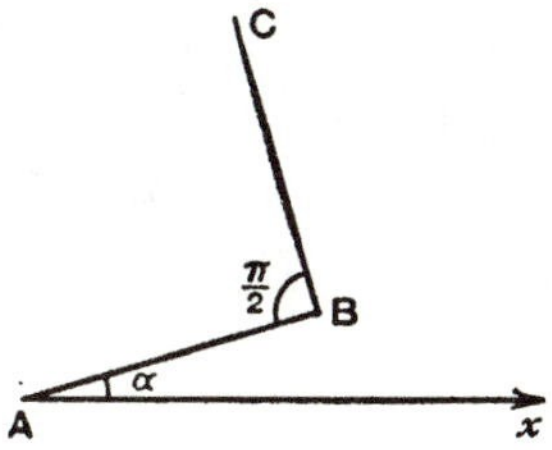

FIG. 63.

2. Use the method of No. 1 to find the maximum value of

$$7\cos\alpha + 24\sin\alpha.$$

3. AD is an altitude of $\triangle$ABC, and Z is the middle point of AB. Prove that the projection of ZC on AD is $\frac{1}{2}c\sin B$.

4. If a and b are given numbers, express $a\cos\theta + b\sin\theta$ in the form $r\sin(\theta+\alpha)$. Give geometrical interpretations of r and α. What are the maximum and minimum values of $a\cos\theta + b\sin\theta$, and for what values of θ do they occur?

5. Prove that $\Sigma\left(\sin\frac{r\pi}{n}\cos\frac{r\pi}{n}\right)$, for $r=1$ to n, is zero.

6. Prove that $\Sigma\cos^2\left(\theta+\frac{2r\pi}{n}\right)$, for $r=1$ to n, is $\frac{n}{2}$.

7. Find $\Sigma(\cos r\theta\cos\overline{r+1}\,\theta)$, for $r=1$ to n.

8. Find $\Sigma\{\sin(\alpha+r\beta)\sin(\alpha+\overline{r+1}\,\beta)\}$, for $r=1$ to n.

9. Sum to n terms:

$$\sin 2\theta\sin^2\theta + \sin 3\theta\sin^2\frac{3\theta}{2} + \sin 4\theta\sin^2 2\theta + \dots.$$

10. Prove that

$$1+2\sum_{r=1}^{n}(\cos r\alpha\cos r\beta) = \frac{\cos n\alpha\cos(n+1)\beta - \cos(n+1)\alpha\cos n\beta}{\cos\beta - \cos\alpha}.$$

11. Prove that $\Sigma(r\sin r\theta)$, for $r=1$ to n, is

$$\frac{(n+1)\sin n\theta - n\sin(n+1)\theta}{2(1-\cos\theta)}.$$

12. Evaluate $\Sigma(r^2\cos r\theta)$, for $r=1$ to n.

13. Prove that $\sin\theta\div\sin\frac{\theta}{2^n} = 2^n\cos\frac{\theta}{2}\cos\frac{\theta}{2^2}\dots\cos\frac{\theta}{2^n}$, and deduce the values of $\Sigma\log\cos\frac{\theta}{2^r}$ and $\Sigma\, 2^{-r}\tan\frac{\theta}{2^r}$, for $r=1$ to n.

14. Sum the series $\log\cos\theta + \log\cos 2\theta + \log\cos 4\theta + \log\cos 8\theta + \dots$ to n terms.

15. Evaluate $\lim\limits_{n\to\infty}\frac{s_1+s_2+\dots+s_n}{n}$ when s_n is equal to the sum to n terms of $1-1+1-1+1-1+\dots.$

16. If $s_n = \Sigma\sin r\theta$, for $r=1$ to n, and if $\theta\neq 2r\pi$, prove that

$$\lim_{n\to\infty}\frac{s_1+s_2+\dots+s_n}{n} = \tfrac{1}{2}\cot\tfrac{1}{2}\theta.$$

Also find the same limit for $\Sigma\cos(2r-1)\theta$, for $r=1$ to n, when $\theta\neq k\pi$.

17. Sum to n terms $\tan\theta + 2\tan 2\theta + 4\tan 4\theta + \dots.$

18. Sum to n terms $\tan^2\theta + 2^2\tan^2 2\theta + 4^2\tan^2 4\theta + \dots$.

19. Prove that $4\Sigma\left\{3^{r-1}\sin^3\left(\frac{\theta}{3^r}\right)\right\}$, for $r=1$ to n, is equal to $3^n \sin\frac{\theta}{3^n} - \sin\theta$. Deduce another result by differentiation.

20. In attempting to draw a regular polygon $A_1A_2 \dots A_n$ a person draws the sides A_1A_2, A_2A_3, ... in order each of length c, but makes each interior angle of the figure too great by α. Prove that the final vertex, A_{n+1}, will be at distance $c\sin\frac{n\alpha}{2}\operatorname{cosec}\left(\frac{\pi}{n}-\frac{\alpha}{2}\right)$ from A_1.

21. If O is the circumcentre of the regular polygon $A_1A_2 \dots A_n$, prove that the sum of the projections of $OA_1, OA_2, \dots OA_n$ on any line is zero.

22. In No. 21, if P is any point, and R is the circumradius, prove that $PA_1^2 + PA_2^2 + \dots + PA_n^2 = n(R^2 + OP^2)$.

23. If P is any point on the minor arc A_1A_{2n+1} of the circumcircle of a regular polygon $A_1A_2 \dots A_{2n+1}$, prove that

$$PA_1 - PA_2 + PA_3 - \dots = 0.$$

24. If O is the centre of the in-circle, radius a, of a regular polygon $A_1A_2 \dots A_n$ and if P is any point, prove that the sum of the squares of the perpendiculars from P to A_1A_2, A_2A_3, ..., $A_{n-1}A_n$, A_nA_1 is

$$n(a^2 + \tfrac{1}{2}OP^2).$$

HARDER MISCELLANEOUS EXAMPLES

EXERCISE VII. f.

1. ABC is a triangle, whose side BC is divided at K in the ratio of $p : q$. If the projections of AB, AC, AK on any given line Ox are denoted by $\overline{AB}$, $\overline{AC}$, $\overline{AK}$, prove that $(p+q)\overline{AK} = q\,.\,\overline{AB} + p\,.\,\overline{AC}$.

Obtain a special result by taking Ox to be the side BC; and deduce that $(p+q)\cot AKC = q\cot B - p\cot C$.

2. OA, OB, OC are concurrent edges and OD is a diagonal of a rectangular box. If OP makes angles α, β, γ and θ with OA, OB, OC and OD, prove that

(i) $OD\cos\theta = OA\cos\alpha + OB\cos\beta + OC\cos\gamma$;

(ii) $\cos^2\alpha + \cos^2\beta + \cos^2\gamma = 1$.

3. In any quadrilateral ABCD, prove that

$$a^2 + b^2 + c^2 - d^2 = 2ab\cos B + 2bc\cos C - 2ac\cos(A+D).$$

4. In any pentagon ABCDE, where $AB = a$, $BC = b$, etc., prove that

$$a^2 + b^2 - c^2 - d^2 - e^2 = 2ab\cos B - 2cd\cos D - 2de\cos E + 2ce\cos(D+E).$$

Sum to n terms the series whose rth terms are :

5. $\sin rx \sin ry \sin rz$.
6. $\cos^2 r\theta \sin^3 r\theta$.
7. $\cos^5 r\theta$.
8. $(n-r+1)\cos(r-1)\theta$.
9. $r\cos(n-r)\theta$.
10. $(n-r+1)\cos^2(r-1)\theta$.

11. Prove that

$$n+2(n-1)\cos\theta+2(n-2)\cos 2\theta+\ldots+2\cos(n-1)\theta=\frac{1-\cos n\theta}{1-\cos\theta}.$$

12. Evaluate $(n-1)\sin\theta+(n-2)\sin 2\theta+\ldots+\sin(n-1)\theta$.

13. What is the product of

$$1+x\cos\theta+x^2\cos 2\theta+\ldots+x^n\cos n\theta \quad \text{and} \quad 1-2x\cos\theta+x^2\ ?$$

14. What is the product of

$$x\sin\theta+x^2\sin 2\theta+\ldots+x^n\sin n\theta \quad \text{and} \quad 1-2x\cos\theta+x^2\ ?$$

15. Prove that

$$\cos\theta\cos\theta+\cos^2\theta\cos 2\theta+\ldots+\cos^n\theta\cos n\theta=\frac{\sin n\theta\cos^{n+1}\theta}{\sin\theta}.$$

16. Prove that

$$1+\cos\phi\sec\phi+\cos 2\phi\sec^2\phi+\ldots+\cos n\phi\sec^n\phi=\frac{\sin(n+1)\phi}{\sin\phi\cos^n\phi}.$$

Sum to n terms the series whose rth terms are:

17. $2^{1-r}\sin^3(2^{r-1}\theta)\cos(2^{r-1}\theta)$.
18. $\sin(2r+1)\theta\sec 2r\theta\sec(2r+2)\theta$.
19. $\sin 4r\theta \operatorname{cosec} r\theta$.
20. $\cos(3^{r-1}\theta)\operatorname{cosec}(3^r\theta)$.

21. Sum to infinity the series :

$$\frac{1}{2^2}\operatorname{th}^2\frac{x}{2}+\frac{1}{4^2}\operatorname{th}^2\frac{x}{4}+\frac{1}{8^2}\operatorname{th}^2\frac{x}{8}+\ldots.$$

22. $A_1A_2\ldots A_n$ is a regular polygon, prove that

$$A_1A_3^2+A_1A_4^2+\ldots+A_1A_{n-1}^2=\tfrac{1}{2}A_1A_2^2\left(n\operatorname{cosec}^2\frac{\pi}{n}-4\right).$$

23. $A_1A_2\ldots A_n$ is a regular polygon inscribed in a circle centre O, radius R ; P is a point near O. Prove that

$$PA_1+PA_2+\ldots+PA_n \simeq n\left(R+\frac{OP^2}{4R}\right).$$

24. A regular polygon of n sides is inscribed in a circle centre O, radius R ; P is a point at distance c from O. Perpendiculars are drawn from P to the sides of the polygon. Prove that the sum of the squares of the sides of the new polygon formed by the feet of these perpendiculars is $n(R^2+c^2)\sin^2\dfrac{2\pi}{n}$.

25. AB is a diameter of a circle centre O ; Q_0 is any point on the circumference ; $Q_1, Q_2, Q_3, \ldots Q_n$ are the middle points of the arcs $AQ_0, AQ_1, \ldots AQ_{n-1}$ respectively. Prove that

$$BQ_1 . BQ_2 . \ldots . BQ_n=\frac{AQ_0}{AQ_n} . OA^n.$$

CHAPTER VIII

COMPLEX NUMBERS

The Idea of Number. In elementary algebra it is found that certain equations have solutions, whereas others, almost of the same form, have none. The idea of number is gradually generalised, and the possibility of the solution of a particular equation may depend on the point to which the process of generalisation has been pushed. Consider the equations:

(i) $2x = 4$;	(ii) $2x = 5$;	(iii) $2x + 3 = 0$;
(iv) $x^2 = 4$;	(v) $x^2 = 2$;	(vi) $x^2 = -1$.

In the algebra of **natural** numbers (i) and (iv) are satisfied by $x = 2$ and the others have no solutions. In the algebra of **fractions** (i) and (iv) are satisfied by $\frac{2}{1}$ and (ii) by $\frac{5}{2}$. In the algebra of **directed** (positive and negative) numbers of the type $\pm\frac{p}{q}$, each of the first three has a solution and (iv) has the two solutions $+\frac{2}{1}$ and $-\frac{2}{1}$, but (v) and (vi) cannot be satisfied.

The most difficult step in the process of generalisation is the introduction of the **real** (rational and irrational) numbers; this will be discussed in the companion volume on Analysis. In real algebra $\sqrt{2}$ is available for the solution of equation (v). For many purposes it is enough to have a number which *approximately* satisfies an equation, and this is the reason why the introduction of irrationals is not urgent.

The rational real numbers have properties exactly analogous to those of the directed numbers $\pm\frac{p}{q}$, in much the same way as the positive directed numbers have properties like the signless fractions. No real number satisfies the equation $x^2 = -1$, even approximately.

In the present Chapter, by introducing the **complex** numbers, we shall carry out a further generalisation of the notion of number, and by suitable new definitions of the meanings of addition, multiplication, etc., shall show that equations like $x^2 = -1$ (in which -1, as well as x, is a complex number) are satisfied by numbers of the new type.

Definition of a Complex Number. Consider a plane and in it two rectangular axes Ox, Oy. The position of a point P in this plane, or the displacement made in moving from the origin O to the point

P, requires for its determination *two* real numbers, e.g. the coordinates of P referred to Ox, Oy.

An **ordered pair of real numbers is called a complex number**; and if a and b, *in that order*, are the real numbers, the complex number is denoted by the symbol $[a, b]$.

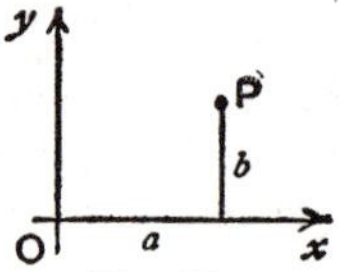

FIG. 64.

Thus there is a unique complex number corresponding to every point P of the plane, viz. the number $[a, b]$ corresponds to the point P whose coordinates are (a, b). Conversely, given a complex number $[a, b]$, there is a unique corresponding point P, with coordinates (a, b), or a unique corresponding displacement, whose components along the axes are a and b.

The statement that the pair of numbers which constitute a complex number is "ordered" means that $[a, b]$ and $[b, a]$ are distinct complex numbers (unless $a = b$). But for this, the one-to-one correspondence between complex numbers and points in a plane would not hold; just as the points (2, 5) and (5, 2) are distinct, so too are the complex numbers [2, 5] and [5, 2].

Definitions of Fundamental Operations. In a logical introduction to the theory of fractions or negative numbers, it is necessary to begin by defining the meanings of the elementary operations as applied to these numbers. In the same way, we must start here by making definitions of equality, addition, etc., for complex numbers.

These definitions simply state what meanings are to be given to the signs $=$, $+$, $-$, $\times$, and $\div$, in this new kind of algebra, which might be called the 'algebra of ordered number-pairs', but is actually called the 'algebra of complex numbers'.

(i) **Equality.** The two complex numbers $[a, b]$ and $[c, d]$ are called **equal** if and only if $a = c$ and $b = d$. In this case, we write $[a, b] = [c, d]$.

This definition secures that the points (and displacements), which correspond to two complex numbers, are the same points (and the same displacements) if and only if the complex numbers are "equal."

(ii) **Addition.** The complex number $[a + c, b + d]$ is called the **sum** of the two complex numbers $[a, b]$ and $[c, d]$; and we write

$$\mathbf{[a, b] + [c, d] = [a + c, b + d].} \quad \text{..........................(1)}$$

If P and Q are the points corresponding to $[a, b]$ and $[c, d]$, see Fig. 65, and if QOPR is a parallelogram, the point R corresponds to $[a+c, b+d]$, for the projection of OR on Ox is equal to the sum of the projections of OP, PR or OP, OQ on Ox; and similarly for projections on Oy.

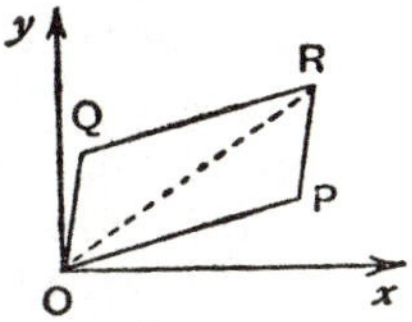

FIG. 65.

In vector notation, $\overline{OR} = \overline{OP} + \overline{OQ}$, so the displacement corresponding to the sum of two complex numbers is the vector sum of the displacements corresponding to the numbers.

(iii) **Subtraction** is defined by the relation,

$$[\mathbf{a}, \mathbf{b}] - [\mathbf{c}, \mathbf{d}] = [\mathbf{a} - \mathbf{c}, \mathbf{b} - \mathbf{d}]. \quad \text{.........................(2)}$$

This is chosen because, by equation (1),

$$[a-c, b-d] + [c, d] = [a-c+c, b-d+d] = [a, b].$$

(iv) **Multiplication** is defined by the relation,

$$[\mathbf{a}, \mathbf{b}] \times [\mathbf{c}, \mathbf{d}] = [\mathbf{ac} - \mathbf{bd}, \mathbf{ad} + \mathbf{bc}]. \quad \text{..................(3)}$$

The reason for this (at first sight peculiar) definition will become apparent in the next few pages, see especially Note 2, p. 142 A complete discussion will be given in the companion volume on Analysis. Any definition is legitimate, if not self-contradictory, and this one happens to be the most convenient and useful one to make.

(v) **Division.** $[a, b] \div [c, d]$ is defined as the complex number $[x, y]$ given by $[x, y] \times [c, d] = [a, b]$, provided that such a number exists.

By (3), this gives $[xc - yd, xd + yc] = [a, b]$;

$$\therefore \; xc - yd = a \quad \text{and} \quad xd + yc = b;$$

$$\therefore \text{ solving,} \quad x = \frac{ac+bd}{c^2+d^2}, \quad y = \frac{bc-ad}{c^2+d^2}, \text{ unless } c = d = 0;$$

$$\therefore \; [a, b] \div [c, d] = \left[\frac{ac+bd}{c^2+d^2}, \; \frac{bc-ad}{c^2+d^2}\right]. \quad \text{...............(4)}$$

This definition of division excludes the divisor $[0, 0]$. The number $[0, 0]$ plays a part in the theory of complex numbers, analogous to that played by 0 in the theory of real numbers. Division by $[0, 0]$ is not defined.

EXERCISE VIII. a.

[The beginner should work through *all* the examples in this Exercise.]

1. What are the values of $[1, 5] + [2, 3]$ and $[2, 3] + [1, 5]$? What general law does this illustrate?

2. What are the values of $[2, 5] \times [3, 4]$ and $[3, 4] \times [2, 5]$? What general law does this illustrate?

3. Verify by a numerical example, or show algebraically, that

$$[a+c, b+d] \times [e, f] = [a, b] \times [e, f] + [c, d] \times [e, f].$$

4. What are the values of:

(i) $[2, 0] + [3, 0]$; (ii) $[a, 0] + [c, 0]$;

(iii) $[2, 0] \times [3, 0]$; (iv) $[a, 0] \times [c, 0]$;

(v) $[a, b] + [a, b] + [a, b] + [a, b] + [a, b]$ and $[a, b] \times [5, 0]$?

5. What are a, b if $[3, 5] + [a, b] = [7, 8]$?

6. What are the values of:

(i) $[3, 5] + [-2, 8]$; (ii) $[-9, 2] + [-7, -1]$;

(iii) $[-6, -4] - [2, -9]$; (iv) $[-6, -4] + [-2, 9]$;

(v) $[6, 1] \times [3, 0]$; (vi) $[6, 1] \times [0, 3]$;

(vii) $[4, 5] \times [-8, 2]$; (viii) $[3, 11] \div [2, 3]$.

7. Simplify (i) $[a, 0] + [0, b]$;

(ii) $[a, 0] \times [1, 0] + [b, 0] \times [0, 1]$.

8. Write down the square of $[a, b]$, i.e. $[a, b] \times [a, b]$. What are the squares of $[1, 0]$, $[-1, 0]$, $[0, 1]$, $[0, -1]$?

9. Simplify $[\cos\theta, \sin\theta] \times [\cos\phi, \sin\phi]$.

10. If $[a, b] \times [c, d] = [0, 0]$, prove that $(a^2+b^2)(c^2+d^2) = 0$, and hence that either $[a, b]$ or else $[c, d]$ must be $[0, 0]$.

Notation. It appears from results such as those of Ex. VIII. a, No. 4, that complex numbers of the special type $[a, 0]$ behave rather like real numbers. *It is therefore convenient to denote* $[a, 0]$ *by the symbol* a, *and in particular* $[0, 0]$ *by* 0. There is a precedent for this in elementary algebra, where, for example, the symbol 2 is used with several different meanings: sometimes it means the natural number 2, sometimes the fraction $\frac{2}{1}$, sometimes the directed number $+\frac{2}{1}$, and later 'the real number 2'; now it is given a further possible meaning, the complex number $[2, 0]$. Complex numbers of the form $[a, 0]$ correspond to points on the x-axis.

It is also convenient to use an abbreviated notation for complex numbers which correspond to points on the y-axis. *The number* $[0, a]$ *is denoted by* ia *or* ai, *and in particular* $[0, 1]$ *by* i.

Further, since $[a, b] = [a, 0] + [0, b]$ and since $[a, 0]$ and $[0, b]$ are denoted by a and bi or ib, *the general complex number* $[a, b]$ *is denoted by* $a+bi$ *or* $a+ib$.

Just as in real algebra $x \times x \times x \times \ldots$ to n factors is denoted by x^n, so, if z is any complex number $a+ib$ and n is a positive integer, the product $z \times z \times z \times \ldots$ to n factors is denoted by z^n. Also $\frac{1}{z^n}$ is written z^{-n} and, in particular, $\frac{1}{z}$ is written z^{-1}.

Mechanical Application of Algebraic Processes. If a, b, c, d, i stand for ordinary real numbers, we have:

$$\text{(i) } (a+bi)+(c+di)=(a+c)+(b+d)i;$$

$$\text{(ii) } (a+bi)\times(c+di)=(ac+bdi^2)+(ad+bc)i.$$

If, on the other hand, $a+bi$ and $c+di$ stand for the complex numbers $[a, b]$ and $[c, d]$), we have

$$\text{(i) } (a+bi)+(c+di)=[a, b]+[c, d]=[a+c, b+d]=(a+c)+(b+d)i;$$

$$\text{(ii) } (a+bi)\times(c+di)=[a, b]\times[c, d]=[ac-bd, ad+bc]$$
$$=(ac-bd)+(ad+bc)i.$$

The two relations (i) are identical in form, as they stand; the two relations (ii) become identical in form, if -1 is written for i^2.

Thus, the correct results of both addition and multiplication are given by a mechanical application of the ordinary processes of algebra to the symbols in their abbreviated form, provided that -1 is written for i^2, wherever it occurs.

The reader may have anticipated this fact from the result obtained in Ex. VIII. a, No. 8, where he found that

$$[0, 1]^2=[0, 1]\times[0, 1]=[-1, 0]$$

or, using the abbreviated notation, $i^2=-1$.

Other operations, subtraction, division, etc., can always be reduced so as ultimately to depend upon addition and multiplication [see the definitions on pp. 138, 139]. The statement in *italics* therefore holds for all the fundamental processes of algebra.

Note 1. The equation, $i^2=-1$, does *not* mean that there exists a number i whose square is the negative number -1. In this equation, i is simply an abbreviation for $[0, 1]$, and -1 is *not* the negative number -1, but an abbreviation for $[-1, 0]$.

Nor does the equation imply that $[0, 1]$ is the only complex number whose square is $[-1, 0]$, and in fact, as was found in Ex. VIII. a, No. 8, $[0, -1]$ is another complex number whose square is $[-1, 0]$. The equation, $x^2=-1$, has two roots if it is an equation in complex algebra, i.e. if x and -1 denote ordered number-pairs. In real algebra, it has no roots. More generally, it will appear later that the equation of the nth degree in complex algebra always has n roots (subject only to the usual conventions of language with respect to " equal " roots). In real algebra it may have n roots or fewer.

Note 2. The advantages of the definitions on p. 139 can now be partly appreciated :

(i) It appears that complex numbers of the form $[n, 0]$ have properties much like those of the real number n. From the law of addition, $[a, b] + [a, b] + \dots$ to n terms, equals $[na, nb]$, and from the law of multiplication $[a, b] \times [n, 0]$ also equals $[na, nb]$.

Therefore, $[a, b] + [a, b] + \dots$ to n terms $= [a, b] \times [n, 0]$.

This would not be true if multiplication was, for example, defined by the relation $[a, b] \times [c, d] = [ac, bd]$.

(ii) It is desirable that fundamental laws should be the same for complex as for real algebra. The reader who has worked Ex. VIII. a, Nos. 1-3, has verified this in some cases and should be able to do so in general.

(iii) The result of Ex. VIII. a, No. 10, shows that the fundamental factor theorem, that if the product of two numbers is zero then one of the numbers is zero, also holds in complex algebra.

We give now some examples of the manipulation of complex numbers.

Example 1. Calculate $[x, y]^3$.

By the principle established on p. 141, we have

$$(x + yi)^3 = x^3 + 3x^2yi + 3xy^2i^2 + y^3i^3 ;$$

$\therefore$ writing -1 for i^2 and $-i$ for $i^3 (= i^2 . i)$, we have

$$(x + yi)^3 = (x^3 - 3xy^2) + (3x^2y - y^3) i ;$$

$$\therefore \quad [x, y]^3 = [x^3 - 3xy^2, 3x^2y - y^3].$$

The reader should show that the same result is obtained by two applications of the law of multiplication for complex numbers.

Example 2. Divide $[a, b]$ by $[c, d]$ when $[c, d] \neq [0, 0]$.

$$\frac{a + bi}{c + di} = \frac{(a + bi)(c - di)}{(c + di)(c - di)} = \frac{(ac + bd) + (bc - ad)i}{c^2 + d^2} ;$$

$\therefore$ as on p. 139, $\quad [a, b] \div [c, d] = \left[\frac{ac + bd}{c^2 + d^2}, \frac{bc - ad}{c^2 + d^2}\right].$

Example 3. Expand $[x, y]^n$, where n is a positive integer.

$$(x + yi)^n = x^n + \binom{n}{1} x^{n-1}(yi) + \binom{n}{2} x^{n-2}(yi)^2 + \dots + (yi)^n$$

$$= \left(x^n - \binom{n}{2} x^{n-2}y^2 + \dots\right) + \left(\binom{n}{1} x^{n-1}y - \binom{n}{3} x^{n-3}y^3 + \dots\right)i.$$

Nomenclature. Complex numbers of the form $[x, 0]$ are sometimes called "real," and those of the form $[0, y]$ are called, in contrast, "pure imaginary." This language, which is a legacy from the time when the theory of complex numbers was imperfectly understood, is most unfortunate and should be avoided. In the same way, when a function of $[x, y]$ has been expressed in the form $[a, b]$, e.g. $[x, y]^3 = [x^3 - 3xy^2, 3x^2y - y^3]$, in Example 1 above, the expressions a and b are often called the real and imaginary parts of the function; this is an equally misleading form of words. They may be called the "first and second parts" of the function.

Complex numbers of the form $[a, b]$, $[a, -b]$ or $a+bi$, $a-bi$ are called **conjugate**. This name is due to *Cauchy*.

If, as in Examples 1-3 above, a function of a complex number has been reduced to the standard form $[a, b]$ of a complex number, we have an equation of the form

$$\phi(x+yi) = \mathsf{X} + \mathsf{Y}i.$$

In this case, it also follows that

$$\phi(x-yi) = \mathsf{X} - \mathsf{Y}i,$$

for the only property of i that is used in the work is $i^2 = -1$, that is $[0, 1]^2 = [-1, 0]$, and this remains true if i is replaced everywhere by $-i$, since $[0, -1]^2 = [-1, 0]$.

Consequently $\phi(x+yi) \,.\, \phi(x-yi) = [\mathsf{X}^2 + \mathsf{Y}^2, 0] = \mathsf{X}^2 + \mathsf{Y}^2.$ ......(5)

Historical Note. The idea of the possibility of dealing with the square root of a negative number is certainly as old as the time of *Diophantus* (c. 245-330) and arose from attempts to solve special equations. Probably *Cardan* (1501-1576) was the first to assume the application of algebraic processes to symbols of the form, $a+\sqrt{-b}$. He discussed the problem: divide 10 into two parts whose product is 40, and gave as the answer $5+\sqrt{-15}$ and $5-\sqrt{-15}$, (not however in this form); he then showed that his answers satisfied the given conditions, if the ordinary rules of algebra were applied.

Complex numbers were used freely by *Euler* (1707-1788), to whom the symbol i for $\sqrt{-1}$ is due, and by many of his contemporaries (*John Bernouilli*, *Cotes*, *De Moivre*, etc.). Their real nature was first made clear by *Wessel* (1797) and *Argand* (1806), who introduced the geometrical interpretation. The name "complex number" is due to *Gauss* (1777-1855), who developed in far greater detail than *Wessel* or *Argand*, but on similar lines, the fundamental principles of the theory and also used them in his investigations of the properties of natural numbers. The work of *Gauss* prepared the way for the discoveries of *Cauchy*, *Riemann* and many others which form the foundation of modern Analysis and play a large part in modern mathematical Physics.

EXERCISE VIII. b.

1. Express the following complex numbers in an alternative form:
$[3, 5]$; $[6, 0]$; $[0, 7]$; $[0, 0]$; $[0, -1]$; $[\alpha, -\beta]$.

2. Express the following complex numbers in the bracket form:
$1+2i$; 5; $3-2i$; $7i$; $-2i$; $-i-2$; $\alpha i+\beta$.

3. Simplify the following:

(i) $[3, 2]+[1, -5]$	(ii) $[4, 6]-[6, 6]$
(iii) $[1, 2]\times[3, 4]$	(iv) $[2, 0]\div[3, 0]$
(v) $(2+3i)-(3-4i)$	(vi) $(1+i)+(1-i)$
(vii) $3i\times 4i$	(viii) $3i\div 4i$
(ix) $[\alpha, \beta]+[0, 0]$	(x) $[\alpha, \beta]\times[0, 0]$
(xi) $[\alpha, \beta]\times[1, 0]$	(xii) $(7-9i)\div(1+i)$.

Express the following, Nos. 4-30, in the form $X+Yi$:

4. $(1+2i)\times 3$.
5. $5i\times(1-i)$.
6. $(1+i)\times(1-i)$.
7. $(2+i)(3-2i)$.
8. $(4+3i)^2$.
9. $(a+bi)^2$.
10. $i(a+bi)$.
11. $ai\times bi$.
12. $a\div i$.
13. $(\cos\theta+i\sin\theta)(\cos\phi+i\sin\phi)$.
14. $(\cos\theta+i\sin\theta)(\cos\theta-i\sin\theta)$.
15. $(\cos\theta+i\sin\theta)(\cos\phi-i\sin\phi)$.
16. $(\cos\theta+i\sin\theta)^2$.
17. $[r\cos\theta, r\sin\theta]\times[s\cos\phi, s\sin\phi]$.
18. $[\cos\theta, \sin\theta]\div[\cos\phi, \sin\phi]$.
19. $(x-\cos\theta-i\sin\theta)(x-\cos\theta+i\sin\theta)$.
20. $(x+\sin\phi+i\cos\phi)(x+\sin\phi-i\cos\phi)$.
21. $(1+i)^2$.
22. $(1-i)^3$.
23. $\dfrac{1+i}{1-i}$.
24. $\dfrac{1+2i}{1-i}$.
25. $\dfrac{2-3i}{1+2i}$.
26. $\dfrac{(1+i)^2}{1-i}$.
27. $(x-iy)^2$.
28. $\dfrac{1}{x-yi}$.
29. $\dfrac{x+iy}{x-iy}$.
30. $\dfrac{1+z}{1-z}$, where $z=[x, y]$.
31. Simplify (i) $(1+i)^{-2}+(1-i)^{-2}$; (ii) $(1+i)^{-4}+(1-i)^{-4}$.
32. If $(2+3i)(3-4i)=a+bi$, find the value of a^2+b^2.
33. Show that the cubes of $\frac{1}{2}(-1\pm i\sqrt{3})$ are each 1.
34. Simplify $(x+yi)(x-yi)$ and factorise $(x-1)^2+(y-2)^2$.

35. If $(x+yi)^n = a+bi$, express a^2+b^2 in terms of x, y.

36. What is the series whose nth term is $(1+i^{2n})(1+i^n)$?

37. Find real numbers x and y such that

$$(2-i)x+(1+3i)y+2=0.$$

38. Prove that $[3, 4]$ is one root of the equation $z^2-6z+25=0$, and find the other root?

39. If $\sqrt{(x+yi)} = \mathrm{A}+\mathrm{B}i$, where $\mathrm{A}>0$, prove that $\mathrm{A}^2-\mathrm{B}^2=x$ and $2\mathrm{AB}=y$. Hence express $\sqrt{(5+12i)}$ in the form $\mathrm{A}+\mathrm{B}i$.
Similarly express $\sqrt{i}$ in the form $\mathrm{A}+\mathrm{B}i$.

40. What is the condition that one root of the equation

$$z^2+2(a+ib)z+c+id=0 \text{ is of the form } [k, 0]\,?$$

The Argand Diagram. The figure, referred to on pp. 137, 138, in which the point P, or the displacement $\overline{\mathrm{OP}}$, corresponding to a complex number $[x, y]$, $=x+yi$, is considered, is called the *Argand Diagram*.

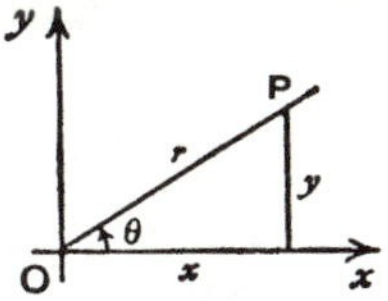

FIG. 66.

Modulus and Amplitude. If the length of OP is r units, and if $\angle(\mathrm{OP}, \mathrm{O}x)=\theta$, then we have

$$\mathbf{x}=\mathbf{r}\cos\boldsymbol{\theta}\,; \quad \mathbf{y}=\mathbf{r}\sin\boldsymbol{\theta}\,; \quad \text{............(6)}$$

also $$\mathbf{r}=+\sqrt{(\mathbf{x}^2+\mathbf{y}^2)}, \quad \text{....................(7)}$$

and $$\cos\boldsymbol{\theta} : \sin\boldsymbol{\theta} : 1 = \mathbf{x} : \mathbf{y} : +\sqrt{(\mathbf{x}^2+\mathbf{y}^2)}. \quad \text{......(8)}$$

r is called the **modulus**, or sometimes the *absolute value*, of the complex number $x+yi$, and is denoted by $|x+yi|$ or by $\mathrm{mod}\,(x+yi)$.

θ is called the **amplitude** of the complex number $x+yi$, and is denoted by $\mathrm{am}\,(x+yi)$.

A complex number is expressed in terms of its modulus and amplitude by the relation

$$x+yi=r\cos\theta+ir\sin\theta, \quad \text{or} \quad x+yi=r(\cos\theta+i\sin\theta)$$

or sometimes, if there is no danger of ambiguity, by $x+yi=(r, \theta)$.

It is customary to denote the complex number $x+yi$ by z and to write $r=|z|$, $\theta=\mathrm{am}\,(z)$. If this is done, z is called the *argument* of the point P in the Argand Diagram. The symbol $|z|$ for $\mathrm{mod}\,z$ is due to *Weierstrass*; its use in connection with real numbers has been explained on p. 78.

The Modulus-Amplitude Form. It is frequently necessary to express a complex number, given in the form $x+yi$, in terms of its modulus and amplitude.

By definition, r is essentially positive and its value is obtained *uniquely* from $r = +\sqrt{(x^2+y^2)}$.

Equation (8) shows that there is also a unique value of θ in the range $-\pi < \theta \leqslant +\pi$; but any value of θ which differs from this value by a multiple of 2π would lead to the same representative point P and is therefore a possible value for $\text{am}\,(x+yi)$. The unique value of θ in the range $-\pi < \theta \leqslant +\pi$ is called the **principal value** of the amplitude.

Since $x = r\cos\theta$, $y = r\sin\theta$, it follows that $\tan\theta = \frac{y}{x}$; but this equation is not sufficient to determine the amplitude, since it gives two values of θ between $-\pi$ and $+\pi$. Of these two values, it is necessary to select that one for which $\cos\theta = \frac{x}{+\sqrt{(x^2+y^2)}}$ and therefore also $\sin\theta = \frac{y}{+\sqrt{(x^2+y^2)}}$.

Example 4. Find the modulus and amplitude of $-2+3i$.

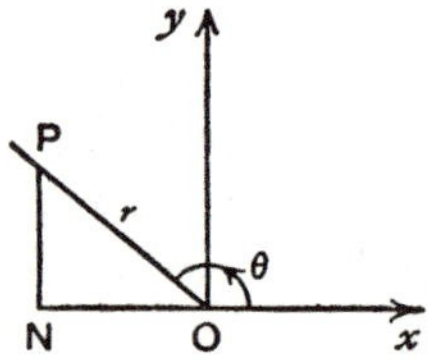

FIG. 67.

From Fig. 67, $r\cos\theta = x = -2$,

$r\sin\theta = y = +3$;

$\therefore\ r = +\sqrt{(x^2+y^2)} = +\sqrt{13}$.

And θ is the angle $\left(\text{between } \frac{\pi}{2} \text{ and } \pi\right)$ given by

$$\sin\theta : \cos\theta : 1 = 3 : -2 : +\sqrt{13}.$$

As is pointed out above, it is not sufficient to say that θ is the angle given by $\tan\theta = -\frac{3}{2}$.

Example 5. Express $1-\cos\theta - i\sin\theta$ in the modulus-amplitude form.

$$1-\cos\theta - i\sin\theta = 2\sin^2\frac{\theta}{2} - 2i\sin\frac{\theta}{2}\cos\frac{\theta}{2}$$

$$= 2\sin\frac{\theta}{2}\left(\sin\frac{\theta}{2} - i\cos\frac{\theta}{2}\right)$$

$$= 2\sin\frac{\theta}{2}(\cos\alpha + i\sin\alpha),$$

where α is chosen so that $\cos\alpha = \sin\frac{\theta}{2}$ and $\sin\alpha = -\cos\frac{\theta}{2}$.

These conditions are satisfied by $\alpha = \frac{\theta-\pi}{2}$.

The result now obtained is of the required form only if $\sin\frac{\theta}{2}$ is positive, i.e. if $4n\pi < \theta < (4n+2)\pi$.

If $\sin\frac{\theta}{2}$ is negative, i.e. if $(4n-2)\pi < \theta < 4n\pi$, we write

$$1-\cos\theta - i\sin\theta = -2\sin\frac{\theta}{2}(-\cos\alpha - i\sin\alpha)$$

$$= -2\sin\frac{\theta}{2}(\cos\overline{\alpha+\pi} + i\sin\overline{\alpha+\pi}),$$

where $\alpha = \frac{\theta-\pi}{2}$, as before.

It is instructive to consider this example geometrically. We shall take the special case, $0<\theta<\pi$. In Fig. 68, the points A and P represent the complex numbers 1 and $\cos\theta + i\sin\theta$.

To find the point corresponding to $1-(\cos\theta+i\sin\theta)$, we take the displacement $\overline{\mathsf{OA}} - \overline{\mathsf{OP}}$, and this is $\overline{\mathsf{PA}}$. Now $\angle \mathsf{AOP} = \theta$, and OA, OP are each of unit length; $\therefore$ the length of PA is $2\sin\frac{\theta}{2}$,

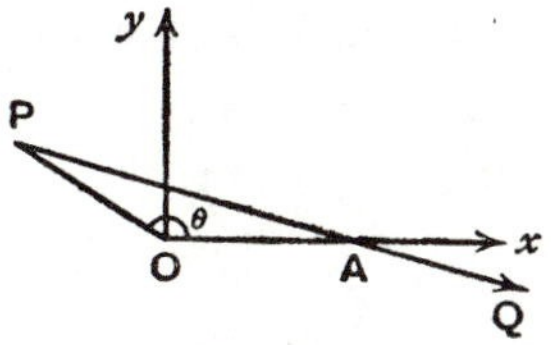

FIG. 68.

and the principal value of the amplitude of $\overline{\mathsf{PA}}$ is measured by the angle from Ox to PA, which is $-\frac{\pi-\theta}{2} = \frac{\theta-\pi}{2}$.

$\therefore$ as before,

$$1-\cos\theta - i\sin\theta = 2\sin\frac{\theta}{2}\left(\cos\frac{\theta-\pi}{2} + i\sin\frac{\theta-\pi}{2}\right).$$

Note. In numerical examples, the modulus and amplitude can often be written down by reference to a figure.

EXERCISE VIII. c.

Draw figures and give the modulus and the principal value of the amplitude of the complex numbers in Nos. 1-16.

1. 1.	**2.** -1.	**3.** i.	**4.** $-i$.
5. $1+i\sqrt{3}$.	**6.** $1-i\sqrt{3}$.	**7.** $-1+i\sqrt{3}$.	**8.** $-1-i\sqrt{3}$.
9. $i-\sqrt{3}$.	**10.** $1+i$.	**11.** $-1+i$.	**12.** $-1-i$.
13. $1-i$.	**14.** $i-1$.	**15.** $\sqrt{3}-i$.	**16.** $-i-\sqrt{3}$.

Express in the modulus-amplitude form:

17. $3+4i$.	**18.** $\sqrt{2}+1-i$.	**19.** $-3+4i$.	**20.** $-3-4i$.
21. $\sqrt{3}-2-i$.		**22.** $\cos\alpha - i\sin\alpha$.	
23. $\sin\alpha - i\cos\alpha$.		**24.** $\sin\alpha + i\cos\alpha$.	

25. What are the modulus and the principal value of the amplitude of $\cos^2 a + i \sin a \cos a$ if

$$\text{(i)}\ -\frac{\pi}{2} < a < 0; \quad \text{(ii)}\ \frac{\pi}{2} < a < \pi; \quad \text{(iii)}\ -\pi < a < -\frac{\pi}{2}?$$

Express in the modulus-amplitude form:

26. $1 + i \tan a$.

27. $1 + i \cot a$.

28. $\tan \beta - i$.

29. $-\sin \beta - i \cos \beta$.

30. $1 + \cos \theta + i \sin \theta$.

31. $1 + \cos \theta - i \sin \theta$.

32. $1 + \sin \theta + i \cos \theta$.

33. $\cos a + i \sin a + \cos \beta + i \sin \beta$.

34. $\cos a - i \sin \beta + i(\sin a + i \cos \beta)$.

35. $1 + r \cos \phi + ir \sin \phi$.

36. Interpret geometrically the relation

$$1 + \left(\cos \frac{2\pi}{3} + i \sin \frac{2\pi}{3}\right) + \left(\cos \frac{4\pi}{3} + i \sin \frac{4\pi}{3}\right) = 0,$$

and generalise the result.

37. Interpret geometrically the relation

$$1 + (\cos \theta + i \sin \theta) = 2 \cos \frac{\theta}{2}\left(\cos \frac{\theta}{2} + i \sin \frac{\theta}{2}\right).$$

38. By using geometrical considerations, find r_1 and r_2, if

$$r_1(\cos \theta_1 + i \sin \theta_1) + r_2(\cos \theta_2 + i \sin \theta_2) = \sin \overline{\theta_2 - \theta_1}\left(\cos \frac{\pi}{2} + i \sin \frac{\pi}{2}\right),$$

where $0 < \theta_1 < \frac{\pi}{2} < \theta_2 < \pi$.

Applications of the Argand Diagram. It was shown on p. 139 that if P and Q are the points of the Argand Diagram corresponding to the complex numbers, $a + bi = z_1$ and $c + di = z_2$, then the point R which corresponds to their *sum*, $z_1 + z_2$, is found by completing the parallelogram QOPR.

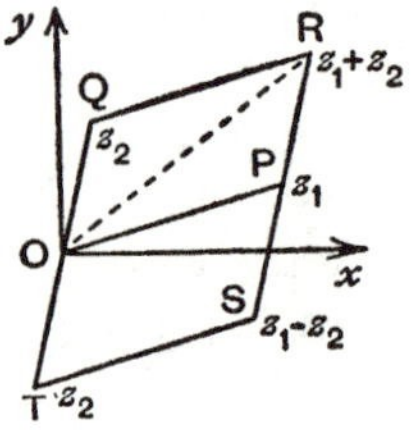

FIG. 69.

Since the length of OR is not greater than the sum of the lengths of OP and OQ, we have, see Fig. 69,

$$|z_1 + z_2| \leqslant |z_1| + |z_2|. \quad \text{.............(9)}$$

The reader should show in a similar way, or deduce from (9) (see Ex. VIII. d, No. 5) that

$$|z_1 + z_2| \geqslant |z_1| - |z_2| \quad \text{.............(10)}$$

and

$$|z_1 - z_2| \geqslant |z_1| - |z_2|. \quad \text{.............(11)}$$

Suppose A and P are the points of the Argand Diagram corresponding to the fixed complex number z_1 and the variable complex number z, see Fig. 70.

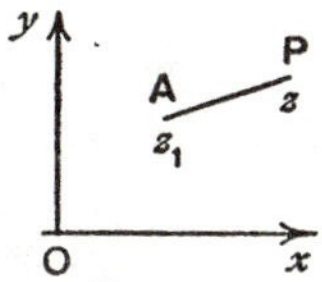

FIG. 70.

Then $z - z_1$ corresponds to the displacement $\overline{AP}$, and $|z - z_1|$ is the length of AP.

If $|z - z_1| = \text{constant} = c$, say, then P moves on a circle, centre A, radius c.

If $\text{am}(z - z_1) = \text{constant} = \alpha$, say, then P moves on the *half-line* AP, such that $\angle(AP, Ox) = \alpha$.

If $z = kz_1$, where k is a positive or negative number, P is a point on the line OA such that $\dfrac{OP}{OA} = k$.

Example 6. Two points P and Q in the Argand Diagram represent complex numbers z and $2z + 3 + i$. If P moves round the circle, centre the origin and radius k, how does Q move ?

If $z' \equiv 2z + 3 + i$, $\qquad |z' - 3 - i| = |2z| = 2|z| = 2k$;

$\therefore$ Q moves on the circle of radius $2k$ whose centre represents the complex number $3 + i$, i.e. the point whose coordinates are (3, 1).

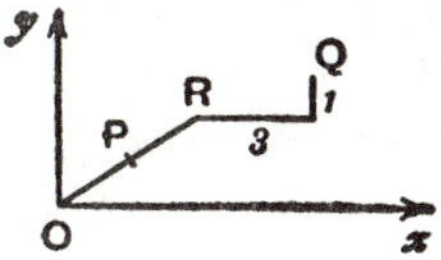

FIG. 71.

Otherwise : the point representing $2z$ is at R in OP produced so that OR $=$ 2OP, see Fig. 71, and the displacement from R to Q consists of 3 units parallel to Ox and 1 unit parallel to Oy. As P describes the given circle, R describes the concentric circle of radius $2k$, and thus Q describes the circle found by displacing this circle a distance 3 units parallel to Ox and 1 unit parallel to Oy.

EXERCISE VIII. d.

1. Given two points A, B in the Argand Diagram, representing the complex numbers α and β, construct the points which represent (i) $\alpha - \beta$; (ii) $\frac{1}{2}(\alpha + \beta)$; (iii) $\alpha + 2\beta$; (iv) $\alpha - 3\beta$.

2. A, B, C are collinear points, such that AB $=$ BC. If A and B represent complex numbers α and β, what does C represent ?

3. A, B, C are collinear points, such that AB $=$ 2BC. If A, B, C represent complex numbers α, β, γ, what is the relation between α, β, γ ?

4. A given point P represents the complex number z. Construct the points which represent (i) $2z$; (ii) $-3z$; (iii) $z + 3$; (iv) $z - 7$; (v $4z + 9$; (vi) $8 - 3z$.

5. Show geometrically that

(i) $|z_1 + z_2| \geqslant |z_1| - |z_2|$;

(ii) $|z_1 - z_2| \geqslant |z_1| - |z_2|$. (See Fig. 69 on p. 148.)

6. If P and Q represent the complex numbers z_1 and z_2, state the geometrical condition for the equality signs in No. 5.

7. If $|z| = 1$, what is the locus of P, when it represents the complex numbers (i) $3z$; (ii) $z+3$; (iii) $4z+9$?

8. If P represents the complex number z, what facts about the position of P are expressed by

(i) $|z| = 5$; (ii) $|z-1| = 2$; (iii) $|z+2| = 3$;
(iv) $|2z-1| = 3$; (v) $|z-2-3i| = 4$; (vi) $\text{am}\,(z) = 0$.

9. Use the modulus notation to express that the point P which represents the complex number z lies

(i) inside the circle, centre (8, 9), radius 7;

(ii) on the circle, centre (a, b), radius c;

(iii) outside the circle, centre $(-1, 0)$, radius 1.

10. What are the greatest and least values of $|z-3|$ if $|z| \leqslant 1$?

11. What are the greatest and least values of $|z+2|$ if $|z| \leqslant 1$?

12. What are the greatest and least values of $|z|$ if $|z-5| \leqslant 2$?

13. What are the greatest and least values of $|z+1|$ if $|z-4| \leqslant 3$?

14. What are the greatest and least values of $|z-4|$ if $|z+3i| \leqslant 1$?

15. A variable point P represents z; what can be said about the position of P if $1 < |z+2-3i| < 2$?

16. If $|z| < 1$, what can be said about the possible positions of the point which represents $1+z$?

17. If $|z| < 1$, prove that the principal value of $\text{am}\,(1+z)$ lies between $-\frac{\pi}{2}$ and $+\frac{\pi}{2}$.

18. If $|z| = \frac{1}{2}$, find the range of principal values of $\text{am}\,(1+z)$.

19. If P_1 and P_2 represent the complex numbers z_1 and z_2 and if m_1 and m_2 are any positive or negative numbers, $(m_1 + m_2 \neq 0)$, explain the significance of the point $\dfrac{m_1 z_1 + m_2 z_2}{m_1 + m_2}$.

20. Generalise No. 19.

Products. Let the numbers corresponding to P and Q, expressed in the modulus-amplitude form, be $r(\cos\theta + i\sin\theta)$ and $s(\cos\phi + i\sin\phi)$; then the polar coordinates of P, Q are (r, θ) and (s, ϕ).

The product of the numbers is $[r \cos \theta, r \sin \theta] \times [s \cos \phi, s \sin \phi]$. By definition, this equals

$$[rs(\cos\theta\cos\phi - \sin\theta\sin\phi), rs(\sin\theta\cos\phi + \cos\theta\sin\phi)]$$
$$= [rs\cos(\theta+\phi), rs\sin(\theta+\phi)]$$
$$= \mathbf{rs}\,\{\cos(\theta+\varphi) + i\sin(\theta+\varphi)\}. \quad \text{......................(12)}$$

This result may be expressed in the form

$$|\mathbf{z}_1 . \mathbf{z}_2| = |\mathbf{z}_1| . |\mathbf{z}_2|; \quad \mathrm{am}(\mathbf{z}_1 . \mathbf{z}_2) = \mathrm{am}\,\mathbf{z}_1 + \mathrm{am}\,\mathbf{z}_2. \quad \text{......(13)}$$

But the second result in (13) is *not necessarily true of the principal values.*

From (12) we see that the product is represented by the point K whose polar coordinates are $(rs, \theta+\phi)$, which is found by taking the point A (1, 0) and making the triangle QOK directly similar to the triangle AOP. For, $OK : OQ = OP : OA$, $\therefore$ $OK = rs$; also

$$\angle xOK = \angle xOQ + \angle QOK = \angle xOQ + \angle AOP$$
$$= \phi + \theta.$$

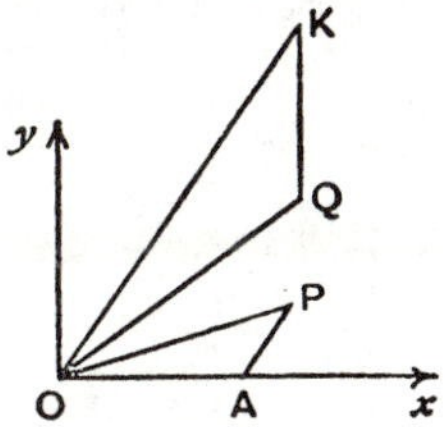
FIG. 72.

Repeated applications of formula (12) give

$$r_1(\cos\theta_1 + i\sin\theta_1) . r_2(\cos\theta_2 + i\sin\theta_2) \,.....$$
$$r_n(\cos\theta_n + i\sin\theta_n) = r_1 r_2 \ldots r_n\{\cos(\Sigma\theta) + i\sin(\Sigma\theta)\}.$$

In particular, if $r_1 = r_2 = \ldots = r_n = 1$ and $\theta_1 = \theta_2 = \ldots = \theta_n = \theta$,

we have $$(\cos\theta + i\sin\theta)^n = \cos n\theta + i\sin n\theta, \quad \text{..................(14)}$$

where n is any positive integer.

This is a special case of an important theorem which will be discussed in Chapter IX.

Quotients. An expression for the result obtained when one complex number is divided by another may be deduced from formula (12) as follows:

$$\frac{r}{s}(\cos\overline{\theta-\phi} + i\sin\overline{\theta-\phi}) \times s(\cos\phi + i\sin\phi)$$

$$= \frac{r}{s} . s(\cos\overline{\theta-\phi+\phi} + i\sin\overline{\theta-\phi+\phi}) = r(\cos\theta + i\sin\theta);$$

$$\therefore \{\mathbf{r}(\cos\theta + i\sin\theta)\} \div \{\mathbf{s}(\cos\varphi + i\sin\varphi)\}$$

$$= \frac{\mathbf{r}}{\mathbf{s}}(\cos\overline{\theta-\varphi} + i\sin\overline{\theta-\varphi}). \quad \text{.....................(15)}$$

This result may be expressed in the form

$$|z_1 \div z_2| = |z_1| \div |z_2|; \quad \text{am}(z_1 \div z_2) = \text{am}\, z_1 - \text{am}\, z_2. \quad \ldots\ldots(16)$$

But the second result in (16) is *not necessarily true of principal values.*

In particular, putting $\theta = 0$, $r = s = 1$, we have

$$\frac{1}{\cos \varphi + i \sin \varphi} = \cos(-\phi) + i \sin(-\phi) = \cos \varphi - i \sin \varphi. \ldots(17)$$

If P, Q represent the complex numbers

$$r(\cos\theta + i \sin\theta), \quad s(\cos\phi + i\sin\phi),$$

the point H which represents $\{r(\cos\theta + i\sin\theta)\} \div \{s(\cos\phi + i\sin\phi)\}$ is obtained by making P in Fig. 73 play the part of K in Fig. 72. Thus, if A is the point (1, 0), construct the triangle AOH directly similar to the triangle QOP; then H represents

$$\frac{r}{s}(\cos\overline{\theta - \phi} + i \sin\overline{\theta - \phi}).$$

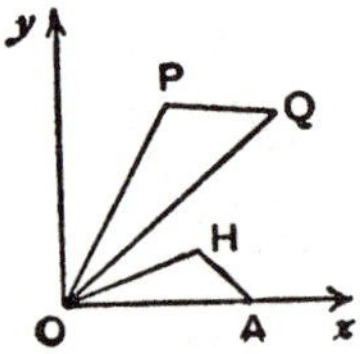

FIG. 73.

Notation. The expression $\cos\theta + i\sin\theta$ is often denoted by **cis θ**, and $(\cos\theta + i\sin\theta)^n$ is denoted by $\text{cis}^n\theta$. Equation (14) may then be written, $\text{cis}^n\theta = \text{cis}\, n\theta$. And since, from (17),

$$\cos(-\theta) + i\sin(-\theta) = \cos\theta - i\sin\theta = (\cos\theta + i\sin\theta)^{-1},$$

we may denote $\cos\theta - i\sin\theta$ either by $\text{cis}(-\theta)$ or by $(\text{cis}\,\theta)^{-1}$.

Example 7. The points B, P, Q in the Argand Diagram represent the complex numbers 2, z, z^2. If P describes the circle on OB as diameter, find the locus of Q.

In Fig. 74, A is the centre of the given circle; the triangles AOP, POQ are similar; therefore, if (r, θ) are the polar coordinates of Q,

$$\text{OP} = \sqrt{(\text{OA} \,.\, \text{OQ})} = \sqrt{r} \quad \text{and} \quad \angle x\text{OP} = \tfrac{1}{2}\theta.$$

Also OB = 2; $\quad \therefore \sqrt{r} = 2\cos\dfrac{\theta}{2}$;

$$\therefore r = 4\cos^2\frac{\theta}{2} = 2(1 + \cos\theta).$$

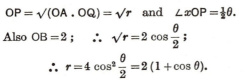

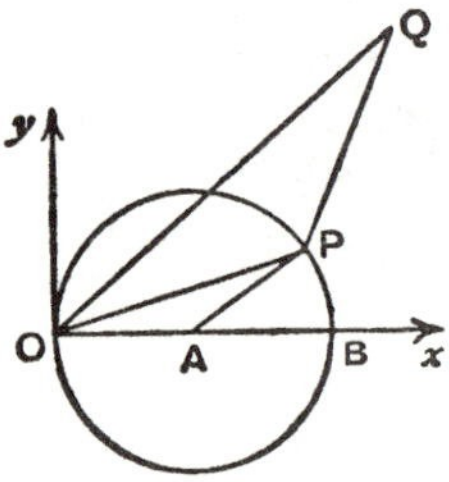

FIG. 74.

This is the polar equation of a cardioid. The reader should sketch the locus.

Example 8. The points B, P, R in the Argand Diagram represent the complex numbers 2, z, $\frac{1}{z}$. If P describes the circle on OB as diameter, find the locus of R.

In Fig. 75, A is the centre of the given circle ; the triangles ROA, AOP are similar, by the construction given above (see Fig. 73). But OA = AP ; ∴ OR = RA ; ∴ the locus of R is the perpendicular bisector of OA, namely the line $x = \frac{1}{2}$.

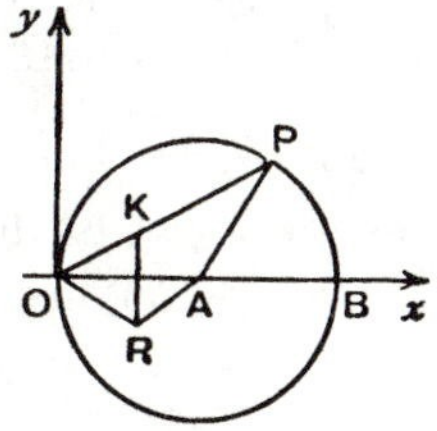

FIG. 75.

Otherwise: if K is the image of R in the x-axis, OP . OK = OP . OR = OA^2 = 1 ; ∴ P, K are inverse points w.r.t. the circle, $|z| = 1$. ∴ the locus of K is the straight line, $x = \frac{1}{2}$. But R is the image of K in Ox ; ∴ the locus of R is also the straight line, $x = \frac{1}{2}$. This application of inversion is important.

It should be noticed that as P moves in an anti-clockwise direction round the circle from B towards O along the upper semi-circle, K moves from the x-axis *upwards* towards $+\infty$, and therefore R moves from the x-axis *downwards* towards $-\infty$. Also when P continues from O to B along the lower semi-circle, K moves *upwards* from $-\infty$ to the x-axis, and so R moves *downwards* from $+\infty$ to the x-axis.

EXERCISE VIII. e.

Simplify the following: (Nos. 1-24).

1. $\dfrac{\cos 2\alpha + i \sin 2\alpha}{\cos \alpha + i \sin \alpha}$.

2. $\dfrac{\cos \beta + i \sin \beta}{\cos \beta - i \sin \beta}$.

3. $\dfrac{1}{\cos 2\theta + i \sin 2\theta}$.

4. $\dfrac{\cos \phi - i \sin \phi}{\cos 2\phi + i \sin 2\phi}$.

5. $\dfrac{\cos 3\alpha + i \sin 3\alpha}{\cos \alpha - i \sin \alpha}$.

6. $\dfrac{\cos 4\theta - i \sin 4\theta}{\cos 2\theta - i \sin 2\theta}$.

7. $\dfrac{(\cos \alpha + i \sin \alpha)(\cos \beta + i \sin \beta)}{\cos \gamma + i \sin \gamma}$.

8. $\dfrac{(\cos \theta + i \sin \theta)^2}{\cos \phi - i \sin \phi}$.

9. $(\cos \theta - i \sin \theta)^3$.

10. $(\cos 2\theta + i \sin 2\theta)^3$.

11. $\dfrac{(\cos \theta - i \sin \theta)^2}{(\cos \theta + i \sin \theta)^3}$.

12. $\dfrac{(\cos 2\theta - i \sin 2\theta)^3}{(\cos 3\theta + i \sin 3\theta)^4}$.

13. $\left(\cos \frac{\pi}{5} + i \sin \frac{\pi}{5}\right)^5$.

14. $\left(\cos \frac{3\pi}{5} + i \sin \frac{3\pi}{5}\right)^5$.

15. $(\sin \theta + i \cos \theta)^3$.

16. $(\sin \theta - i \cos \theta)^5$.

17. $(1+\cos\theta+i\sin\theta)^3$.

18. $(1+i\sin\theta-\cos\theta)^4$.

19. $\dfrac{1+\cos 2\theta+i\sin 2\theta}{\cos 2\theta+i\sin 2\theta}$.

20. $\dfrac{1-\cos 2\theta+i\sin 2\theta}{1+\cos 2\theta-i\sin 2\theta}$.

21. $\dfrac{(1+\sin 2\theta+i\cos 2\theta)^4}{(1+\sin 2\theta-i\cos 2\theta)^4}$.

22. $\dfrac{1}{(1-\sin\theta-i\cos\theta)^3}$.

23. $\dfrac{\text{cis}^3\, 5\theta\,(\text{cis}\,\theta)^{-3}}{\text{cis}^5\, 2\theta\ \text{cis}^2\, 3\theta}$.

24. $\dfrac{(1-\text{cis}\,\theta)^3}{\{1+\text{cis}\,(-\theta)\}^3}$.

25. Write down the cubes of $\text{cis}\,\dfrac{\theta}{3}$, $\text{cis}\,\dfrac{\theta+2\pi}{3}$, $\text{cis}\,\dfrac{\theta+4\pi}{3}$. What inference can be drawn from the results ?

26. Write down the values of

$$\text{cis}^4 a,\quad \text{cis}^4\left(a+\frac{\pi}{2}\right),\quad \text{cis}^4(a+\pi),\quad \text{cis}^4\left(a+\frac{3\pi}{2}\right).$$

What inference can be drawn ?

27. Simplify cis A cis B cis C, if $A+B+C=\pi$.

28. Simplify $(\text{cis}\,\theta)^n+(\text{cis}\,\theta)^{-n}$.

29. If $z=\cos\theta+i\sin\theta$, express in terms of θ

$$\text{(i) } z+\frac{1}{z};\quad \text{(ii) } z-\frac{1}{z};\quad \text{(iii) } z^n+\frac{1}{z^n};\quad \text{(iv) } z^n-\frac{1}{z^n}.$$

30. If $u=\text{cis}\,\theta$, $v=\text{cis}\,\phi$, express $\dfrac{u}{v}+\dfrac{v}{u}$ in terms of θ, ϕ.

31. If n is a positive integer, prove that

$$(\sin\theta+i\cos\theta)^n=\left(\frac{1+\sin\theta+i\cos\theta}{1+\sin\theta-i\cos\theta}\right)^n=\text{cis}\left(\frac{n\pi}{2}-n\theta\right).$$

32. If the complex number z is represented by the given point P, and if $|z|=1$, show how to construct the points which represent

$$\text{(i) } 2z^2;\quad \text{(ii) } z^3;\quad \text{(iii) } z^{-1}.$$

33. Given the point P which represents any complex number z, construct the points which represent

$$\text{(i) } z^2;\quad \text{(ii) } z^2+3;\quad \text{(iii) } (z+1)^2;\quad \text{(iv) } -z^{-1}.$$

34. If the point P which represents the complex number z moves along the x-axis from $x=-1$ to $x=+1$, describe the corresponding motions of the points which represent $z+a+bi$, az, iz, and $(a+bi)z$.

35. If the point P in No. 34 moves with uniform speed, describe the corresponding motions of the points representing z^2 and $1/z$.

36. If $z_1=1/z$, $z_2=1/(1-z_1)$, $z_3=1/(1-z_2)$, and the point which represents z moves along the x-axis from $x=-1$ to $x=+1$, find the corresponding motions of the points representing z_1, z_2, z_3.

37. If the point which represents the complex number z moves round the circle $|z|=1$ in the anti-clockwise direction starting from the point $(1, 0)$, describe the motions of the points which represent the complex numbers

$$\text{(i) } 2z^2\text{; (ii) } \frac{1}{z}\text{; (iii) } \frac{1}{z-2}\text{; (iv) } (z-1)^2.$$

38. Answer the same question as in No. 37 for

$$\text{(i) } z-1\text{; (ii) } \frac{1}{z-1}\text{; (iii) } \frac{2}{z-1}\text{; (iv) } \frac{z+1}{z-1}.$$

39. Answer the same question as in No. 37 for

$$\text{(i) } z+2\text{; (ii) } \frac{3}{z+2}\text{; (iii) } \frac{2z+1}{z+2}\text{; (iv) } \frac{az+b}{cz+d},$$

where a, b, c, d are real numbers.

40. Answer the same question as in No. 37 for

$$\text{(i) } iz\text{; (ii) } i(z+1)\text{; (iii) } \frac{i}{z+1}\text{; (iv) } \frac{i}{z+i}.$$

41. If $z_1=(1-iz)/(z-i)$ and the point representing z moves from -1 to $+1$ along the x-axis, how does the point which represents z_1 move?

42. With the notation of No. 41, if

$$z_2=(1-iz_1)/(z_1-i), \quad z_3=(1-iz_2)/(z_2-i), \text{ etc.},$$

describe the motions of the points which represent $z_2, z_3, z_4, \ldots$.

43. If P is a given point on the circle $|z-1|=1$, state a construction for the point Q, such that the complex number represented by P is the square of that represented by Q. Find the locus of Q when P moves round the circle.

Principal Values. When a function, $f(x)$, such as $\sin^{-1}x$, $\tan^{-1}x$, or $\text{am}(x+3i)$ has more than one real value for a given value, x_1, of x, the numerically least of these values is called the *principal value* of $f(x)$ corresponding to $x=x_1$; and if there are two numerically equal least values, the positive one is called the principal value.

Thus $\tan^{-1}(-1)$ has values $\frac{3\pi}{4}, \frac{7\pi}{4}, \ldots, -\frac{\pi}{4}, -\frac{5\pi}{4}, \ldots$ and of these we call $-\frac{\pi}{4}$ the principal value. Again $\cos^{-1}\frac{1}{2}$ has values $2n\pi \pm \frac{\pi}{3}$, which include both $+\frac{\pi}{3}$ and $-\frac{\pi}{3}$; of these, we call $+\frac{\pi}{3}$ the principal value.

The aggregate of these selected principal values of $f(x)$ is called the range of principal values of $f(x)$.

There is no recognised standard notation for distinguishing principal values from general values. We shall henceforward usually mean by $\sin^{-1}x$ the principal value of the function, and, if we wish to call special attention to the fact that the general value is intended we shall write $\text{Sin}^{-1}x$; similarly we shall use $\text{Tan}^{-1}x$, $\text{Am}\,z$, etc., to denote general values. But, if the context is such as to remove any possibility of ambiguity, $\sin^{-1}x$, $\cos^{-1}x$, etc., may be used to represent general values.

The beginner is advised not to omit any of the examples which are marked with an asterisk in Ex. VIII. f. No. 8 is particularly important.

EXERCISE VIII. f.

Verify the results of Nos. 1-4.

1.* $-\frac{\pi}{2} \leqslant \sin^{-1}x \leqslant \frac{\pi}{2}$.

2.* $0 \leqslant \cos^{-1}x \leqslant \pi$.

3.* $-\frac{\pi}{2} < \tan^{-1}x < \frac{\pi}{2}$.

4.* $-\pi < \text{am}(x+iy) \leqslant \pi$.

5.* Draw rough graphs of $\sin^{-1}x$, $\cos^{-1}x$, and $\tan^{-1}x$.

6. Find the values of (i) $\sin^{-1}x + \cos^{-1}x$; (ii) $\tan^{-1}x + \cot^{-1}x$.

7. For what values of x is $\sin^{-1}x$ equal to

(i) $\cos^{-1}\sqrt{(1-x^2)}$; (ii) $\cos^{-1}\{-\sqrt{(1-x^2)}\}$; (iii) $-\cos^{-1}\sqrt{(1-x^2)}$?

8.* From the definition $-\pi < \text{am}(x+iy) \leqslant \pi$, prove that

(i) if $x > 0$, $\text{am}(x+iy) = \tan^{-1}\frac{y}{x}$;

(ii) if $x < 0 < y$, $\text{am}(x+iy) = \pi + \tan^{-1}\frac{y}{x}$;

(iii) if $x < 0$ and $y < 0$, $\text{am}(x+iy) = -\pi + \tan^{-1}\frac{y}{x}$.

9. What are the values of $\text{am}(x+yi)$, (i) when $y=0$, (ii) when $x=0$? Draw the graphs of $\text{am}\,x$ and $\text{am}\,ix$.

10. Draw the graphs of (i) $\text{am}(x+ix)$; (ii) $\text{am}(x-ix)$.

11. If m is positive, for what values of n is $\tan^{-1}m + \tan^{-1}n > \frac{\pi}{2}$? Answer the same question when m is negative.

Can $\tan^{-1}m + \tan^{-1}n$ be $< -\frac{\pi}{2}$?

12.* Prove that the value of k in the formula

$$\tan^{-1}m + \tan^{-1}n = k\pi + \tan^{-1}\left(\frac{m+n}{1-mn}\right)$$

is zero unless $mn > 1$, and that if $mn > 1$, k is $+1$ or -1, according as m and n are positive or negative.

13.* If r_1, θ_1 and r_2, θ_2 are the moduli, and principal values of the amplitudes, of two complex numbers, and r, θ of their product, prove that

(i) if $\theta_1+\theta_2 \leqslant -\pi$, $\theta=\theta_1+\theta_2+2\pi$;

(ii) if $-\pi<\theta_1+\theta_2 \leqslant \pi$, $\theta=\theta_1+\theta_2$;

(iii) if $\pi<\theta_1+\theta_2$, $\theta=\theta_1+\theta_2-2\pi$.

14. Draw the graph of $2\tan^{-1}x-\tan^{-1}\dfrac{2x}{1-x^2}$.

15. Draw the graph of $2\cos^{-1}x-\cos^{-1}(2x^2-1)$.

16.* What meanings must be assigned to the many valued functions $\text{Sin}^{-1}x$, $\text{Cos}^{-1}x$, $\text{Tan}^{-1}x$ in order that the following relations may be true ?

(i) If $|x|<1$, $\dfrac{d}{dx}(\text{Sin}^{-1}x)=\dfrac{1}{\sqrt{(1-x^2)}}$;

(ii) if $|x|<1$, $\dfrac{d}{dx}(\text{Cos}^{-1}x)=\dfrac{-1}{\sqrt{(1-x^2)}}$;

(iii) for all values of x, $\dfrac{d}{dx}(\text{Tan}^{-1}x)=\dfrac{1}{1+x^2}$.

EASY MISCELLANEOUS EXAMPLES

EXERCISE VIII. g.

1. Express in the modulus-amplitude form (i) $\dfrac{1+i}{\sqrt{2}}$; (ii) $(1+i)^n$.

2. Evaluate $(1+i\sqrt{3})^8+(1-i\sqrt{3})^8$.

3. If $z=\text{cis}\,\theta$, prove that $\dfrac{2}{1+z}=1-i\tan\dfrac{\theta}{2}$, and that

$$\frac{1+z}{1-z}=i\cot\frac{\theta}{2}.$$

4. If n is a positive integer, prove that

$$(1+i)^n+(1-i)^n=2^{\frac{n}{2}+1}.\cos\frac{n\pi}{4}.$$

5. If A, B, C are the vertices of a triangle in the Argand Diagram representing the complex numbers α, β, γ, what are the points which represent (i) $\frac{1}{2}(\beta+\gamma)$; (ii) $\frac{1}{3}(\alpha+\beta+\gamma)$; (iii) $k\alpha+(1-k)\beta$, where k is a real number ?

6. ABCD is a quadrilateral in the Argand Diagram; E, F, G, H, P, Q are the middle points of AB, BC, CD, DA, AC, BD. If A, B, C, D represent the complex numbers α, β, γ, δ, what numbers are represented by E, G and the middle points of EG, FH, PQ ? What conclusion can be drawn from these results ?

7. Four points represent the complex numbers a, β, γ, δ. Interpret geometrically the condition $a+\gamma=\beta+\delta$.

8. Two fixed points A, B and a variable point P represent the complex numbers a, β, z. Find the locus of P if

(i) $|z-a|=|\beta|$; (ii) $|z-a|=|z-\beta|$; (iii) $|z-a|=3|z-\beta|$.

9. Verify that $\frac{1}{2}\sqrt{2}.(\pm 1 \pm i)$ are four fourth roots of -1, and deduce two quadratic factors of x^4+1.

10. If $a=\text{cis}\,2a$, $b=\text{cis}\,2\beta$, $c=\text{cis}\,2\gamma$, and $d=\text{cis}\,2\delta$, express in the modulus-amplitude form

(i) $a+b$; (ii) $a-b$; (iii) $(a-c)(b+d)$.

11. If $a=\text{cis}\,\theta$ and $b=\text{cis}\,\phi$, prove that $\cos(\theta+\phi)=\frac{1}{2}\left(ab+\frac{1}{ab}\right)$.

12. Interpret geometrically the relation $z=a+t(\beta-a)$, where a and β are fixed complex numbers and t is a real variable.

13. If $|z|=1$ and $\text{am}\,z=\theta$, find the values of

$$\text{(i)}\left|\frac{2}{1-z^2}\right|;\quad \text{(ii)}\ \text{am}\left(\frac{2}{1-z^2}\right),$$

where $\text{am}\,w$ denotes the principal value of the amplitude of w.

14. If $|z_1-z_2|=|z_1+z_2|$, prove that $\text{am}\,z_1$ and $\text{am}\,z_2$ differ by $\frac{\pi}{2}$ or $\frac{3\pi}{2}$.

15. Two fixed points, A and B, and a variable point P, represent the complex numbers a, β, and z. Find the locus of P if

(i) $\text{am}(z-a)=\text{am}\,\beta$; (ii) $\text{am}(z-a)-\text{am}(z-\beta)=\frac{\pi}{6}$.

16. In No. 15, if β becomes variable and P describes a given curve Σ, what is the locus of the point that represents β, if (i) $\beta=a+z$, (ii) $\beta=az$, and (iii) $\beta=|a|\div z$.

17. The transformations $z_1=\frac{a+bz}{c+dz}$, $z_2=\frac{a+bz_1}{c+dz_1}$ give z_2 in terms of z. Find a, b, c, d so that the resulting single transformation may be $z_2=1-\frac{1}{z}$.

18. The transformation $z=\frac{1+Zi}{Z+i}$ when repeated a second time leads to $\frac{1}{z}$. Do any other transformations of the form $z=\frac{a+bZ}{c+dZ}$ have this property ?

19. What is the condition that the equations $3x+4y=p$, $x^2+y^2=c^2$ can only be solved in the algebra of complex numbers ?

20. If $\omega^3 = 1$, but ω is not 1, prove that

(i) $1 + \omega + \omega^2 = 0$;

(ii) ω^2 is a root of $z^3 = 1$ and of $z^2 + z + 1 = 0$;

(iii) $a^2 - ab + b^2 \equiv (\omega a + \omega^2 b)(\omega^2 a + \omega b)$.

21. With the notation of No. 20, find the values of

(i) $(1 + \omega)^3$;

(ii) $(1 + 2\omega + 3\omega^2)(1 + 3\omega + 2\omega^2)$;

(iii) $1 + \omega + \omega^2 + \omega^3 + \ldots$ to n terms.

22. With the notation of No. 20, expand

(i) $(a - b\omega)(a - b\omega^2)(a - b)$;

(ii) $(a + b + c)(a + b\omega + c\omega^2)(a + b\omega^2 + c\omega)$.

HARDER MISCELLANEOUS EXAMPLES.

EXERCISE VIII. h.

1. Simplify

$$\{(\cos \alpha - \cos \beta) + i(\sin \alpha - \sin \beta)\}^n + \{(\cos \alpha - \cos \beta) - i(\sin \alpha - \sin \beta)\}^n$$

(i) for n even; (ii) for n odd.

2. Prove that $\{\text{cis}\, 2\theta + \text{cis}\,(-2\phi)\}\, \text{cis}\, \phi = 2\, \text{cis}\, \theta \cos(\theta + \phi)$.

3. Express $\dfrac{3}{2 + \cos\theta + i\sin\theta}$ in the form $x + yi$, and prove that $x^2 + y^2 = 4x - 3$.

4. If $p + iq = x\, \text{cis}\, \alpha\, \text{cis}\, \theta + y\, \text{cis}\, \beta\, \text{cis}\, \theta$, obtain a relation independent of θ.

5. If $\text{cis}\, \alpha = a$ and $\text{cis}\, \beta = b$, prove that $\sin(\alpha - \beta) = \dfrac{(b^2 - a^2)i}{2ab}$.

6. If $a = \text{cis}\, 2\alpha$, $b = \text{cis}\, 2\beta$, $c = \text{cis}\, 2\gamma$, and $d = \text{cis}\, 2\delta$, express in the modulus-amplitude form

(i) $a^2 - b^2$; (ii) $ab - cd$; (iii) $abcd - \dfrac{1}{abcd}$.

7. If $(1 + \text{cis}\, \theta)(1 + \text{cis}\, 2\theta) = u + iv$, prove that

(i) $v = u \tan \dfrac{3\theta}{2}$; (ii) $u^2 + v^2 = 16 \cos^2\theta \cos^2 \dfrac{\theta}{2}$.

8. If $(1 + ix_1)(1 + ix_2)(1 + ix_3) = \mathsf{A} + \mathsf{B}i$, prove that

$$\mathsf{A} \tan(\Sigma \tan^{-1} x) = \mathsf{B}.$$

9. The complex numbers a, z, $z - a$ are represented by points A, P, Q. If A is fixed and P describes a given curve, what is the locus of Q ?

10. If the point P, which represents the complex number z, describes the circle centre (1, 0) and radius 1, what locus is described by the point representing $\dfrac{1}{az+b+ci}$, where a, b, c are real?

11. If $|z^2-1|=2$, prove that the point P, which represents z, moves so that PA . PB is constant, where A and B are certain fixed points.

12. If the point which represents z moves on the unit circle, centre O, what curve is described by the point representing $2z+z^2$?

13. If $z=x+yi$, where y is positive, prove that $\left|\dfrac{z-i}{z+i}\right|<1$.

14. If $z=x+yi$, $\mathrm{Z}=\dfrac{e^x \operatorname{cis} y-1}{e^x \operatorname{cis} y+1}$ and $-\dfrac{\pi}{2}<y<\dfrac{\pi}{2}$, prove that $|\mathrm{Z}|<1$.

15. The fixed points A, B and the variable point P represent the complex numbers α, β, z. What is the locus of P, if

(i) $\operatorname{am}\left(\dfrac{z-\beta}{z-\alpha}\right)=\dfrac{\pi}{3}$; (ii) $\operatorname{am}\left(\dfrac{z-\beta}{z-\alpha}\right)=-\dfrac{\pi}{3}$; (iii) $\operatorname{am}\left(\dfrac{z-\beta}{z-\alpha}\right)=-\dfrac{2\pi}{3}$?

16. If $z_1(a_2-a_3)+z_2(a_3-a_1)+z_3(a_1-a_2)=0$, and a_1, a_2, a_3 are real, prove that the points which represent the complex numbers z_1, z_2, z_3, are collinear.

17. If A, B, C are points which represent complex numbers α, β, γ, and if $a=|\beta-\gamma|$, $b=|\gamma-\alpha|$, $c=|\alpha-\beta|$, prove that the in-centre of ABC represents $\dfrac{a\alpha+b\beta+c\gamma}{a+b+c}$. What numbers do the e-centres represent?

18. Interpret the relation $\dfrac{\alpha-\gamma}{\beta-\gamma}=\operatorname{cis}\dfrac{\pi}{3}$ between the complex numbers α, β, γ, and prove that $\alpha^2+\beta^2+\gamma^2=\beta\gamma+\gamma\alpha+\alpha\beta$ follows from it.

19. Interpret the relation $\dfrac{\alpha-\gamma}{\alpha-\beta}=\dfrac{\lambda-\nu}{\lambda-\mu}$ between the six complex numbers $\alpha, \beta, \gamma, \lambda, \mu, \nu$, and show that it can be written

$$\begin{vmatrix} \alpha & \beta & \gamma \\ \lambda & \mu & \nu \\ 1 & 1 & 1 \end{vmatrix}=0.$$

20. If a and b are real, prove geometrically that

$$\left\{\left(a\cos\frac{2\pi}{n}-b\sin\frac{2\pi}{n}\right)+i\left(a\sin\frac{2\pi}{n}+b\cos\frac{2\pi}{n}\right)\right\}^n=(a+ib)^n.$$

21. An ellipse in the Argand Diagram has foci $(\pm d, 0)$ and z_1, z_2 are complex numbers which correspond to the ends of conjugate semi-diameters, prove that $z_1^2+z_2^2=d^2$.

22. Draw the graphs of (i) $\operatorname{cosec}^{-1}x$; (ii) $\sec^{-1}x$; (iii) $\cot^{-1}x$; for principal values only.

23. Draw the graph of $\sec x + \sec y + 2 = 0$.

24. Prove that, unless $x = (2n+1)\pi$,

$$\frac{x - \operatorname{am}(\operatorname{cis} x)}{2\pi} = \left[\frac{x+\pi}{2\pi}\right].\quad \text{(See footnote, p. 46.)}$$

25. Prove that the relation $z = (1 + Zi)/(Z + i)$ transforms the part of the axis of x between the points $z = -1$ and $z = +1$ into a semicircle passing through the points $Z = 1$ and $Z = -1$. Find all the figures that can be found from the originally selected part of the axis of x by successive applications of this transformation.

26. If $\omega^3 = 1$, but $\omega \neq 1$, prove that $\omega^n + \omega^{2n} = 2$ or -1, where n is a positive integer.

27. Prove that $x^2 + xy + y^2$ is a factor of $(x+y)^n - x^n - y^n$, where n is odd and not divisible by 3.

28. Prove that $x^2 + y^2 + z^2 - yz - zx - xy$ is a factor of

$$(y-z)^n + (z-x)^n + (x-y)^n, \text{ if } n \text{ is not divisible by 3.}$$

29. Prove that $(x^2 + y^2 + z^2 - yz - zx - xy)^2$ is a factor of

$$(y-z)^n + (z-x)^n + (x-y)^n, \text{ if } n \equiv 1 \pmod{3}.$$

30. Write down the product of $x + y\omega + z\omega^2$, $x + y\omega^2 + z\omega$ where $\omega = \operatorname{cis}\dfrac{2\pi}{3}$; hence express $x^3 + y^3 + z^3 - 3xyz$ and

$$(x^2 - yz)^3 + (y^2 - zx)^3 + (z^2 - xy)^3 - 3(x^2 - yz)(y^2 - zx)(z^2 - xy),$$

each as the product of three factors. Prove that the first expression is the square root of the second.

31. If $f_n(x)$ is the sum to n terms of the series whose rth term is $a_r x^r$, what do the expressions,

$$\text{(i) } f_n(x) + f_n(\omega x) + f_n(\omega^2 x), \quad \text{(ii) } f_n(x) + \omega f_n(\omega x) + \omega^2 f_n(\omega^2 x),$$

represent when $\omega^3 = 1$ but $\omega \neq 1$?

CHAPTER IX

DE MOIVRE'S THEOREM AND APPLICATIONS

Definition of a^n, where a is complex and n is a rational number. For integral values of n, the definition has been given on pp. 140, 141. If n is fractional, it is equal to $\frac{p}{q}$, where p and q are integers, and there is no loss of generality in supposing that q is positive; and then any value of z which satisfies the equation, $z^q = a^p$, is called a value of a^n. We reserve the notation, $\sqrt[q]{a^p}$, for the principal value of a^n, as defined on p. 165.

De Moivre's Theorem. *If n is any rational number,*

$$\cos n\theta + i \sin n\theta \text{ is a value of } (\cos\theta + i\sin\theta)^n. \quad \text{..........(1)}$$

(i) First, suppose that n is a positive integer.

We have proved on p. 151, by actual multiplication, that

$$(\cos\theta_1 + i\sin\theta_1)(\cos\theta_2 + i\sin\theta_2)\ldots(\cos\theta_n + i\sin\theta_n)$$
$$= \cos(\Sigma\theta) + i\sin(\Sigma\theta).$$

Putting $\theta_1 = \theta_2 = \ldots = \theta_n = \theta$, we have

$$(\cos\theta + i\sin\theta)^n = \cos n\theta + i\sin n\theta.$$

(ii) Next, suppose that n is a negative integer. Put $n = -m$.

Then $(\cos\theta + i\sin\theta)^n = (\cos\theta + i\sin\theta)^{-m}$, and this, by definition,

$$= \frac{1}{(\cos\theta + i\sin\theta)^m} = \frac{1}{\cos m\theta + i\sin m\theta}, \text{ by (i).}$$

But $(\cos m\theta + i\sin m\theta)(\cos m\theta - i\sin m\theta) = \cos^2 m\theta - i^2\sin^2 m\theta = 1$;

$$\therefore \frac{1}{\cos m\theta + i\sin m\theta} = \cos m\theta - i\sin m\theta$$
$$= \cos(-m\theta) + i\sin(-m\theta);$$

$$\therefore (\cos\theta + i\sin\theta)^n = \cos n\theta + i\sin n\theta.$$

Therefore, if n is a positive or negative integer, there is only one value of $(\cos\theta + i\sin\theta)^n$, and this value is $\cos n\theta + i\sin n\theta$.

(iii) Next, suppose that n is a fraction. Put $n = \frac{p}{q}$, where p, q are integers and q is positive.

In this case $(\cos\theta + i\sin\theta)^n$ is many-valued, and we shall prove presently that it has q values. At the moment, we merely wish to show that $\cos n\theta + i\sin n\theta$ is one value of $(\cos\theta + i\sin\theta)^n$.

By (i), $$\left(\cos\frac{p\theta}{q}+i\sin\frac{p\theta}{q}\right)^q=\cos p\theta+i\sin p\theta.$$

Also by (i) or (ii), $\cos p\theta+i\sin p\theta=(\cos\theta+i\sin\theta)^p$;

$$\therefore\ \left(\cos\frac{p\theta}{q}+i\sin\frac{p\theta}{q}\right)^q=(\cos\theta+i\sin\theta)^p;$$

$\therefore$ by the definition of a^n, given above, it follows that

$$\cos\frac{p\theta}{q}+i\sin\frac{p\theta}{q}\text{ is a value of }(\cos\theta+i\sin\theta)^{\frac{p}{q}}.$$

The theorem is therefore proved for all rational values of n.

Writing $-\theta$ for θ, we see that $\cos n\theta-i\sin n\theta$ is a value of $(\cos\theta-i\sin\theta)^n$, for all rational values of n.

The values of $(\cos\theta+i\sin\theta)^{\frac{p}{q}}$, where p, q are integers and q is positive. De Moivre's Theorem states that $\cos\frac{p\theta}{q}+i\sin\frac{p\theta}{q}$ is *one* value of $(\cos\theta+i\sin\theta)^{\frac{p}{q}}$. Suppose that $s(\cos\phi+i\sin\phi)$ represents *any* value of $(\cos\theta+i\sin\theta)^{\frac{p}{q}}$.

Then, by definition,

$$s^q(\cos\phi+i\sin\phi)^q=(\cos\theta+i\sin\theta)^p;$$

$$\therefore\ s^q(\cos q\phi+i\sin q\phi)=\cos p\theta+i\sin p\theta;$$

$$\therefore\ s^q=1;\quad \cos q\phi=\cos p\theta;\quad \sin q\phi=\sin p\theta.$$

But s is positive, since it is the modulus of a complex number;

$\therefore$ $s^q=1$ requires that $s=1$; since, if $s>1$, $s^q>1$ and if $0\leqslant s<1$, $s^q<1$.

Also the other equations require that $q\phi=p\theta+2r\pi$, where r is an integer or zero.

Taking, in succession, $r=0, 1, 2, \dots, (q-1)$, we obtain the q values,

$$\operatorname{cis}\frac{p\theta}{q},\ \operatorname{cis}\left(\frac{p\theta}{q}+\frac{2\pi}{q}\right),\ \operatorname{cis}\left(\frac{p\theta}{q}+\frac{4\pi}{q}\right),\ \dots\ \operatorname{cis}\left(\frac{p\theta}{q}+\frac{2(q-1)\pi}{q}\right).$$

These q values are all distinct, because the angles given by $r=r_1$, $r=r_2$ differ by $\frac{2(r_1\sim r_2)\pi}{q}$, which is less than 2π, since $|r_1-r_2|<q$.

Also, *no further values are given by other values of r,* because any other value of r must differ from one of the numbers $0, 1, 2, \dots, (q-1)$ by a multiple of q.

If p, q are prime to one another, the same results may be written, but in a different order, as

$$\text{cis}\,\frac{p\theta}{q},\ \text{cis}\,\frac{p(\theta+2\pi)}{q},\ \text{cis}\,\frac{p(\theta+4\pi)}{q},\ \dots,\ \text{cis}\,\frac{p[\theta+2(q-1)\pi]}{q},$$

because the numbers $0, p, 2p, 3p, \dots, (q-1)p$ are congruent (mod. q) to $0, 1, 2, \dots, (q-1)$, in some order.

If p is not prime to q, the function $(\cos\theta+i\sin\theta)^{\frac{p}{q}}$ is taken to mean $\{(\cos\theta+i\sin\theta)^p\}^{\frac{1}{q}}$ and has therefore q distinct values, viz., the q values of $(\cos p\theta+i\sin p\theta)^{\frac{1}{q}}$; these may be written,

$$\text{cis}\left(\frac{p\theta}{q}+\frac{2r\pi}{q}\right),\ r=0, 1, 2, \dots, (q-1);$$

but in this case, the expression $\text{cis}\left\{\frac{p(\theta+2s\pi)}{q}\right\}$, where s is any integer, does not assume q distinct values and therefore does not represent all the values of $(\cos\theta+i\sin\theta)^{\frac{p}{q}}$. Thus, the function $(\cos\theta+i\sin\theta)^{\frac{4}{8}}$ has the 8 values given by $(\cos 4\theta+i\sin 4\theta)^{\frac{1}{8}}$, and is distinct from the two-valued function $(\cos\theta+i\sin\theta)^{\frac{1}{2}}$; its 8 values are represented by

$\text{cis}\left(\frac{4\theta}{8}+\frac{2r\pi}{8}\right)$, $r=0, 1, 2, \dots, 7$; but the expression $\text{cis}\left\{\frac{4(\theta+2s\pi)}{8}\right\}$, where s is any integer, has only 2 distinct values. Cf. Ex. IX. a, No. 8.

Principal Value of $(\cos\theta+i\sin\theta)^{\frac{p}{q}}$. The principal value of $(\cos\theta+i\sin\theta)^{\frac{p}{q}}$ is taken to be $\text{cis}\,\frac{p\theta}{q}$, only if $-\pi<\theta\leqslant\pi$.

Otherwise, if k is the (positive or negative) integer such that $-\pi<\theta+2k\pi\leqslant\pi$, the principal value of $(\cos\theta+i\sin\theta)^{\frac{p}{q}}$ is taken to be

$$\text{cis}\left\{\frac{p}{q}(\theta+2k\pi)\right\}=\text{cis}\left(\frac{p\theta}{q}+\frac{2pk\pi}{q}\right).$$

For any given value of k, this can of course be reduced to the form, $\text{cis}\left(\frac{p\theta}{q}+\frac{2r\pi}{q}\right)$, where r is some integer less than q.

This definition of principal value holds whether p is prime to q or not.

The reader should notice that the principal value of $(\cos\theta + i\sin\theta)^{\frac{p}{q}}$, where $-\pi < \theta \leqslant \pi$, is not the same as the principal value of $(\cos p\theta + i\sin p\theta)^{\frac{1}{q}}$, unless also $-\pi < p\theta \leqslant \pi$.

Further, the principal value of the qth root of $(\cos\theta + i\sin\theta)^p$ or the principal qth root of $(\cos\theta + i\sin\theta)^p$ is taken to mean the principal value of $(\cos\theta + i\sin\theta)^{\frac{p}{q}}$, as defined above. Thus the principal value of the 8th root of $(\cos\theta + i\sin\theta)^4$ means the principal value of $(\cos\theta + i\sin\theta)^{\frac{4}{8}}$, and this is defined above as

$$\text{cis}\,\{\tfrac{4}{8}(\theta + 2k\pi)\} = \text{cis}\,\{\tfrac{1}{2}(\theta + 2k\pi)\},$$

where k is the integer given by $-\pi < \theta + 2k\pi \leqslant \pi$. The principal value of $(\cos\theta + i\sin\theta)^{\frac{4}{8}}$ is therefore the same as the principal value of $(\cos\theta + i\sin\theta)^{\frac{1}{2}}$, but it is not the same as the principal value of $(\cos 4\theta + i\sin 4\theta)^{\frac{1}{8}}$, unless, with the same notation as before, $-\pi < 4(\theta + 2k\pi) \leqslant \pi$. Cf. Ex. IX. a, No. 10.

Values of $\mathbf{z^{\frac{p}{q}}}$. *Definition.* If r is any real positive number, and if p and q are integers, q being positive, the symbol $\sqrt[q]{(r^p)}$ denotes the (unique) positive qth root of r^p.

Every complex number, z, can be written in the form,

$$r(\cos\theta + i\sin\theta),$$

where $$r = |z| \quad \text{and} \quad -\pi < \theta \leqslant \pi.$$

$\therefore$ the values of $z^{\frac{p}{q}}$ may be written

$$\sqrt[q]{(r^p)}\ \text{cis}\left(\frac{p\theta}{q} + \frac{2s\pi}{q}\right),$$

where $s = 0, 1, 2, \ldots, (q-1)$.

Thus there are q values, whether p is prime to q or not; and of these, since $-\pi < \theta \leqslant \pi$, the *principal value* is

$$\sqrt[q]{(r^p)}\ \text{cis}\,\frac{p\theta}{q}.$$

Geometrical Representation of Powers and Roots. Fig. 76 represents the circle $|z| = 1$ in the Argand Diagram. The point P_1 which represents the complex number $(\cos\theta + i\sin\theta)$ lies on the circle, and the arc AP_1, measured from the point A, (1, 0), is of length θ.

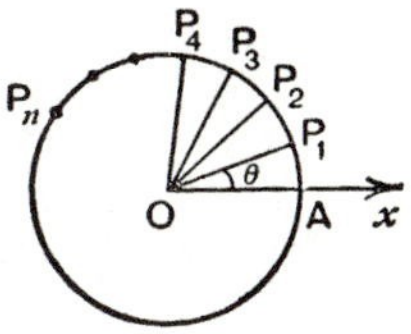

FIG. 76.

To apply the geometrical method of construction, given on p. 151, for the points representing the numbers

$$(\cos\theta + i\sin\theta)^2,\ (\cos\theta + i\sin\theta)^3,\ \dots,\ (\cos\theta + i\sin\theta)^n,$$

we construct in succession the triangles P_1OP_2, P_2OP_3, ..., $P_{n-1}OP_n$, each similar to $\triangle AOP_1$. The points $P_2, P_3, \dots, P_n$, being on the circle at arcual distances $2\theta, 3\theta, \dots, n\theta$ from A, represent the numbers

$$(\cos 2\theta + i\sin 2\theta),\ (\cos 3\theta + i\sin 3\theta),\ \dots,\ (\cos n\theta + i\sin n\theta).$$

This illustrates part (i) of de Moivre's Theorem.

Suppose now, see Fig. 77, that Q is the point on the circle which represents $\cos\alpha + i\sin\alpha$, and that we want to represent geometrically an nth root of that number. We shall have to find a point P on the circle such that the arc AQ $= n$. arc AP; but as the arcual distance of Q from A can be regarded as α or $\alpha + 2\pi$ or $\alpha + 4\pi$ or ... or $\alpha + 2r\pi$, where r is any integer, the arc AP may be taken as $\dfrac{\alpha + 2r\pi}{n}$, where r is any integer. This gives n points P, say $P_1, P_2, \dots P_n$, representing the n nth roots of $\cos\alpha + i\sin\alpha$. $P_1P_2 \dots P_n$ is a regular polygon inscribed in the circle.

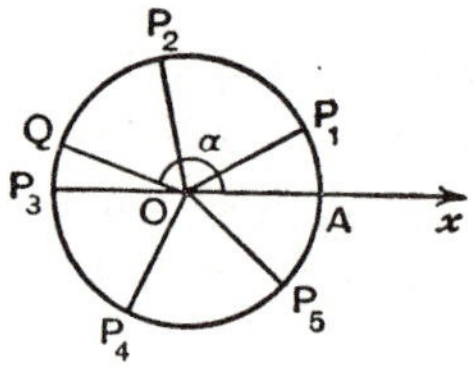

FIG. 77.

Note. The q values of $(\cos\alpha + i\sin\alpha)^{\frac{p}{q}}$ can also be represented in a similar way; similarly the nth roots of $c(\cos\alpha + i\sin\alpha)$ can be represented by the corners of a regular n-sided polygon inscribed in the circle, centre the origin, radius $\sqrt[n]{c}$.

Example 1. What is the principal value of $(1-i)^{\frac{3}{5}}$ and what are its other values?

$$1 - i = \sqrt{2}\left(\frac{1}{\sqrt{2}} - \frac{i}{\sqrt{2}}\right) = \sqrt{2}\left\{\cos\left(-\frac{\pi}{4}\right) + i\sin\left(-\frac{\pi}{4}\right)\right\};$$

$\therefore$ the principal value of $(1-i)^{\frac{3}{5}}$ is

$$(\sqrt[10]{2})^3\left\{\cos\left(-\frac{3\pi}{20}\right) + i\sin\left(-\frac{3\pi}{20}\right)\right\}$$

$$= \sqrt[10]{8}\left\{\cos\frac{3\pi}{20} - i\sin\frac{3\pi}{20}\right\}.$$

Also any value of $(1-i)^{\frac{3}{5}}$ can be expressed in the form,

$$\sqrt[10]{8}\left\{\cos\left(-\frac{3\pi}{20} + \frac{2r\pi}{5}\right) + i\sin\left(-\frac{3\pi}{20} + \frac{2r\pi}{5}\right)\right\},$$

where $r = 0, 1, 2, 3, 4$.

Example 2. Solve $(x-1)^n = x^n$.

Take the nth root of each side;

$$\therefore\ x - 1 = x\left(\cos\frac{2r\pi}{n} + i\sin\frac{2r\pi}{n}\right), \text{ where } r = 0, 1, 2, \ldots, (n-1);$$

$$\therefore\ x\left(1 - \cos\frac{2r\pi}{n} - i\sin\frac{2r\pi}{n}\right) = 1;$$

$$\therefore\ x\left(2\sin^2\frac{r\pi}{n} - 2i\sin\frac{r\pi}{n}\cos\frac{r\pi}{n}\right) = 1;$$

$$\therefore\ 2x\sin\frac{r\pi}{n}\left(\sin\frac{r\pi}{n} - i\cos\frac{r\pi}{n}\right) = 1.$$

Multiplying each side by $\sin\frac{r\pi}{n} + i\cos\frac{r\pi}{n}$,

$$2x\sin\frac{r\pi}{n} = \sin\frac{r\pi}{n} + i\cos\frac{r\pi}{n};$$

$$\therefore\ x = \tfrac{1}{2}\left(1 + i\cot\frac{r\pi}{n}\right),$$

where $r = 1, 2, \ldots, (n-1)$.

Example 3. Points P and Q in the Argand Diagram represent the complex numbers z and w. If $|z| = 1$ and if am z steadily increases from $-\pi$ to $+\pi$, describe the corresponding motion of Q if

(i) $w = z^{\frac{1}{3}}$; (ii) $w = \sqrt[3]{z}$.

If $z = \text{cis}\,\theta$, $z^{\frac{1}{3}} = \text{cis}\,\frac{\theta}{3}$ or $\text{cis}\,\frac{\theta + 2\pi}{3}$ or $\text{cis}\,\frac{\theta - 2\pi}{3}$.

(i) For each position of P, there are 3 positions of Q, say Q_1, Q_2, Q_3, which move continuously along the circle $|z| = 1$, anti-clockwise. Q_1 moves from $\theta = -\frac{\pi}{3}$ to $\theta = +\frac{\pi}{3}$; and at the same time, Q_2 moves from $\theta = +\frac{\pi}{3}$ to $\theta = \pi$, and Q_3 moves from $\theta = -\pi$ to $\theta = -\frac{\pi}{3}$.

(ii) $\sqrt[3]{z} = \text{cis}\,\frac{\theta}{3}$ for $-\pi < \theta < \pi$; $\therefore$ there is only one position of Q for each position of P, and Q moves from $\theta = -\frac{\pi}{3}$ to $\theta = +\frac{\pi}{3}$.

EXERCISE IX. a.

1. Write down the square roots of

(i) $\cos 2\theta + i \sin 2\theta$; (ii) $\cos 3\theta - i \sin 3\theta$;

(iii) $\sin \theta + i \cos \theta$; (iv) i; (v) $-i$.

2. Write down the cube roots of (i) $\cos 3\theta + i \sin 3\theta$; (ii) 1; (iii) i; (iv) $-i$; (v) $\cos \theta - i \sin \theta$; (vi) $\sin \theta - i \cos \theta$.

3. Write down the values of (i) $(-1)^{\frac{2}{3}}$; (ii) $(-i)^{\frac{3}{4}}$; (iii) $(1+i)^{\frac{1}{2}}$; (iv) $(1 - i\sqrt{3})^{\frac{2}{5}}$; (v) $128^{\frac{1}{7}}$.

4. Write down all the roots of (i) $x^5 = 1$; (ii) $x^4 + 1 = 0$.

5. Represent in the Argand Diagram (i) the cube roots of -1; (ii) the fourth roots of i; (iii) the fifth roots of 32; (iv) $(-5 - 12i)^3$.

6. Simplify $\left(\cos\frac{\pi}{8} + i \sin\frac{\pi}{8}\right)^4 \div \left(\cos\frac{\pi}{4} - i \sin\frac{\pi}{4}\right)^2$.

7. Simplify $\sqrt{\left(\cos\frac{\pi}{6} - i \sin\frac{\pi}{6}\right)^{11}} \div \sqrt{\left(\cos\frac{\pi}{6} + i \sin\frac{\pi}{6}\right)}$.

8. Give (i) the two values of $\left(\cos\frac{2\pi}{3} + i \sin\frac{2\pi}{3}\right)^{\frac{1}{2}}$; (ii) the eight values of $\left(\cos\frac{2\pi}{3} + i \sin\frac{2\pi}{3}\right)^{\frac{4}{8}}$; state which are the principal values.

9. Give the principal values of

(i) $\left(\cos\frac{3\pi}{4} + i \sin\frac{3\pi}{4}\right)^{\frac{7}{3}}$; (ii) $\left(\cos\frac{5\pi}{4} + i \sin\frac{5\pi}{4}\right)^{\frac{1}{3}}$.

10. Give (i) the principal 8th root of $\cos\frac{4\pi}{3} + i \sin\frac{4\pi}{3}$;

(ii) the principal square root of $\cos\frac{\pi}{3} + i \sin\frac{\pi}{3}$.

11. Find (i) the product, (ii) the sum, of the five values of $(\cos \pi + i \sin \pi)^{\frac{2}{5}}$. Of what equation are these five values the roots ?

12. Use the result of No. 4 (i) to write down factors of $x^5 - 1$, and deduce that $x^5 - 1 = (x - 1)\{x^2 + \frac{1}{2}(\sqrt{5} + 1)x + 1\}\{x^2 - \frac{1}{2}(\sqrt{5} - 1)x + 1\}$.

13. Solve $x^n = 1$. What is the sum of the roots ? If a is a root other than unity, prove that $1 + a + a^2 + \ldots + a^{n-1} = 0$.

14. Find the roots of $x^6 = 1$ which do *not* satisfy $x^2 + x + 1 = 0$.

15. Solve $(x + i)^6 + (x - i)^6 = 0$.

16. Solve $(1 + x)^n = (1 - x)^n$.

17. Solve $x^{2n} - 2x^n \cos na + 1 = 0$.

18. Solve $x^9 + x^5 - x^4 = 1$.

19. If $z = \text{cis}\,\theta$, express $\sqrt{\dfrac{1+z}{1-z}}$ in the form $\mathsf{A}+\mathsf{B}i$, (i) for $0<\theta<\pi$, (ii) for $\pi<\theta<2\pi$.

20. How many values are there of (i) $i^{\frac{1}{3}}+(-i)^{\frac{1}{3}}$; (ii) $i^{\frac{1}{3}}\times(-i)^{\frac{1}{3}}$?

21. Find the principal value of $(1+\cos\theta+i\sin\theta)^{\frac{3}{4}}$

(i) if $-\pi<\theta<\pi$; (ii) if $\pi<\theta<3\pi$.

22. A regular hexagon is inscribed in the circle $|z|=1$ in the Argand Diagram, and one vertex represents $\cos\alpha+i\sin\alpha$. What do the others represent?

23. Two points P and Q in the Argand Diagram represent complex numbers z and w, the modulus of z being unity. P moves so that am z steadily increases from $-\pi$ to $+\pi$.

Describe the corresponding motion of Q when

(i) $w=2z+3$; (ii) $w=iz+2$; (iii) $w=3z^2$.

24. Answer the same question as in No. 23 for

(i) $w=z^3$; (ii) $w=z^3+1$; (iii) $w=z^{\frac{1}{2}}$.

25. Answer the same question as in No. 23 for

(i) $w=\dfrac{1}{\sqrt{z}}$; (ii) $w=(z+1)^2$; (iii) $w=z^2+2z$.

26. Answer the same question as in No. 23 for

(i) $w=\sqrt{(z+1)}$; (ii) $w=\dfrac{1}{\sqrt{(z+1)}}$; (iii) $w=(z+1)^{\frac{1}{2}}$.

27. By using substitutions of the form $a=\text{cis}\,2\alpha$ in the identity,

$$(a_1-a_2)(a_3-a_4)+(a_2-a_3)(a_1-a_4)+(a_3-a_1)(a_2-a_4)=0,$$

prove that $\sin(\theta_1-\theta_2)\sin(\theta_3-\theta_4)+\ldots+\ldots=0$.

Expression of Powers of cos θ and sin θ in terms of Multiple Angles. The student no doubt realises already the great importance of formulae, such as $\sin^2\theta=\frac{1}{2}(1-\cos 2\theta)$, $\cos^3\theta=\frac{1}{4}(3\cos\theta+\cos 3\theta)$, etc. Besides being essential for dealing with all kinds of identities, they are wanted, for example, for the integration of powers of $\sin\theta$ and $\cos\theta$, and for the summation of series like $\Sigma(\sin^2 n\theta)$, $\Sigma\cos^2(\theta+n\phi)$, etc. We shall now investigate expressions for

$$\cos^n\theta,\quad \sin^n\theta,\quad \cos^p\theta\sin^q\theta.$$

If $\cos\theta+i\sin\theta=z$, then $\cos\theta-i\sin\theta=\dfrac{1}{z}$;

$$\therefore\ 2\cos\theta=z+\frac{1}{z};\quad 2i\sin\theta=z-\frac{1}{z};\quad \ldots\ldots\ldots\ldots(2)$$

also $\quad\cos n\theta+i\sin n\theta=z^n$ and $\cos n\theta-i\sin n\theta=\dfrac{1}{z^n}$;

$$\therefore\ 2\cos n\theta=z^n+\frac{1}{z^n};\quad 2i\sin n\theta=z^n-\frac{1}{z^n}.\quad \ldots\ldots\ldots(3)$$

By means of (2) and the binomial theorem for a positive integral index, the functions $\cos^n\theta$, $\sin^n\theta$, $\cos^p\theta \sin^q\theta$, etc., can be expressed in terms of z and expanded in powers of z and $1/z$.

By means of (3), the expression in powers of z and $1/z$ can be replaced by cosines or sines of multiple angles.

Example 4. Express $\cos^5\theta$ in terms of multiple angles.

From (2), $$(2\cos\theta)^5 = \left(z+\frac{1}{z}\right)^5 = z^5 + 5z^3 + 10z + \frac{10}{z} + \frac{5}{z^3} + \frac{1}{z^5}$$
$$= \left(z^5 + \frac{1}{z^5}\right) + 5\left(z^3 + \frac{1}{z^3}\right) + 10\left(z + \frac{1}{z}\right)$$
$$= 2\cos 5\theta + 5(2\cos 3\theta) + 10(2\cos\theta), \text{ from (3)};$$
$$\therefore\ \cos^5\theta = \tfrac{1}{16}(\cos 5\theta + 5\cos 3\theta + 10\cos\theta).$$

Check by putting $\theta = 0$.

Example 5. Express $\sin^5\theta$ in terms of multiple angles.

From (2), $$(2i\sin\theta)^5 = \left(z-\frac{1}{z}\right)^5 = z^5 - 5z^3 + 10z - \frac{10}{z} + \frac{5}{z^3} - \frac{1}{z^5}$$
$$= \left(z^5 - \frac{1}{z^5}\right) - 5\left(z^3 - \frac{1}{z^3}\right) + 10\left(z - \frac{1}{z}\right)$$
$$= 2i\sin 5\theta - 5(2i\sin 3\theta) + 10(2i\sin\theta), \text{ from (3)};$$
$$\therefore\ \sin^5\theta = \tfrac{1}{16}(\sin 5\theta - 5\sin 3\theta + 10\sin\theta).$$

Check by putting $\theta = \frac{\pi}{2}$.

Example 6. Express $\cos^3\theta \sin^4\theta$ in terms of multiple angles.

From (2),
$$(2\cos\theta)^3(2i\sin\theta)^4 = \left(z+\frac{1}{z}\right)^3\left(z-\frac{1}{z}\right)^4 = \left(z^2 - \frac{1}{z^2}\right)^3\left(z - \frac{1}{z}\right)$$
$$= \left(z^6 - 3z^2 + \frac{3}{z^2} - \frac{1}{z^6}\right)\left(z - \frac{1}{z}\right)$$
$$= \left(z^7 + \frac{1}{z^7}\right) - \left(z^5 + \frac{1}{z^5}\right) - 3\left(z^3 + \frac{1}{z^3}\right) + 3\left(z + \frac{1}{z}\right)$$
$$= 2\cos 7\theta - 2\cos 5\theta - 3(2\cos 3\theta) + 3(2\cos\theta);$$
$$\therefore\ \cos^3\theta\sin^4\theta = \tfrac{1}{64}(\cos 7\theta - \cos 5\theta - 3\cos 3\theta + 3\cos\theta).$$

Check by putting $\theta = \frac{\pi}{4}$.

Note. It is instructive to consider the form of the expansion in the general case. It is evident that $\cos^n\theta$ can always be expressed as a sum of *cosines* of multiples of θ, since it depends on the binomial expansion of $\left(z+\frac{1}{z}\right)^n$. But $\sin^n\theta$ depends on the expansion of $\left(z-\frac{1}{z}\right)^n$, which involves terms like $z^r+\frac{1}{z^r}$ if n is even, and terms like $z^r-\frac{1}{z^r}$ if n is odd. Therefore $\sin^n\theta$ can be expanded in terms of *cosines or sines* of multiples of θ according as n is even or odd. For the details, see Ex. IX. b, Nos. 15-18.

EXERCISE IX. b.

1. If $z=\cos\theta+i\sin\theta$, write down the values of $z^3+\frac{1}{z^3}$ and $z^4-\frac{1}{z^4}$ in terms of θ.

2. Express $\cos 7\theta$ and $\sin 6\theta$ in terms of $\cos\theta+i\sin\theta$.

Use the general method to express the following in terms of cosines or sines of multiples of θ.

3. $\cos^3\theta$. **4.** $\cos^4\theta$. **5.** $\cos^7\theta$. **6.** $\sin^4\theta$.

7. $\sin^7\theta$. **8.** $\sin^3\theta\cos\theta$. **9.** $\cos^4\theta\sin^3\theta$. **10.** $\cos^5\theta\sin^4\theta$.

11. Prove that $16\cos^5\theta-\cos 5\theta=5\cos\theta(1+2\cos 2\theta)$.

12. Evaluate (i) $\int\sin^5\theta\,d\theta$; (ii) $\int\cos^7\theta\,d\theta$; (iii) $\int\cos^5\theta\sin^4\theta\,d\theta$;

by means of expansions in terms of multiples of θ.

13. Evaluate $\int\sin^4\theta\cos^6\theta\,d\theta$.

14. Explain *how to find* the value of

$$\sin^4 A\cos^6 A+\sin^4 2A\cos^6 2A+\dots \text{ to } n \text{ terms.}$$

15. Prove that, if n is even, $2^{n-1}\cos^n\theta=\cos n\theta+n\cos(n-2)\theta+\dots$ and that there are $\frac{n}{2}+1$ terms, of which the last is $\frac{n!}{2\{(\frac{1}{2}n)!\}^2}$. Show that the others after the first are the values of

$$\binom{n}{r}\cos(n-2r)\theta,\text{ for } r=1, 2, 3, \dots, \tfrac{1}{2}(n-2).$$

16. Prove that, if n is odd,

$$2^{n-1}\cos^n\theta=\cos n\theta+\Sigma\binom{n}{r}\cos(n-2r)\theta,\text{ for } r=1, 2, \dots, \tfrac{1}{2}(n-1).$$

17. Prove that $2^{2n-1}(-1)^n \sin^{2n}\theta$

$$=\cos 2n\theta - 2n\cos(2n-2)\theta + \dots + \frac{(-1)^n(2n)!}{2(n!)^2},$$

and give the general term.

18. Prove that $2^{2n}(-1)^n \sin^{2n+1}\theta$

$$=\sin(2n+1)\theta - (2n+1)\sin(2n-1)\theta + \dots + \frac{(-1)^n(2n+1)!\sin\theta}{n!(n+1)!},$$

and give the general term.

19. If $\cos^3\theta\sin^4\theta$ is expressed in the form

$$A_1\cos\theta + A_3\cos 3\theta + A_5\cos 5\theta + A_7\cos 7\theta,$$

deduce by differentiation that $A_1 + 9A_3 + 25A_5 + 49A_7 = 0$, and find the value of $A_1 + 3^4A_3 + 5^4A_5 + 7^4A_7$. Verify the result by means of Example 6, p. 170.

Expansions of cos nθ, sin nθ, and tan nθ, where n is any positive integer. We have

$$\cos n\theta + i\sin n\theta = (\cos\theta + i\sin\theta)^n \equiv (c+is)^n, \text{ say,}$$

$$= c^n + \binom{n}{1}c^{n-1}is + \binom{n}{2}c^{n-2}i^2s^2 + \binom{n}{3}c^{n-3}i^3s^3 + \binom{n}{4}c^{n-4}i^4s^4 + \dots$$

$$= \left\{c^n - \binom{n}{2}c^{n-2}s^2 + \binom{n}{4}c^{n-4}s^4 - \dots\right\} + i\left\{\binom{n}{1}c^{n-1}s - \binom{n}{3}c^{n-3}s^3 + \dots\right\},$$

and so, by equating the first and second parts of the two complex numbers,

$$\cos n\theta = c^n - \binom{n}{2}c^{n-2}s^2 + \binom{n}{4}c^{n-4}s^4 - \dots. \qquad \text{..............(4)}$$

$$\sin n\theta = \binom{n}{1}c^{n-1}s - \binom{n}{3}c^{n-3}s^3 + \dots. \qquad \text{..................(5)}$$

Also $\cos n\theta + i\sin n\theta = \cos^n\theta\,(1 + i\tan\theta)^n \equiv \cos^n\theta\,(1+it)^n$, say, which gives the same results as before in the form

$$\cos n\theta = \cos^n\theta\left\{1 - \binom{n}{2}t^2 + \binom{n}{4}t^4 - \dots\right\},$$

$$\sin n\theta = \cos^n\theta\left\{\binom{n}{1}t - \binom{n}{3}t^3 + \dots\right\}.$$

By division,

$$\tan n\theta = \frac{\binom{n}{1}t - \binom{n}{3}t^3 + \dots}{1 - \binom{n}{2}t^2 + \binom{n}{4}t^4 - \dots}. \qquad \text{..........................(6)}$$

Expansion of tan $(\theta_1+\theta_2+\ldots+\theta_n)$.

Similarly, $\cos(\theta_1+\theta_2+\ldots+\theta_n)+i\sin(\theta_1+\theta_2+\ldots+\theta_n)$

$=(\cos\theta_1+i\sin\theta_1)(\cos\theta_2+i\sin\theta_2)\ldots(\cos\theta_n+i\sin\theta_n)$

$=\cos\theta_1\cos\theta_2\ldots\cos\theta_n(1+it_1)(1+it_2)\ldots(1+it_n)$, where $t_r\equiv\tan\theta_r$

$=\cos\theta_1\cos\theta_2\ldots\cos\theta_n(1+i\Sigma_1+i^2\Sigma_2+i^3\Sigma_3+\ldots)$,

where Σ_r denotes the sum of the products of $\tan\theta_1$, $\tan\theta_2$, ... taken r at a time.

Equating the first and second parts of the complex numbers,

$$\cos(\theta_1+\theta_2+\ldots+\theta_n)=\cos\theta_1\cos\theta_2\ldots\cos\theta_n(1-\Sigma_2+\Sigma_4-\ldots);$$

$$\sin(\theta_1+\theta_2+\ldots+\theta_n)=\cos\theta_1\cos\theta_2\ldots\cos\theta_n(\Sigma_1-\Sigma_3+\ldots);$$

$$\therefore\ \tan(\theta_1+\theta_2+\ldots+\theta_n)=\frac{\Sigma_1-\Sigma_3+\Sigma_5-\ldots}{1-\Sigma_2+\Sigma_4-\Sigma_6+\ldots}. \quad\ldots\ldots(7)$$

Formula (7) is easily remembered; it includes (6) as the special case when θ_1, θ_2, ... θ_n are equal.

Formula (6) expresses $\tan n\theta$ in terms of $\tan\theta$. Formulae (4), (5) can be transformed by means of the identity $\sin^2\theta+\cos^2\theta=1$, so that, for example, $\cos n\theta$ can be expressed entirely in terms of $\cos\theta$ as in the example below. The general results will be discussed on p. 178.

Example 7. Express $\cos 6\theta$ in terms of $\cos\theta$.

We have

$$\begin{aligned}\cos 6\theta &= c^6-\binom{6}{2}c^4s^2+\binom{6}{4}c^2s^4-s^6\\ &= c^6-15c^4(1-c^2)+15c^2(1-c^2)^2-(1-c^2)^3\\ &= 32\cos^6\theta-48\cos^4\theta+18\cos^2\theta-1.\end{aligned}$$

EXERCISE IX. c.

From formulae (4) and (5) find expressions for

1. $\sin 5\theta$ in terms of $\sin\theta$.

2. $\cos 5\theta$ in terms of $\cos\theta$.

3. $\dfrac{\sin 6\theta}{\sin\theta}$ in terms of $\cos\theta$.

4. $\cos 6\theta$ in terms of $\sin\theta$.

5. Give the formulae for $\tan 4\theta$ and $\tan 5\theta$ in terms of $\tan\theta$.

6. What equation is satisfied by $\tan\theta$ if $\tan 6\theta=0$?

7. What equation is satisfied by $\tan\theta$ if $7\theta=\frac{\pi}{2}$?

8. Give the expansion of $\tan(\theta_1+\theta_2-\theta_3)$.

9. Give the relations holding between the tangents if
(i) $\theta_1+\theta_2+\theta_3=\pi$; (ii) $\theta_1+\theta_2+\theta_3+\theta_4=2\pi$; (iii) $\theta_1+\theta_2+\theta_3=\frac{3\pi}{2}$.

10. What results can be deduced from

$$(\sin\theta+i\cos\theta)^n=\left\{\cos\left(\frac{\pi}{2}-\theta\right)+i\sin\left(\frac{\pi}{2}-\theta\right)\right\}^n=\operatorname{cis}\left(\frac{n\pi}{2}-n\theta\right),$$

where n is a positive integer ?

11. Give the last terms in the formulae (4) and (5), (i) if n is even, (ii) if n is odd.

12. Give the last terms of the numerator and denominator of the formula for $\tan n\theta$, (i) if n is even, (ii) if n is odd.

13. Show that the coefficient of c^n in

$$c^n-\binom{n}{2}c^{n-2}(1-c^2)+\binom{n}{4}c^{n-4}(1-c^2)^2-\dots \text{ is } 2^{n-1}.$$

14. Prove that $2^{\frac{n}{2}}\cos\frac{n\pi}{4}=1-\binom{n}{2}+\binom{n}{4}-\dots$, and give the last term.

15. Prove that $\sec\theta\cos 5\theta=1-12\sin^2\theta+16\sin^4\theta$.

16. In any triangle ABC, prove that

$$c^3=a^3\cos 3B+3a^2b\cos(A-2B)+3ab^2\cos(2A-B)+b^3\cos 3A.$$

17. Find the equation whose roots are $\pm\tan\frac{\pi}{7}$, $\pm\tan\frac{2\pi}{7}$, $\pm\tan\frac{3\pi}{7}$.

18. If $\tan\theta_1$, $\tan\theta_2$, $\tan\theta_3$, $\tan\theta_4$ are the roots of the equation $t^4+bt^3+ct^2+et+f=0$, find the value of $\tan(\theta_1+\theta_2+\theta_3+\theta_4)$.

Summation of Series. *Sum the series*

$$C\equiv 1+x\cos\theta+x^2\cos 2\theta+\dots+x^{n-1}\cos(n-1)\theta,$$

$$S\equiv \quad x\sin\theta+x^2\sin 2\theta+\dots+x^{n-1}\sin(n-1)\theta.$$

Put $\cos\theta+i\sin\theta=z$.

Then

$$C+iS=1+xz+x^2z^2+\dots+x^{n-1}z^{n-1}$$

$$=\frac{1-x^nz^n}{1-xz}=\frac{(1-x^nz^n)\left(1-\frac{x}{z}\right)}{(1-xz)\left(1-\frac{x}{z}\right)}$$

$$=\frac{1-\frac{x}{z}-x^nz^n+x^{n+1}z^{n-1}}{1-x\left(z+\frac{1}{z}\right)+x^2}.$$

But $\dfrac{1}{z}=\cos\theta-i\sin\theta$ and $z+\dfrac{1}{z}=2\cos\theta$;

$$\therefore\ \mathsf{C}+i\mathsf{S}=\frac{1-x(\cos\theta-i\sin\theta)-x^n(\cos n\theta+i\sin n\theta)+x^{n+1}(\cos\overline{n-1}\theta+i\sin\overline{n-1}\theta)}{1-2x\cos\theta+x^2}$$

$$=\frac{1-x\cos\theta-x^n\cos n\theta+x^{n+1}\cos\overline{n-1}\theta+i\{x\sin\theta-x^n\sin n\theta+x^{n+1}\sin\overline{n-1}\theta\}}{1-2x\cos\theta+x^2};$$

∴ by equating the first and second parts, we have

$$\mathsf{C}=\frac{1-x\cos\theta-x^n\cos n\theta+x^{n+1}\cos(n-1)\theta}{1-2x\cos\theta+x^2}\qquad\ldots\ldots\ldots\ldots(8)$$

and
$$\mathsf{S}=\frac{x\sin\theta-x^n\sin n\theta+x^{n+1}\sin(n-1)\theta}{1-2x\cos\theta+x^2}\qquad\ldots\ldots\ldots\ldots\ldots(9)$$

These results may also be obtained by multiplying the given series by $1-2x\cos\theta+x^2$ and showing that in the product all the terms disappear except a few at the beginning and end.

Note. If $|x|<1$, since $\lim_{r\to\infty} x^r=0$, we see that

$\dfrac{1-x\cos\theta}{1-2x\cos\theta+x^2}$ is the sum to infinity of $1+x\cos\theta+x^2\cos 2\theta+\ldots$ (10)

and

$\dfrac{x\sin\theta}{1-2x\cos\theta+x^2}$ is the sum to infinity of $x\sin\theta+x^2\sin 2\theta+\ldots$ ...(11)

Example 8. Sum the series

$$\cos a+\binom{n}{1}\cos(a+\beta)+\binom{n}{2}\cos(a+2\beta)+\ldots+\cos(a+n\beta).$$

Put $\cos a+i\sin a=a$, $\cos\beta+i\sin\beta=b$.
Then the given series is the first part of the complex number

$$a+\binom{n}{1}ab+\binom{n}{2}ab^2+\ldots+\binom{n}{n}ab^n=a(1+b)^n;$$

but
$$1+b=1+\cos\beta+i\sin\beta=2\cos^2\frac{\beta}{2}+2i\sin\frac{\beta}{2}\cos\frac{\beta}{2}$$

$$=2\cos\frac{\beta}{2}\left(\cos\frac{\beta}{2}+i\sin\frac{\beta}{2}\right);$$

$$\therefore\ a(1+b)^n = \text{cis}\,\alpha\left(2\cos\frac{\beta}{2}\,\text{cis}\,\frac{\beta}{2}\right)^n = \left(2\cos\frac{\beta}{2}\right)^n \text{cis}\,\alpha\,\text{cis}\,\frac{n\beta}{2}$$

$$= \left(2\cos\frac{\beta}{2}\right)^n \text{cis}\left(\alpha+\frac{n\beta}{2}\right);$$

$$\therefore \text{ the given series} = \left(2\cos\frac{\beta}{2}\right)^n \cos\left(\alpha+\frac{n\beta}{2}\right).$$

EXERCISE IX. d.

1. Sum to n terms,

$$\cos\theta+\tfrac{1}{2}\cos 2\theta+\tfrac{1}{4}\cos 3\theta+\tfrac{1}{8}\cos 4\theta+\dots.$$

Deduce the sum to infinity.

2. (i) Sum to n terms,

$$\cos\theta\cos\theta+\cos^2\theta\cos 2\theta+\cos^3\theta\cos 3\theta+\dots.$$

(ii) Deduce the sum to infinity if θ is not a multiple of π.

3. (i) Sum to n terms,

$$\sin\theta\sin\theta+\sin^2\theta\sin 2\theta+\sin^3\theta\sin 3\theta+\dots.$$

(ii) Deduce the sum to infinity if θ is not an odd multiple of $\frac{\pi}{2}$.

4. If n is a positive integer, express $\dfrac{a(1-b^n)}{1-b}$ in the form $A+Bi$, when $a=\text{cis}\,\alpha$, $b=\text{cis}\,\beta$.

What results can be deduced from the identity,

$$\frac{a(1-b^n)}{1-b}\equiv a+ab+ab^2+\dots+ab^{n-1}?$$

5. If n is odd, prove that

$$1-\binom{n}{1}\cos 2\theta+\binom{n}{2}\cos 4\theta-\dots-\cos 2n\theta=(-1)^{\frac{n-1}{2}}\sin n\theta\,(2\sin\theta)^n.$$

6. If n is even, sum the series

$$\binom{n}{1}\sin 2\theta-\binom{n}{2}\sin 4\theta+\dots-\sin 2n\theta.$$

7. Sum to $n+1$ terms,

$$(2\cos\theta)^n-\binom{n}{1}(2\cos\theta)^{n-1}\cos\theta+\binom{n}{2}(2\cos\theta)^{n-2}\cos 2\theta-\dots.$$

8. Sum to $n+1$ terms,

$$\sin^n\phi\cos n\theta+\binom{n}{1}\sin^{n-1}\phi\cos(n-1)\theta\sin(\theta-\phi)$$
$$+\binom{n}{2}\sin^{n-2}\phi\cos(n-2)\theta\sin^2(\theta-\phi)+\dots.$$

9. Prove that, in any triangle ABC, if $b < c$, the sum to infinity of $\sin A + \frac{b}{c}\sin 2A + \frac{b^2}{c^2}\sin 3A + \dots$ is $\frac{c\sin C}{a}$.

10. Use the identity,

$$\sin\theta(\sin\phi - i\cos\phi) \equiv \cos(\theta-\phi) - \cos\phi(\cos\theta + i\sin\theta),$$

to obtain an expansion for $\sin^n\theta\cos\left(n\phi - \frac{n\pi}{2}\right)$.

11. Prove that, in any triangle ABC, $c^n = a^n\cos nB +$

$$\binom{n}{1}a^{n-1}b\cos\{(n-1)B - A\} + \binom{n}{2}a^{n-2}b^2\cos\{(n-2)B - 2A\} + \dots$$

to $n+1$ terms.

12. If $|x| < 1$, find the coefficients of x^n in the expansions in powers of x of

$$\text{(i)}\ \frac{\cos\theta - x}{1 - 2x\cos\theta + x^2};\quad \text{(ii)}\ \frac{1 - x^2}{1 - 2x\cos\theta + x^2}.$$

13. If θ is not a multiple of π, find the sum to infinity of

$$\sin\alpha + \cos\theta\sin(\alpha+\theta) + \cos^2\theta\sin(\alpha+2\theta) + \cos^3\theta\sin(\alpha+3\theta) + \dots.$$

14. Sum to $n+1$ terms,

$$1 + nx\cos\theta + \binom{n}{2}x^2\cos 2\theta + \binom{n}{3}x^3\cos 3\theta + \dots.$$

15. Sum to $n+1$ terms,

$$\cos n\theta + nx^2\cos(n-2)\theta + \binom{n}{2}x^4\cos(n-4)\theta + \dots.$$

16. What results can be deduced by writing cis θ for z in

$$1 + 2z + 3z^2 + \dots + nz^{n-1} \equiv \frac{1 - (n+1)z^n + nz^{n+1}}{(1-z)^2}?$$

17. Sum to n terms,

$$\binom{2n}{1}\cos\theta + \binom{2n}{3}\cos 3\theta + \binom{2n}{5}\cos 5\theta + \dots.$$

18. Prove that, if $|x| < 1$,

$$\frac{(1-x)\cos\theta}{1 - 2x\cos 2\theta + x^2} = \sum_0^\infty x^n\cos(2n+1)\theta.$$

Deduce that $\cos(2n+1)\theta\sec\theta = \sum\frac{(n+r)!}{(n-r)!}\frac{(-4\sin^2\theta)^r}{(2r)!}$,

for $r = 0$ to n, where $0!$ is taken to mean unity.

19. If $ac > b^2$ and $\left|x\sqrt{\frac{a}{c}}\right| < 1$, prove that $\frac{a}{ax^2 - 2bx + c}$ can be expanded in powers of x and that the coefficient of x^n is

$$\left(\frac{a}{c}\right)^{\frac{1}{2}n+1} \sin (n+1)\theta \operatorname{cosec} \theta,$$

where

$$\cos \theta \sqrt{(ac)} = b.$$

20. Prove that $$\frac{\sin x}{(n-1)!\,(n+1)!} + \frac{2 \sin 2x}{(n-2)!\,(n+2)!} + \dots + \frac{n \sin nx}{(2n)!}$$

$$= \frac{2^{n-2}}{(2n-1)!} (1 + \cos x)^{n-1} \sin x.$$

Expansions of cos nθ, sin nθ in terms of cos θ, sin θ separately. We suppose that n is a positive integer, and we write $\cos \theta = c$, $\sin \theta = s$.

Forms of the Expansions.

(i) From $\cos n\theta = c^n - \binom{n}{2} c^{n-2}s^2 + \binom{n}{4} c^{n-4}s^4 - \dots,$

putting $$s^2 = 1 - c^2, \quad s^4 = (1 - c^2)^2, \dots$$

we see that **cos nθ is a polynomial in cos θ of degree n,**

viz. $$\cos n\theta = a_n c^n + a_{n-2} c^{n-2} + \dots + a_{n-2r} c^{n-2r} + \dots, \qquad \text{......(12)}$$

the last term being $a_1 c$ if n is odd, and a_0 if n is even.

(ii) Differentiation w.r.t. θ gives (cf. foot of p. 128)

$n \sin n\theta$

$$= \sin \theta \{ n a_n c^{n-1} + (n-2) a_{n-2} c^{n-3} + \dots + (n-2r) a_{n-2r} c^{n-2r-1} + \dots \};$$

$\therefore$ **$\frac{\sin n\theta}{\sin \theta}$ is a polynomial in cos θ of degree n − 1,**

viz. $$\frac{\sin n\theta}{\sin \theta} = b_{n-1} c^{n-1} + b_{n-3} c^{n-3} + \dots + b_{n-2r-1} c^{n-2r-1} + \dots, \qquad \text{...(13)}$$

the last term being b_0 if n is odd, and $b_1 c$ if n is even.

(iii) Changing θ to $\frac{\pi}{2} - \theta$, we get

if n is even, $$(-1)^{\frac{n}{2}} \cos n\theta = a_n s^n + a_{n-2} s^{n-2} + \dots + a_0, \qquad \text{.........(14)}$$

and $$(-1)^{\frac{n}{2}-1} \frac{\sin n\theta}{\cos \theta} = b_{n-1} s^{n-1} + b_{n-3} s^{n-3} + \dots + b_1 s; \qquad (15)$$

and, if n is odd, $$(-1)^{\frac{n-1}{2}} \sin n\theta = a_n s^n + a_{n-2} s^{n-2} + \dots + a_1 s, \quad \text{........(16)}$$

and $$(-1)^{\frac{n-1}{2}} \frac{\cos n\theta}{\cos\theta} = b_{n-1} s^{n-1} + b_{n-3} s^{n-3} + \dots + b_0. \quad \text{...(17)}$$

The forms of the results are easily recalled by means of the special cases when $n=2, 3, \dots$; thus

$$\cos 2\theta, \quad \cos 3\theta, \quad \frac{\sin 2\theta}{\sin\theta}, \quad \frac{\sin 3\theta}{\sin\theta}$$

can be expressed as polynomials in $\cos\theta$; and

$$\cos 2\theta, \quad \frac{\cos 3\theta}{\cos\theta}, \quad \sin 3\theta, \quad \frac{\sin 2\theta}{\cos\theta}$$

can be expressed as polynomials in $\sin\theta$.

Relation between Consecutive Coefficients.

From $$\cos n\theta = a_n c^n + a_{n-2} c^{n-2} + \dots \equiv \Sigma a_r c^r,$$

by differentiating twice with respect to θ, (cf. foot of p. 128), we get

$$n^2 \cos n\theta = \frac{d}{d\theta}\{\Sigma r a_r c^{r-1} s\} = \Sigma\{r a_r c^r - r(r-1) a_r c^{r-2}(1-c^2)\};$$

$$\therefore \; n^2 \Sigma a_r c^r \equiv \Sigma\{r^2 a_r c^r - r(r-1) a_r c^{r-2}\}.$$

Equate coefficients of c^r;

$$\therefore \; n^2 a_r = r^2 a_r - (r+2)(r+1) a_{r+2};$$

$$\therefore \; a_{r+2} = -\frac{n^2 - r^2}{(r+1)(r+2)} a_r. \quad \text{........................(18)}$$

A similar result to (18) can be found, by the same process, for each of the expansions. Such results enable us to calculate all the coefficients, if one coefficient is known. In applying this method we begin by finding the first or the last coefficient of the expansion (12-17), according as we wish to have the result arranged in descending or ascending powers.

When one expansion has been obtained, any other expansion can be deduced rapidly from it by one of the methods given on p. 178.

The First and Last Coefficients.

(i) From $\cos n\theta = c^n - \binom{n}{2} c^{n-2}(1-c^2) + \binom{n}{4} c^{n-4}(1-c^2)^2 - \dots,$

it follows that

$$a_n = 1 + \binom{n}{2} + \binom{n}{4} + \dots = \tfrac{1}{2}\{(1+1)^n + (1-1)^n\} = 2^{n-1}.$$

Also, if n is even, putting $\theta=\frac{\pi}{2}$ in (12), we have

$$a_0=\cos\frac{n\pi}{2}=(-1)^{\frac{n}{2}}.$$

And, if n is odd, $a_1=\lim_{\theta\to\frac{\pi}{2}}\frac{\cos n\theta}{\cos\theta}=\lim_{\phi\to 0}\frac{\cos n\left(\frac{\pi}{2}-\phi\right)}{\cos\left(\frac{\pi}{2}-\phi\right)}$

$$=\lim_{\phi\to 0}\frac{\sin\frac{n\pi}{2}\sin n\phi}{\sin\phi}=n(-1)^{\frac{n-1}{2}}.$$

(ii) For the expansion of $\frac{\sin n\theta}{\sin\theta}$, by differentiating as on p. 178, we see that, in (13), $b_{n-1}=a_n=2^{n-1}$.

Also, if n is even, $b_1=\lim_{\theta\to\frac{\pi}{2}}\frac{\sin n\theta}{\sin\theta\cos\theta}=\lim_{\phi\to 0}\frac{\sin n\left(\frac{\pi}{2}-\phi\right)}{\cos\left(\frac{\pi}{2}-\phi\right)}$

$$=\lim_{\phi\to 0}\frac{-\cos\frac{n\pi}{2}\sin n\phi}{\sin\phi}=-n(-1)^{\frac{n}{2}}.$$

And, if n is odd, $b_0=\frac{\sin\frac{n\pi}{2}}{\sin\frac{\pi}{2}}=(-1)^{\frac{n-1}{2}}.$

(iii) The first and last coefficients in the expansions (14) to (17) can be obtained in the same way or may be deduced from the results of (i) and (ii) by writing $\frac{\pi}{2}-\theta$ for θ.

Note. The numerical value of the coefficient of the term of highest degree is in every case 2^{n-1}. As regards the sign to be attached, this can be conjectured from special cases. Thus, $\frac{\sin 2\theta}{\cos\theta}=2\sin\theta$ and $\frac{\sin 4\theta}{\cos\theta}=-8\sin^3\theta+4\sin\theta$ suggest that the leading coefficient of $\frac{\sin n\theta}{\cos\theta}$, when n is even, is $(-1)^{\frac{n}{2}-1}2^{n-1}$.

For convenience of reference, the various expansions are given in Ex. IX. e, Nos. 12-25. An example is added below to illustrate the method which should be followed in any required special case.

We have assumed that n is a positive integer. It is natural to enquire whether the expansions in Ex. IX. e, Nos. 14-17, 22-25 can be interpreted if n is fractional, when the series involved are evidently no longer finite. This involves a discussion which is beyond the range of this book; see Bromwich, *Infinite Series*, 1st ed., Ch. IX. § 68.

Example 9. Discuss, ab initio, the expansion of $\dfrac{\sin n\theta}{\cos\theta}$, where n is even, in ascending powers of $\sin\theta$.

From

$$\cos n\theta + i\sin n\theta = (c+is)^n = \left\{c^n - \binom{n}{2}c^{n-2}s^2 + \ldots\right\} + i\left\{\binom{n}{1}c^{n-1}s - \ldots\right\}$$

it follows that

$$\sin n\theta = \binom{n}{1}c^{n-1}s - \binom{n}{3}c^{n-3}s^3 + \ldots;$$

$$\therefore\ \frac{\sin n\theta}{\cos\theta} = \binom{n}{1}c^{n-2}s - \binom{n}{3}c^{n-4}s^3 + \ldots.$$

Since n is even, by using $c^2 = 1 - s^2$, we can express c^{n-2}, c^{n-4}, etc., as polynomials in s.

Therefore $\dfrac{\sin n\theta}{\cos\theta}$ *can* be expressed as a polynomial of degree $n-1$ in $\sin\theta$. [It cannot be expressed as a polynomial in $\cos\theta$.]

We may therefore write, for n even,

$$\sin n\theta \equiv c(k_{n-1}s^{n-1} + k_{n-3}s^{n-3} + \ldots + k_r s^r + \ldots + k_1 s).$$

Differentiating, $\quad n\cos n\theta \equiv \Sigma(rk_r s^{r-1}c^2 - k_r s^{r+1})$

$$\equiv \Sigma\{rk_r s^{r-1} - (r+1)k_r s^{r+1}\}.$$

Differentiating again,

$$-n^2\sin n\theta \equiv \Sigma\{r(r-1)k_r s^{r-2}c - (r+1)^2 k_r s^r c\};$$

$$\therefore\ -n^2\,\Sigma(k_r s^r) \equiv \Sigma\{r(r-1)k_r s^{r-2} - (r+1)^2 k_r s^r\}.$$

Equating coefficients of s^r,

$$-n^2 k_r = (r+2)(r+1)k_{r+2} - (r+1)^2 k_r;$$

$$\therefore\ k_{r+2} = -\frac{n^2-(r+1)^2}{(r+1)(r+2)}k_r.$$

But $\quad k_1 = \lim\limits_{\theta\to 0}\dfrac{\sin n\theta}{\sin\theta\cos\theta} = n;\quad \therefore\ k_3 = -\dfrac{n^2-2^2}{2\,.\,3}\,.\,n;$

$$\therefore\ k_5 = \frac{(n^2-4^2)(n^2-2^2)}{2\,.\,3\,.\,4\,.\,5}\,.\,n;$$

$$\therefore\ k_7 = -\frac{(n^2-6^2)(n^2-4^2)(n^2-2^2)}{7!}\,.\,n;\ \text{etc.};$$

$\therefore$ if n is even,

$$\frac{\sin n\theta}{\cos\theta} = n\sin\theta - \frac{n(n^2-2^2)}{3!}\sin^3\theta + \frac{n(n^2-2^2)(n^2-4^2)}{5!}\sin^5\theta - \dots .$$

EXERCISE IX. e.

Which of the functions in Nos. 1-6 can be expressed as polynomials in $\sin\theta$, and which of them as polynomials in $\cos\theta$?

In each case, use the methods of pp. 179, 180 to obtain (a) the term of lowest degree, (b) the term of highest degree.

1. $\sin n\theta$, if n is odd.

2. $\cos n\theta$, if n is even.

3. $\dfrac{\sin n\theta}{\cos\theta}$, if n is even.

4. $\dfrac{\cos n\theta}{\cos\theta}$, if n is odd.

5. $\dfrac{\sin n\theta}{\sin\theta}$, if n is odd.

6. $\dfrac{\sin n\theta}{\sin\theta\cos\theta}$, if n is even.

7. Assuming that $\cos n\theta = a_0 + a_2\sin^2\theta + \dots + a_n\sin^n\theta$, where n is even, find the value of a_0. Then find by differentiation the other coefficients.

8. Assuming that $\sin n\theta = a_1\sin\theta + a_3\sin^3\theta + \dots + a_n\sin^n\theta$, where n is odd, find the value of a_1. Then find by differentiation the other coefficients.

9. Prove by differentiation that the constants in equation (13), p. 178, are connected by the relation,

$$b_{r+2} = -\frac{n^2-(r+1)^2}{(r+1)(r+2)}b_r.$$

10. Prove that $\dfrac{\sin 7\theta}{\sin\theta} = y^3 + y^2 - 2y - 1$, where $y = 2\cos 2\theta$.

11. Prove that $\dfrac{\sin 9\theta}{\sin\theta} = (x^2-1)(x^6-6x^4+9x^2-1)$, where $x = 2\cos\theta$.

12. $$2\cos n\theta = (2c)^n - n(2c)^{n-2} + \frac{n(n-3)}{2!}(2c)^{n-4} - \frac{n(n-4)(n-5)}{3!}(2c)^{n-6} + \dots .$$

13. $$\frac{\sin n\theta}{\sin\theta} = (2c)^{n-1} - (n-2)(2c)^{n-3} + \frac{(n-3)(n-4)}{2!}(2c)^{n-5} - \frac{(n-4)(n-5)(n-6)}{3!}(2c)^{n-7} + \dots .$$

14. n even, $$\cos n\theta = (-1)^{\frac{n}{2}}\left\{1 - \frac{n^2}{2!}c^2 + \frac{n^2(n^2-2^2)}{4!}c^4 - \frac{n^2(n^2-2^2)(n^2-4^2)}{6!}c^6 + \dots\right\}.$$

15. n odd, $$\cos n\theta = (-1)^{\frac{n-1}{2}}\left\{nc - \frac{n(n^2-1^2)}{3!}c^3 + \frac{n(n^2-1^2)(n^2-3^2)}{5!}c^5 - \dots\right\}.$$

16. n even, $$\frac{\sin n\theta}{\sin\theta} = (-1)^{\frac{n}{2}+1}\left\{nc - \frac{n(n^2-2^2)}{3!}c^3 + \frac{n(n^2-2^2)(n^2-4^2)}{5!}c^5 - \dots\right\}.$$

17. n odd, $$\frac{\sin n\theta}{\sin\theta} = (-1)^{\frac{n-1}{2}}\left\{1 - \frac{n^2-1^2}{2!}c^2 + \frac{(n^2-1^2)(n^2-3^2)}{4!}c^4 - \dots\right\}.$$

18. n even, $$2\cos n\theta = (-1)^{\frac{n}{2}}\left\{(2s)^n - n(2s)^{n-2} + \frac{n(n-3)}{2!}(2s)^{n-4} - \frac{n(n-4)(n-5)}{3!}(2s)^{n-6} + \dots\right\}.$$

19. n odd, $$2\sin n\theta = (-1)^{\frac{n-1}{2}}\left\{(2s)^n - n(2s)^{n-2} + \frac{n(n-3)}{2!}(2s)^{n-4} - \frac{n(n-4)(n-5)}{3!}(2s)^{n-6} + \dots\right\}.$$

20. n even, $$\frac{\sin n\theta}{\cos\theta} = (-1)^{\frac{n}{2}-1}\left\{(2s)^{n-1} - (n-2)(2s)^{n-3} + \frac{(n-3)(n-4)}{2!}(2s)^{n-5} - \dots\right\}.$$

21. n odd, $$\frac{\cos n\theta}{\cos\theta} = (-1)^{\frac{n-1}{2}}\left\{(2s)^{n-1} - (n-2)(2s)^{n-3} + \frac{(n-3)(n-4)}{2!}(2s)^{n-5} - \dots\right\}.$$

22. n even, $$\cos n\theta = 1 - \frac{n^2}{2!}s^2 + \frac{n^2(n^2-2^2)}{4!}s^4 - \dots + (-1)^{\frac{n}{2}}\,.\,2^{n-1}s^n.$$

23. n odd, $$\frac{\cos n\theta}{\cos\theta} = 1 - \frac{n^2-1^2}{2!}s^2 + \frac{(n^2-1^2)(n^2-3^2)}{4!}s^4 - \dots + (-1)^{\frac{n-1}{2}}\,.\,2^{n-1}s^{n-1}.$$

24. n even, $$\frac{\sin n\theta}{\cos\theta} = ns - \frac{n(n^2-2^2)}{3!}s^3 + \frac{n(n^2-2^2)(n^2-4^2)}{5!}s^5 - \dots + (-1)^{\frac{n}{2}+1}\,.\,2^{n-1}s^{n-1}.$$

25. n odd, $\sin n\theta = ns - \frac{n(n^2-1^2)}{3!}s^3$

$$+\frac{n(n^2-1^2)(n^2-3^2)}{5!}s^5 - \ldots + (-1)^{\frac{n-1}{2}} . 2^{n-1} . s^n.$$

26. What are the coefficients of c^{n-8} and c^{n-10} in the expansion of $\cos n\theta$ in powers of $\cos\theta$?

27. What result can be deduced from No. 24 by writing $2p$ for n and $\frac{\pi}{4}$ for θ ?

28. Prove that

$$2^n - (n-1)2^{n-1} + \frac{(n-2)(n-3)}{2!}2^{n-2} - \frac{(n-3)(n-4)(n-5)}{3!}2^{n-3} + \ldots$$

equals $2^{\frac{n+1}{2}} \sin\frac{(n+1)\pi}{4}$.

29. Prove that $1 - \frac{n^2}{2!} + \frac{n^2(n^2-1^2)}{4!} - \frac{n^2(n^2-1^2)(n^2-2^2)}{6!} + \ldots = \cos\frac{n\pi}{3}$.

30. Prove that $1 - \frac{n^2-1^2}{3!} + \frac{(n^2-1^2)(n^2-2^2)}{5!} - \ldots = \frac{1}{n}\sin\frac{n\pi}{3}\operatorname{cosec}\frac{\pi}{3}$.

31. By writing $x = \cos\theta + i\sin\theta$, show how to express $x^n + \frac{1}{x^n}$ as a polynomial of degree n in $x + \frac{1}{x}$.

32. Express $x^9 + \frac{1}{x^9}$ in terms of $x + \frac{1}{x}$.

33. Express $\left(x^7 - \frac{1}{x^7}\right) \div \left(x - \frac{1}{x}\right)$ as a cubic in $\left(x - \frac{1}{x}\right)^2$.

34. Verify that the coefficient of c^{2r+1} in the expansion in No. 15 is the same as the coefficient of c^{2r+1} in the expansion of $\cos n\theta$ in No. 12, if n is odd.

35. Verify that the coefficient of c^{2r} in the expansion in No. 14 is the same as the coefficient of c^{2r} in the expansion of $\cos n\theta$ in No. 12, if n is even.

36. If $y = \sin n\theta$ and $\theta = \sin^{-1}x$, show that

$$\cos\theta\frac{dy}{dx} = n\cos n\theta \quad \text{and} \quad (1-x^2)\frac{d^2y}{dx^2} - x\frac{dy}{dx} + n^2y = 0.$$

Differentiate this k times by Leibnitz' Theorem and deduce that $y_{k+2} = (k^2 - n^2)y_k$, where y_r is the value for $x=0$ of $\frac{d^ry}{dx^r}$. What are the values of y_0, y_1, y_2, y_k ? Obtain the result of No. 25 by assuming Maclaurin's Theorem.

37. If $y = \cos n\theta$ and $\theta = \sin^{-1}x$, show that

$$(1-x^2)\frac{d^2y}{dx^2} - x\frac{dy}{dx} + n^2y = 0.$$

Hence obtain the result of No. 22 by the method of No. 36.

EASY MISCELLANEOUS EXAMPLES

EXERCISE IX. f.

1. Solve $x^6+x^5+x^4+x^3+x^2+x+1=0$.

2. Solve $x^{12}-x^6+1=0$.

3. Solve $(ax-b)^n=(a-bx)^n$.

4. Solve $(1-xi)^n+i(1+xi)^n=0$.

5. Find which roots of $x^{10}=1$ make $x^4+x^3+x^2+x+1=0$.

6. If $x^5=1$, prove that $x^4+x-x^3-x^2=0$ or $\pm\sqrt{5}$.

7. Given that $\tan\alpha=2$, find $\tan 3\alpha$. Use the result to find the cube roots of $88+16i$.

8. Expand $\frac{1}{2}(1-i)\{(1-xi)^n+i(1+xi)^n\}$ in powers of x.

9. If $\omega=\text{cis}\,\dfrac{2\pi}{3}$, expand in powers of x:

$$\text{(i) } (1+x)^{3n}+(1+\omega x)^{3n}+(1+\omega^2x)^{3n};$$

$$\text{(ii) } (1+x)^{3n}+\omega(1+\omega x)^{3n}+\omega^2(1+\omega^2x)^{3n};$$

$$\text{(iii) } (1+x)^{3n}+\omega^2(1+\omega x)^{3n}+\omega(1+\omega^2x)^{3n}.$$

10. If $x^{13}=1$, $x\neq 1$, and if $y_1=x+x^3+x^9$, $y_2=x^2+x^5+x^6$, $y_3=x^4+x^{10}+x^{12}$, $y_4=x^7+x^8+x^{11}$, prove that

$$\text{(i) } y_1^2=y_2+2y_3;\quad \text{(ii) } y_1y_2=y_1+y_2+y_4;\quad \text{(iii) } y_1y_4+y_2=-1.$$

11. Solve $\text{cis}\,r\theta=\text{cis}\,s\theta$ where r, s are unequal positive integers.

12. Show that the roots of $(1-z)^n=z^n$ are of the form $\frac{1}{2}+\text{B}i$.

13. Prove that the points which represent the roots of $z^n=(z+1)^n$ in the Argand Diagram are collinear.

14. Prove that if $|z|=1$, $z\neq 1$, the points representing $\sqrt{\dfrac{1+z}{1-z}}$ lie on an orthogonal line-pair.

15. Prove that the points representing 1, -1, $a+bi$, and $\dfrac{1}{a+bi}$ are concyclic.

16. If the point which represents z moves on the circle $|z|=1$, find the loci of the points which represent

$$\text{(i) } \sqrt{(2z-3)};\quad \text{(ii) } (z+1)^3.$$

17. If α, β are the roots of $t^2-2t+2=0$, prove that

$$\frac{(x+\alpha)^n-(x+\beta)^n}{\alpha-\beta}=\sin n\phi\,\text{cosec}^n\,\phi,\ \text{where } \cot\phi=x+1.$$

18. Sum to $n+1$ terms, $1+\binom{n}{1}\cos\theta+\binom{n}{2}\cos 2\theta+\ldots$.

19. Prove that

$$\cos 4n\theta + \binom{4n}{2}\cos(4n-4)\theta + \binom{4n}{4}\cos(4n-8)\theta + \dots$$
$$+ \binom{4n}{2n-2}\cos 4\theta + \tfrac{1}{2}\binom{4n}{2n} = 2^{4n-2}(\cos^{4n}\theta + \sin^{4n}\theta).$$

20. (i) Prove that $\left(x^{2n+1} - \dfrac{1}{x^{2n+1}}\right) \div \left(x - \dfrac{1}{x}\right)$ can be expressed as a polynomial in $\left(x - \dfrac{1}{x}\right)^2$ of degree n. Find the coefficient of $\left(x - \dfrac{1}{x}\right)^{2n}$ and the constant term.

(ii) Also show that $\sin^3\theta$ is a factor of $(2n+1)\sin\theta - \sin(2n+1)\theta$.

21. If $u_n = (n+1)\sin n\theta - n\sin(n+1)\theta$, find the value of $u_n - u_{n-1}$, and prove that $1 - \cos\theta$ is a factor of u_n.

22. Prove geometrically that, if $z = \text{cis}\,\theta$, and $-\pi < \theta < \pi$, then $n\{\sqrt[n]{z} - 1\} \to i\theta$ as $n \to \infty$.

23. If $\cos^3\theta\sin^4\theta \equiv A_1\cos\theta + A_3\cos 3\theta + A_5\cos 5\theta + A_7\cos 7\theta$, prove that

$$A_1 - \tfrac{1}{3}A_3 + \tfrac{1}{5}A_5 - \tfrac{1}{7}A_7 = \tfrac{2}{35}.$$

24. Use the identity,

$$\frac{1}{(x-a_1)(x-a_2)} \equiv \frac{1}{(a_1-a_2)(x-a_1)} - \frac{1}{(a_1-a_2)(x-a_2)},$$

to show that

$$\sin(a_1 - a_2)\cos(2\theta + a_1 + a_2)$$
$$\equiv \sin(\theta - a_2)\cos(\theta + 2a_1 + a_2) - \sin(\theta - a_1)\cos(\theta + a_1 + 2a_2).$$

25. Use the identity,

$$abc + (b+c)(c+a)(a+b) \equiv (a+b+c)(bc+ca+ab)$$

to show that

$$\cos(\Sigma\alpha)\,.\left\{1 + 8\cos\frac{\beta-\gamma}{2}\cos\frac{\gamma-\alpha}{2}\cos\frac{\alpha-\beta}{2}\right\}$$
$$\equiv \Sigma\cos\alpha\,.\,\Sigma\cos(\beta+\gamma) - \Sigma\sin\alpha\,.\,\Sigma\sin(\beta+\gamma).$$

26. Prove that

$$2\sin^2 n\theta = \frac{n^2}{2!}(2s)^2 - \frac{n^2(n^2-1^2)}{4!}(2s)^4 + \frac{n^2(n^2-1^2)(n^2-2^2)}{6!}(2s)^6 - \dots,$$

where $s \equiv \sin\theta$ and n is a positive integer.

HARDER MISCELLANEOUS EXAMPLES

EXERCISE IX. g.

1. Express $x^7 + 1$ as the product of four factors.

2. Express $x^8 + x^7 + x^6 + \dots + x + 1$ as the product of four factors.

3. If $a = \text{cis}\,\dfrac{2r\pi}{n}$, and if r and p are prime to n, prove that

$$1 + a^p + a^{2p} + \ldots + a^{(n-1)p} = 0.$$

4. If p, q are real and $(x+p)^2 + q^2 \equiv (x+a)(x+\beta)$, prove that $\dfrac{(x+a)^n - (x+\beta)^n}{a-\beta} = q^{n-1} \sin n\theta \operatorname{cosec}^n \theta$, where $\tan\theta = \dfrac{q}{x+p}$.

5. If $x = 2\cos\theta$, prove that $\dfrac{1+\cos 9\theta}{1+\cos\theta} = (x^4 - x^3 - 3x^2 + 2x + 1)^2$.

6. If $u_n - 2u_{n+1}\cos\theta + u_{n+2} = 0$, and also $u_1 = p\sin\theta + q\cos\theta$ and $u_2 = p\sin 2\theta + q\cos 2\theta$, prove that $u_n = p\sin n\theta + q\cos n\theta$.

7. Deduce trigonometrical identities from the relation

$$\Sigma (b-c)^4 \equiv 2\Sigma\{(a-b)^2(a-c)^2\}.$$

8. If $(1+x)^n \equiv a_0 + \Sigma a_r x^r$, where n is a positive integer, prove that

$$a_0 + a_4 + a_8 + \ldots = 2^{n-2} + 2^{\frac{1}{2}n-1}\cos\frac{n\pi}{4}.$$

9. If $r^2 = a^2 + b^2 + c^2$ and $z = \dfrac{b+ic}{r+a}$, prove that

(i) $\dfrac{c+ia}{r+b} = \dfrac{i(1-z)}{1+z}$; (ii) $\dfrac{a+ib}{r+c} = \dfrac{1+iz}{1-iz}$.

10. If $\Sigma\cos\theta_r = 0 = \Sigma\sin\theta_r$, prove that

(i) $\Sigma\cos 4\theta_r = 2\Sigma\Sigma\cos 2(\theta_r + \theta_s)$;

(ii) $\Sigma\sin 4\theta_r = 2\Sigma\Sigma\sin 2(\theta_r + \theta_s)$;

where r and s are 1, 2, 3, 4, 5 and are unequal.

11. Prove that, if $2(ad+bc) = (a+d)(b+c)$, the four points representing the complex numbers a, b, c, d are concyclic and form a harmonic range on the circle.

12. If a, b, c, A, B, C are real and $ac > b^2$, $AC > B^2$, prove that the points representing the roots of $az^2 + 2bz + c = 0$, $Az^2 + 2Bz + C = 0$ are concyclic with the origin if $bC = cB$.

13. If w, z are complex numbers such that $|z| = 1$ and $\dfrac{1}{w} = 1 - z + z^2$, and if they are represented by the points P, Q prove that PQ passes through the point (1, 0) and that the x-axis bisects an angle between OP and OQ.

14. If a_1, a_2, a_3, a_4, a_5 are the fifth roots of unity, prove that $\Sigma\tan^{-1}\dfrac{1}{a_r+1}$, for $r = 1$ to 5, equals $\tan^{-1}\frac{4}{3} + n\pi$, where n is an integer or zero.

15. Use the relation $\dfrac{1}{x^2+1} \equiv \dfrac{i}{2}\left(\dfrac{1}{x+i} - \dfrac{1}{x-i}\right)$ to prove that

$$\frac{d^n}{dx^n}\left(\frac{1}{x^2+1}\right) = (-1)^n \,.\, n! \sin(n+1)\theta \sin^{n+1}\theta,$$

where $x = \cot\theta$.

16. Prove that $\frac{d^n}{dx^n}\left(\frac{x}{x^2+1}\right)=(-1)^n . n! \cos(n+1)\theta \sin^{n+1}\theta$, where $x=\cot\theta$.

17. Prove that

$$1-(n-1)+\frac{(n-2)(n-3)}{2!}-\frac{(n-3)(n-4)(n-5)}{3!}+\ldots$$
$$=(-1)^n \sin\frac{2(n+1)\pi}{3}\operatorname{cosec}\frac{2\pi}{3}.$$

18. Prove that

$$1-3n+\frac{(3n-1)(3n-2)}{2!}-\frac{(3n-2)(3n-3)(3n-4)}{3!}+\ldots=(-1)^n.$$

19. Prove that

$$1-n+\frac{n(n-3)}{2!}-\frac{n(n-4)(n-5)}{3!}+\ldots=2\cos\frac{n\pi}{3}.$$

20. Prove that

$$\cos^2 n\theta=\tfrac{1}{2}\{1+(-1)^n\}-\tfrac{1}{2}(-1)^n\left\{\frac{n^2}{2!}(2c)^2-\frac{n^2(n^2-1^2)}{4!}(2c)^4+\ldots\right\}.$$

21. Find the coefficient of x^n in $(1+2x\cos\theta+x^2)^n$ and deduce that $1+c_1{}^2\cos 2\theta+\ldots+c_n{}^2\cos 2n\theta$ equals

$$n!\cos n\theta\left\{\frac{(2\cos\theta)^n}{n!}+\frac{(2\cos\theta)^{n-2}}{1^2.(n-2)!}+\frac{(2\cos\theta)^{n-4}}{1^2.2^2.(n-4)!}+\ldots\right\}, \text{ where } c_r=\binom{n}{r}.$$

22. If c_r is the coefficient of x^r in $(x^{-n}+\ldots+x^{-1}+1+x+\ldots+x^n)^4$, prove that

$$\left\{\frac{\sin(2n+1)\theta}{\sin\theta}\right\}^4=c_0+2c_1\cos 2\theta+2c_2\cos 4\theta+\ldots+2c_{4n}\cos 8n\theta,$$

and deduce that

$$\int_0^{\frac{\pi}{2}}\sin^4(2n+1)\theta\operatorname{cosec}^4\theta\, d\theta=\frac{\pi}{6}(2n+1)(8n^2+8n+3).$$

23. Prove that $2\operatorname{cosec}\alpha\,\Sigma\{\sin r\alpha\cos(n-r)\beta\}$, for $r=1$ to $n-1$, is equal to

$$\frac{\cos n\alpha-\cos n\beta}{\cos\alpha-\cos\beta}-\frac{\sin n\alpha}{\sin\alpha}.$$

24. If p, q, r are any positive integers or zero, subject to $p+q+r=n$, prove that

$$\sum_{p,q,r}\cos(p\alpha+q\beta+r\gamma)=\tfrac{1}{4}\sum_{\alpha,\beta,\gamma}\frac{\cos\{(n+1)\alpha-\frac{1}{2}(\beta+\gamma)\}}{\sin\frac{\gamma-\alpha}{2}\sin\frac{\alpha-\beta}{2}}.$$

CHAPTER X

ONE-VALUED FUNCTIONS OF A COMPLEX VARIABLE

The meaning of the word "convergent" as applied to series of real numbers has been given in Ch. V., p. 77. A general discussion of the principles of convergence is reserved for the companion volume on Analysis. We give here only what is necessary for the extension of the idea to series of complex numbers.

Absolutely Convergent Series. If the series, $\Sigma|x_n|$, of positive real numbers

$$|x_1|+|x_2|+|x_3|+\ldots+|x_n|+\ldots$$

is convergent, we say that the series, Σx_n, of real numbers

$$x_1+x_2+x_3+\ldots+x_n+\ldots$$

is **absolutely convergent.**

Thus the series,

$$\frac{1}{1^2}-\frac{1}{2^2}+\frac{1}{3^2}-\frac{1}{4^2}+\ldots,$$

is absolutely convergent, because it is shown on p. 210 that the series, $\frac{1}{1^2}+\frac{1}{2^2}+\frac{1}{3^2}+\frac{1}{4^2}+\ldots$, is convergent.

If a series is absolutely convergent, it is necessarily convergent. This may be seen by expressing it as the "difference" between two convergent series; thus,

$$x_1+x_2+\ldots+x_n \equiv \{x_1+|x_1|+x_2+|x_2|+\ldots+x_n+|x_n|\} \\ -\{|x_1|+|x_2|+\ldots+|x_n|\}.$$

Since $\Sigma|x_n|$ is convergent, the second bracket tends to a limit, say V, when $n \to \infty$; also $0 \leqslant (x_n+|x_n|) \leqslant 2|x_n|$; $\therefore$ the first bracket never decreases as n increases and is always less than 2V; it therefore tends to a limit $\leqslant 2V$, when $n \to \infty$; $\therefore$ Σx_n is convergent.

Thus the convergence of $\Sigma\frac{1}{n^2}$ implies the convergence of

$$\Sigma\frac{(-1)^{n+1}}{n^2}.$$

But obviously a series may be convergent without being absolutely convergent; for example, $1-\frac{1}{2}+\frac{1}{3}-\frac{1}{4}+\ldots$ is convergent, with sum to infinity log 2; but the series, $1+\frac{1}{2}+\frac{1}{3}+\frac{1}{4}+\dots$ is divergent, see p. 69. A series which is convergent, but not absolutely convergent, is called *conditionally convergent* or sometimes *semi-convergent.*

Series of Complex Terms. Consider the series, Σz_n, of complex numbers

$$z_1+z_2+z_3+\dots+z_n+\dots,$$

where $z_n \equiv x_n+iy_n$, and x_n, y_n are real numbers.

This series is called *convergent* if the series Σx_n and Σy_n are both convergent. If X and Y are the sums to infinity of these series, so that

$$\mathsf{X}=\lim_{n\to\infty}(x_1+x_2+\dots+x_n) \quad \text{and} \quad \mathsf{Y}=\lim_{n\to\infty}(y_1+y_2+\dots+y_n),$$

$\mathsf{X}+i\mathsf{Y}$ is called the *sum to infinity* of the series Σz_n.

For example, since

$$\lim_{n\to\infty}\left(\frac{1}{2}+\frac{1}{2^2}+\dots+\frac{1}{2^n}\right)=1 \quad \text{and} \quad \lim_{n\to\infty}\left(\frac{1}{3}+\frac{1}{3^2}+\dots+\frac{1}{3^n}\right)=\tfrac{1}{2},$$

we say that the series

$$(\tfrac{1}{2}+\tfrac{1}{3}i)+(\tfrac{1}{4}+\tfrac{1}{9}i)+(\tfrac{1}{8}+\tfrac{1}{27}i)+\dots$$

is convergent, with sum to infinity $1+\frac{1}{2}i$.

More shortly, we may say that Σz_n is convergent if

$$\sigma_n(\equiv z_1+z_2+\dots+z_n)$$

tends to a limit Z when $n\to\infty$, for this only means that

$$(x_1+x_2+\dots+x_n) \quad \text{and} \quad (y_1+y_2+\dots+y_n)$$

tend to limits X and Y and that $\mathsf{X}+i\mathsf{Y}$ is denoted by Z.

Absolute Convergence of Series of Complex Terms. If

$$|z_n| \equiv +\sqrt{(x_n^2+y_n^2)}$$

and if the series, $\Sigma|z_n|$, of positive real numbers

$$|z_1|+|z_2|+|z_3|+\dots+|z_n|+\dots$$

is convergent, we say that the series, Σz_n, of complex numbers

$$z_1+z_2+z_3+\dots+z_n+\dots$$

is absolutely convergent, and it is easy to see that it is then necessarily "convergent," in the sense defined above.

For $|x_n| \leqslant +\sqrt{(x_n^2+y_n^2)}=|z_n|$ and similarly $|y_n| \leqslant |z_n|$; therefore Σx_n and Σy_n are absolutely convergent; $\therefore$ they are convergent; $\therefore$ Σz_n is convergent.

The Geometric Progression Consider the series

$$1+z+z^2+z^3+\dots+z^n+\dots$$

where $z \equiv r(\cos\theta + i\sin\theta)$.

If $\phi_n(z) \equiv 1+z+z^2+\dots+z^{n-1}$, we have, as for the G.P. of real numbers, $\phi_n(z) = \dfrac{1}{1-z} - \dfrac{z^n}{1-z}$, provided $z \neq 1$.

Hence this series is convergent, and has $\dfrac{1}{1-z}$ for sum to infinity, if $\lim\limits_{n\to\infty} \dfrac{z^n}{1-z} = 0$, i.e. if $\lim\limits_{n\to\infty} z^n = 0$, since $z \neq 1$.

Now $z^n = r^n(\cos n\theta + i\sin n\theta)$ and r is positive. Also if $r<1$, $\lim r^n = 0$, see p. 78; but $|r^n \cos n\theta| \leqslant r^n$ and $|r^n \sin n\theta| \leqslant r^n$, $\therefore$ if $r<1$, $\lim(r^n \cos n\theta) = 0$ and $\lim(r^n \sin n\theta) = 0$; $\therefore$ $\lim z^n = 0$.

Thus the geometric progression of complex numbers is convergent when $|z|<1$, that is, if it is absolutely convergent.

The values of $\phi_n(z)$ and $\dfrac{1}{1-z}$ in terms of r and θ were virtually obtained in Ch. IX., p. 174, where x was used instead of r.

The Exponential Series. Consider the series

$$1+\frac{z}{1!}+\frac{z^2}{2!}+\frac{z^3}{3!}+\dots+\frac{z^n}{n!}+\dots, \quad \dots\dots\dots\dots(1)$$

where $z \equiv x+iy = r(\cos\theta + i\sin\theta)$ and $r>0$.

It has been proved in Ch. V that the series of positive numbers

$$1+\frac{r}{1!}+\frac{r^2}{2!}+\frac{r^3}{3!}+\dots+\frac{r^n}{n!}+\dots \quad \dots\dots\dots\dots(2)$$

is convergent for all values of r. Thus the series (1) is absolutely convergent and therefore it is convergent.

In other words, if

$$\mathsf{E}_n(z) \equiv 1+\frac{z}{1!}+\frac{z^2}{2!}+\dots+\frac{z^n}{n!} \quad \dots\dots\dots\dots(3)$$

then $\lim\limits_{n\to\infty}\{\mathsf{E}_n(z)\}$ exists, for all values of z. This limit may be denoted by $\mathsf{E}(z)$, thus

$$\mathsf{E}(z) = \lim_{n\to\infty}\{\mathsf{E}_n(z)\}. \quad \dots\dots\dots\dots(4_1)$$

If $z \equiv x+0i$, the function $\mathsf{E}(z)$ corresponds to the function $\exp(x)$ of the real variable x, see Ch. V., p. 90, in just the same way that the complex number $x+0i$ corresponds to the real number x.

For this reason, the function $\mathsf{E}(z)$ is usually written exp (z), **but it** should be understood that exp (z), so used, acquires its meaning *by definition* as a function of complex algebra. In fact, the definition of equation (4_1) is replaced by the following definition:

$$\exp(\mathbf{z}) = \lim_{n\to\infty} \{\mathbf{E}_n(\mathbf{z})\}. \quad \ldots\ldots\ldots\ldots\ldots\ldots(4_2)$$

Proofs of the properties of exp (z) must be based on this definition; it is not permissible to *assume* that the properties established for exp (x) hold also for exp (z).

Functional Law for exp (z). If $z \equiv 0 + yi$, the exponential series is

$$1 + \frac{(iy)}{1!} + \frac{(iy)^2}{2!} + \frac{(iy)^3}{3!} + \ldots + \frac{(iy)^n}{n!} + \ldots;$$

$$\therefore\ \mathsf{E}_n(yi) = \left\{1 - \frac{y^2}{2!} + \frac{y^4}{4!} - \ldots\right\} + i\left\{y - \frac{y^3}{3!} + \frac{y^5}{5!} - \ldots\right\},$$

where the series in brackets are finite.

But, by pp. 80, 81, when $n \to \infty$, the expressions in these two brackets tend to $\cos y$ and $\sin y$;

$$\therefore\ \exp(yi) = \lim_{n\to\infty} \mathsf{E}_n(yi) = \cos y + i \sin y;$$

$$\therefore\ \exp(y_1 i)\exp(y_2 i)$$
$$= (\cos y_1 + i \sin y_1)(\cos y_2 + i \sin y_2) = \cos(y_1 + y_2) + i \sin(y_1 + y_2)$$
$$= \exp\{i(y_1 + y_2)\}.$$

If then z_1, z_2 are each of the form $0 + iy$, the exponential function, exp (z), satisfies the functional law,

$$\exp(z_1)\exp(z_2) = \exp(z_1 + z_2).$$

We proceed to prove that this result is true for all complex values of z.

Let $\quad z_1 \equiv x_1 + iy_1,\ z_2 \equiv x_2 + iy_2,$ and $|z_1| = r_1,\ |z_2| = r_2.$

$$\mathsf{E}_n(z_1) \cdot \mathsf{E}_n(z_2) = \left\{1 + \frac{z_1}{1!} + \frac{z_1^2}{2!} + \ldots + \frac{z_1^n}{n!}\right\}\left\{1 + \frac{z_2}{1!} + \frac{z_2^2}{2!} + \ldots + \frac{z_2^n}{n!}\right\}$$

$$= 1 + \frac{z_1}{1!} + \frac{z_1^2}{2!} + \ldots + \frac{z_1^n}{n!}$$

$$+ \frac{z_2}{1!} + \frac{z_1 z_2}{1!\,1!} + \frac{z_1^2 z_2}{2!\,1!} + \ldots + \frac{z_1^n z_2}{n!\,1!}$$

$$+ \frac{z_2^2}{2!} + \frac{z_1 z_2^2}{1!\,2!} + \frac{z_1^2 z_2^2}{2!\,2!} + \ldots + \frac{z_1^n z_2^2}{n!\,2!}$$

$$\cdot \quad \cdot \quad \cdot \quad \cdot \quad \cdot \quad \cdot \quad \cdot \quad \cdot \quad \cdot \quad \cdot \quad \cdot \quad \cdot$$

$$+ \frac{z_2^n}{n!} + \frac{z_1 z_2^n}{1!\,n!} + \frac{z_1^2 z_2^n}{2!\,n!} + \ldots + \frac{z_1^n z_2^n}{n!\,n!}.$$

Add up by diagonals; the terms of order s, where $s \leqslant n$, give

$$\frac{z_1^s}{s!}+\frac{z_1^{s-1}z_2}{(s-1)!\,1!}+\frac{z_1^{s-2}z_2^2}{(s-2)!\,2!}+\ldots+\frac{z_2^s}{s!}$$

$$=\frac{1}{s!}\left\{z_1^s+\frac{s}{1!}z_1^{s-1}z_2+\frac{s(s-1)}{2!}z_1^{s-2}z_2^2+\ldots+z_2^s\right\}$$

$$=\frac{(z_1+z_2)^s}{s!},$$

by the binomial theorem for a positive integral index.

$$\therefore\ \mathsf{E}_n(z_1)\mathsf{E}_n(z_2)-\mathsf{E}_n(z_1+z_2)=\sum_{p,q}\frac{z_1^p z_2^q}{p!\,q!}, \qquad \ldots\ldots\ldots\ldots(5)$$

where the summation extends to all values of p, q such that $p+q>n$, $p \leqslant n$, $q \leqslant n$.

In precisely the same way, we have

$$\mathsf{E}_n(r_1)\mathsf{E}_n(r_2)-\mathsf{E}_n(r_1+r_2)=\sum_{p,q}\frac{r_1^p r_2^q}{p!\,q!}. \qquad \ldots\ldots\ldots\ldots\ldots\ldots(6)$$

Now from Ch. V., p. 91, when $n \to \infty$, $\mathsf{E}_n(r_1)$, $\mathsf{E}_n(r_2)$, $\mathsf{E}_n(r_1+r_2)$ tend to the limits e^{r_1}, e^{r_2}, $e^{r_1+r_2}$; but $e^{r_1}e^{r_2}=e^{r_1+r_2}$;

$$\therefore\ \mathsf{E}_n(r_1)\mathsf{E}_n(r_2)-\mathsf{E}_n(r_1+r_2)\to 0 \text{ when } n\to\infty.$$

Now each term of the Σ expression in (6) is the modulus of the corresponding term of the Σ expression in (5); also the modulus of a sum $\leqslant$ the sum of the moduli;

$$\therefore\ |\mathsf{E}_n(z_1)\mathsf{E}_n(z_2)-\mathsf{E}_n(z_1+z_2)| \leqslant \mathsf{E}_n(r_1)\mathsf{E}_n(r_2)-\mathsf{E}_n(r_1+r_2);$$

$$\therefore\ |\mathsf{E}_n(z_1)\mathsf{E}_n(z_2)-\mathsf{E}_n(z_1+z_2)| \to 0 \text{ when } n\to\infty.$$

But, when $n\to\infty$, $\mathsf{E}_n(z_1)$, $\mathsf{E}_n(z_2)$, $\mathsf{E}_n(z_1+z_2)$ tend respectively to the limits $\exp(z_1)$, $\exp(z_2)$, $\exp(z_1+z_2)$;

$$\therefore\ \exp(\mathbf{z}_1)\,.\,\exp(\mathbf{z}_2)=\exp(\mathbf{z}_1+\mathbf{z}_2), \qquad \ldots\ldots\ldots\ldots\ldots\ldots(7)$$

where z_1, z_2 are any two complex numbers.

This proof of the functional law for $\exp(z)$ suggests an alternative method for developing the theory of the exponential function of a real variable.

We start by proving from first principles that, for all real values of x,

$$1+x+\frac{x^2}{2!}+\frac{x^3}{3!}+\ldots$$

is an absolutely convergent series. Denote its sum to infinity by $\mathsf{E}(x)$.

From equation (6) above, we have, for $r_1>0$, $r_2>0$,

$$\mathsf{E}_n(r_1+r_2)<\mathsf{E}_n(r_1)\mathsf{E}_n(r_2). \qquad \ldots\ldots\ldots\ldots\ldots\ldots\ldots\ldots(8)$$

Similarly, the product set out above shows that

$$E_{2n}(r_1+r_2) - E_n(r_1)\,E_n(r_2)$$

equals the sum of a number of *positive* terms ;

$$\therefore\ E_n(r_1)E_n(r_2) < E_{2n}(r_1+r_2)\ ; \qquad \text{.........................(9)}$$

$$\therefore\ \text{from (8) and (9),}\quad E_n(r_1+r_2) < E_n(r_1)\,E_n(r_2) < E_{2n}(r_1+r_2).$$

But when $n \to \infty$, $E_n(r_1)$, $E_n(r_2)$, $E_n(r_1+r_2)$, $E_{2n}(r_1+r_2)$ tend respectively to the limits $E(r_1)$, $E(r_2)$, $E(r_1+r_2)$, $E(r_1+r_2)$;

$$\therefore\ E(r_1+r_2) \leqslant E(r_1)\,E(r_2) \leqslant E(r_1+r_2)\ ;$$

$$\therefore\ E(r_1+r_2) = E(r_1)\,E(r_2).$$

The properties of the exponential function of a real variable can then be deduced from this functional law.

Expression of exp (z) in the Modulus-Amplitude Form. By equation (7), $\exp(z) \equiv \exp(x+iy) = \exp(x+0i)\exp(0+iy)$.

But
$$\exp(x+0i) = 1 + \frac{x}{1!} + \frac{x^2}{2!} + \ldots + \frac{x^n}{n!} + \ldots$$
$$= e^x, \text{ see Ch. V., p. 90.}$$

Also, by p. 192, $\qquad \exp(0+iy) = \cos y + i \sin y$;

$\therefore$ exp (z) is a complex number, with modulus e^x and amplitude $2n\pi + y$, and we write

$$\mathbf{exp\,(z) \equiv exp\,(x+iy) = e^x(cos\,y + i\,sin\,y).} \qquad \text{.........(10)}$$

Thus we see that the function exp (z) is periodic, with period $2\pi i$. The principal value of the amplitude of exp $(x+iy)$ is obtained by choice of n such that $-\pi < 2n\pi + y \leqslant +\pi$.

The special relations,

$$\exp(iy) = \cos y + i \sin y\ ;\ \exp(-iy) = \cos y - i \sin y$$

give important forms for $\cos y$ and $\sin y$:

$$\mathbf{cos\,y = \tfrac{1}{2}\{exp\,(iy) + exp\,(-iy)\},} \qquad \text{.......................(11)}$$

$$\mathbf{sin\,y = \frac{1}{2i}\{exp\,(iy) - exp\,(-iy)\}.} \qquad \text{..................(12)}$$

These forms, however, are merely alternative ways of writing equations (6) and (5) in Ch. V. (pp. 80, 81).

Example 1. Express $\exp(i\alpha)\exp(x \operatorname{cis}\beta)$, (i) in the modulus-amplitude form, (ii) as a power series in x.

What conclusions can be drawn by comparing the two results ?

(i)
$$\exp(i\alpha)\exp(x \operatorname{cis}\beta) = \exp\{i\alpha + x(\cos\beta + i\sin\beta)\}$$
$$= \exp(x\cos\beta)\exp\{i(\alpha + x\sin\beta)\}$$
$$= e^{x\cos\beta}\operatorname{cis}(\alpha + x\sin\beta).$$

(ii) $\exp(i\alpha)\exp(x \operatorname{cis}\beta)$

$$=\operatorname{cis}\alpha\left\{1+x\operatorname{cis}\beta+\frac{x^2}{2!}\operatorname{cis}2\beta+\ldots+\frac{x^n}{n!}\operatorname{cis}n\beta+\ldots\right\}$$

$$=\operatorname{cis}\alpha+x\operatorname{cis}(\alpha+\beta)+\frac{x^2}{2!}\operatorname{cis}(\alpha+2\beta)+\ldots+\frac{x^n}{n!}\operatorname{cis}(\alpha+n\beta)+\ldots.$$

It follows that

$$e^{x\cos\beta}\cos(\alpha+x\sin\beta)=\cos\alpha+x\cos(\alpha+\beta)+\ldots+\frac{x^n}{n!}\cos(\alpha+n\beta)+\ldots$$

and

$$e^{x\cos\beta}\sin(\alpha+x\sin\beta)=\sin\alpha+x\sin(\alpha+\beta)+\ldots+\frac{x^n}{n!}\sin(\alpha+n\beta)+\ldots.$$

Example 2. Discuss the convergence of the series

$$\cos\theta\sin\theta+\frac{\cos^2\theta}{2!}\sin 2\theta+\ldots+\frac{\cos^n\theta}{n!}\sin n\theta+\ldots$$

and find the sum to infinity.

Consider the series

$$1+\cos\theta\operatorname{cis}\theta+\frac{\cos^2\theta}{2!}\operatorname{cis}2\theta+\ldots+\frac{\cos^n\theta}{n!}\operatorname{cis}n\theta+\ldots. \qquad \ldots\text{(i)}$$

Since $|\operatorname{cis} n\theta|=1$, in the series of moduli the sum to n terms is

$$\leqslant 1+|\cos\theta|+\frac{|\cos\theta|^2}{2!}+\ldots+\frac{|\cos\theta|^{n-1}}{(n-1)!}<e^{|\cos\theta|};$$

$\therefore\ \Sigma\dfrac{\cos^n\theta\operatorname{cis}n\theta}{n!}$ is absolutely convergent. Hence it is convergent, and this implies the convergence of the given series.

Also series (i)

$$=1+(\cos\theta\operatorname{cis}\theta)+\frac{(\cos\theta\operatorname{cis}\theta)^2}{2!}+\ldots+\frac{(\cos\theta\operatorname{cis}\theta)^n}{n!}+\ldots$$

$$=\exp(\cos\theta\operatorname{cis}\theta)=\exp(\cos^2\theta+i\cos\theta\sin\theta)$$

$$=e^{\cos^2\theta}\exp\{i(\cos\theta\sin\theta)\}$$

$$=e^{\cos^2\theta}\{\cos(\cos\theta\sin\theta)+i\sin(\cos\theta\sin\theta)\};$$

$\therefore$ the sum to infinity of the given series $=e^{\cos^2\theta}\sin(\cos\theta\sin\theta)$.

Note. A direct proof of the absolute convergence of the given series is as follows: $|\sin n\theta|\leqslant 1$, $\therefore$ the sum to n terms of the series of moduli

$$\leqslant\frac{|\cos\theta|}{1!}+\frac{|\cos\theta|^2}{2!}+\ldots+\frac{|\cos\theta|^n}{n!}<e^{|\cos\theta|}-1.$$

EXERCISE X. a.

Express the following in the form, $a+ib$:

1. $\exp(1+i\pi)$.
2. $\exp(i)+\exp(-i)$.
3. $\exp\left(-1+\frac{i\pi}{3}\right)$.
4. $\exp(\cos\theta+i\sin\theta)$.
5. $\exp(a+ib)\exp(a-ib)$.
6. $\exp(\log r+i\theta)$.
7. $\exp\{\sec a\exp(ia)\}$.
8. $\exp(x\operatorname{cis}\theta)\exp(y\operatorname{cis}\phi)$.

Give simplified values of the following:

9. $\exp(i\pi)$.
10. $\exp(-i\pi)$.
11. $\exp(\operatorname{cis}\theta)+\exp\{\operatorname{cis}(-\theta)\}$.
12. $\exp(\operatorname{cis}\theta\tan\theta)$.
13. $\exp(i\operatorname{cis}\theta)-\exp\{-i\operatorname{cis}(-\theta)\}$.
14. $\exp\{\exp(\operatorname{cis}\theta)\}$.
15. Prove that $\exp(-\theta-i\phi)=(\operatorname{ch}\theta-\operatorname{sh}\theta)(\cos\phi-i\sin\phi)$.

16. If $X+iY=\exp(x+iy)$, find the relation between X and Y, (i) if x is constant and equal to c; (ii) if y is constant and equal to m.

17. Find u, v if $\exp\left(\frac{x-a+iy}{x+a+iy}\right)=u+iv$.

18. Find u if $(1-a^2\cos 2\theta-ia^2\sin 2\theta)^{-1}\exp(i\theta)=u+iv$.

19. The complex numbers z, z' are represented by the points P, P′, where $z'=\exp(z)$. Discuss the movement of P, (i) if P′ describes the unit circle, centre the origin, clockwise, starting from the point $(-1, 0)$, (ii) if P′ describes the negative half of the y'-axis, starting from the origin.

20. Find real numbers a and b such that

$$\exp(a+ib)=\exp(2b+ia).$$

21. Show that the equation, $\exp x=x+a$, where a is real, has no solution of the form, $x=iv$, where v is real.

If it has a solution, $x=u+iv$, where $v\neq 0$, prove that u is positive.

22. What results can be obtained by equating the first and second parts of the complex numbers in the relation,

$$\exp(z)=1+\frac{z}{1!}+\frac{z^2}{2!}+\ldots+\frac{z^n}{n!}+\ldots;$$

(i) if $z=\cos a+i\sin a$; (ii) if $z=1+i\tan\beta$?

23. Prove that $\exp[(a+ib)x]-\exp[(a-ib)x]=2i\,e^{ax}\sin bx$.

Use this relation to find the coefficient of x^n when $e^{x\cos\theta}\sin(x\sin\theta)$ is expanded in powers of x.

24. Express $2e^{\theta}\cos\theta$ in the form $\exp(u+iv)+\exp(u-iv)$.

Hence find the coefficient of θ^n in the expansion of $e^{\theta}\cos\theta$ in powers of θ.

25. Prove that $\exp\{\exp(\theta i)\} - \exp\{-\exp(\theta i)\} = a + ib$, where $a = 2\cos(\sin\theta)\,\mathrm{sh}(\cos\theta)$ and $b = 2\sin(\sin\theta)\,\mathrm{ch}(\cos\theta)$.

26. Expand $e^{x\sin\alpha}\sin(x\cos\alpha)$ in ascending powers of x.

Sum to infinity the following series :

27. $x\cos\theta + x^2\cos 2\theta + x^3\cos 3\theta + \dots$; $(-1 < x < 1)$.

28. $x\sin\alpha + x^2\sin(\alpha+\beta) + x^3\sin(\alpha+2\beta) + \dots$; $(-1 < x < 1)$.

29. $1 + \cos\theta + \dfrac{\cos 2\theta}{2!} + \dfrac{\cos 3\theta}{3!} + \dots$.

30. $\sin\theta - \dfrac{\sin 2\theta}{2!} + \dfrac{\sin 3\theta}{3!} - \dots$.

31. $\cos\alpha - x\cos(\alpha+\beta) + \dfrac{x^2}{2!}\cos(\alpha+2\beta) - \dfrac{x^3}{3!}\cos(\alpha+3\beta) + \dots$.

32. $1 + \cos\theta\tan\theta + \dfrac{1}{2!}\cos 2\theta\tan^2\theta + \dfrac{1}{3!}\cos 3\theta\tan^3\theta + \dots$.

33. $$\cos\alpha + \cos\beta\cos(\alpha+\beta) + \frac{\cos^2\beta}{2!}\cos(\alpha+2\beta) + \frac{\cos^3\beta}{3!}\cos(\alpha+3\beta) + \dots .$$

34. $\cos\alpha + \dfrac{\cos 3\alpha}{3!} + \dfrac{\cos 5\alpha}{5!} + \dots$.

35. If $\mathrm{C} = \dfrac{\cos 2\theta}{2!} + \dfrac{\cos 4\theta}{4!} + \dfrac{\cos 6\theta}{6!} + \dots$

and $\mathrm{S} = \dfrac{\sin 2\theta}{2!} + \dfrac{\sin 4\theta}{4!} + \dfrac{\sin 6\theta}{6!} + \dots$,

prove that $\mathrm{C}^2 + \mathrm{S}^2 = \{\mathrm{ch}(\cos\theta) - \cos(\sin\theta)\}^2$.

The Generalised Circular Functions. If z is any complex number, $\cos z$ and $\sin z$ are *defined* by the relations :

$$\cos z = \tfrac{1}{2}\{\exp(iz) + \exp(-iz)\}, \quad \text{.................}(13)$$

$$\sin z = \frac{1}{2i}\{\exp(iz) - \exp(-iz)\}. \quad \text{.................}(14)$$

Further, we write

$$\tan z = \frac{\sin z}{\cos z}, \ \operatorname{cosec} z = \frac{1}{\sin z}, \text{ etc.}$$

The definitions are equivalent to

$$\cos z + i\sin z = \exp(iz)\,; \quad \cos z - i\sin z = \exp(-iz) \quad \text{......}(15)$$

and to the forms,

$$\cos z = 1 - \frac{z^2}{2!} + \frac{z^4}{4!} - \frac{z^6}{6!} + \dots, \qquad \dots\dots\dots\dots(16)$$

$$\sin z = \frac{z}{1!} - \frac{z^3}{3!} + \frac{z^5}{5!} - \frac{z^7}{7!} + \dots . \qquad \dots\dots\dots\dots(17)$$

The definitions are of course chosen so that formulae established for circular functions of a real variable (defined geometrically) hold also for the generalised functions. The reason that they hold is indicated below :

Results such as $\cos 0 = 1$, $\sin 0 = 0$, $\cos(-z) = \cos z$, $\sin(-z) = -\sin z$, are immediately deduced from the definitions, (13), (14) and (4_2) above.

Suppose it is required to prove that

$$\cos(z_1 + z_2) = \cos z_1 \cos z_2 - \sin z_1 \sin z_2.$$

Using relations (13), (14) it is necessary to prove that

$$\tfrac{1}{2}\{\exp[i(z_1 + z_2)] + \exp[-i(z_1 + z_2)]\} \equiv \tfrac{1}{4}\{\exp(iz_1) + \exp(-iz_1)\}\{\exp(iz_2) + \exp(-iz_2)\} + \tfrac{1}{4}\{\exp(iz_1) - \exp(-iz_1)\}\{\exp(iz_2) - \exp(-iz_2)\}. \quad \dots\dots(18)$$

Now we know that, for real values of y_1, y_2,

$$\cos(y_1 + y_2) \equiv \cos y_1 \cos y_2 - \sin y_1 \sin y_2$$

and ∴ from relations (11), (12), p. 194,

$$\tfrac{1}{2}\{\exp[i(y_1 + y_2)] + \exp[-i(y_1 + y_2)]\} \equiv \tfrac{1}{4}\{\exp(iy_1) + \exp(-iy_1)\}\{\exp(iy_2) + \exp(-iy_2)\} + \tfrac{1}{4}\{\exp(iy_1) - \exp(-iy_1)\}\{\exp(iy_2) - \exp(-iy_2)\}. \quad \dots\dots(19)$$

Since the result of simplifying the right side of (19) gives the left side, and since the process of simplification of the right side of (18) corresponds precisely to that of (19), the truth of (19) implies the truth of (18). Hence to every general formula in the trigonometry of the real angle, there corresponds a similar formula for the generalised circular functions of a complex variable.

The Generalised Hyperbolic Functions. If z is any complex number, ch z and sh z are *defined* by the relations :

$$\operatorname{ch} z = \tfrac{1}{2}\{\exp(z) + \exp(-z)\}, \qquad \dots\dots\dots\dots(20)$$

$$\operatorname{sh} z = \tfrac{1}{2}\{\exp(z) - \exp(-z)\}. \qquad \dots\dots\dots\dots(21)$$

Further, we write $\operatorname{th} z = \dfrac{\operatorname{sh} z}{\operatorname{ch} z}$, $\operatorname{cosech} z = \dfrac{1}{\operatorname{sh} z}$, etc.

These definitions are equivalent to the forms,

$$\operatorname{ch} z = 1 + \frac{z^2}{2!} + \frac{z^4}{4!} + \frac{z^6}{6!} + \dots, \qquad \dots\dots\dots\dots(22)$$

$$\operatorname{sh} z = \frac{z}{1!} + \frac{z^3}{3!} + \frac{z^5}{5!} + \frac{z^7}{7!} + \dots . \qquad \dots\dots\dots\dots(23)$$

The definitions are of course chosen so that formulae established for hyperbolic functions of a real variable (see p. 105) hold also for the generalised functions; this fact may be established by the same method as has just been used for the generalised circular functions.

There is a simple connection between the generalised circular and hyperbolic functions.

$$\cos(iz) = \tfrac{1}{2}\{\exp(i \, . \, iz) + \exp(-i \, . \, iz)\} = \tfrac{1}{2}\{\exp(-z) + \exp(z)\},$$

$$\therefore \ \cos iz = \operatorname{ch} z. \quad \text{.........................(24)}$$

Also

$$\sin(iz) = \frac{1}{2i}\{\exp(i \, . \, iz) - \exp(-i \, . \, iz)\} = \frac{i}{2}\{\exp(z) - \exp(-z)\}$$

$$\therefore \ \sin iz = i \operatorname{sh} z. \quad \text{.........................(25)}$$

Relations (24), (25) justify the rule given on p. 105 for deducing formulae connecting hyperbolic functions from the corresponding formulae for the circular functions.

For example:

$$\operatorname{sh}(A+B) = \frac{1}{i}\sin(iA+iB)$$

$$= \frac{1}{i}\{\sin iA \cos iB + \cos iA \sin iB\}$$

$$= \frac{1}{i}\{i \operatorname{sh} A \operatorname{ch} B + i \operatorname{ch} A \operatorname{sh} B\}$$

$$= \operatorname{sh} A \operatorname{ch} B + \operatorname{ch} A \operatorname{sh} B.$$

Also a product of sines introduces i^2 and so leads to a change of sign in the corresponding formula for hyperbolic functions.

The generalised hyperbolic functions are periodic; we have from (24) and (25)

$$\operatorname{ch}(z + 2n\pi i) = \operatorname{ch} z; \ \operatorname{sh}(z + 2n\pi i) = \operatorname{sh} z; \ \operatorname{th}(z + n\pi i) = \operatorname{th} z, \quad \text{...(26)}$$

where n is any integer.

Expression of Generalised Circular Functions in the Form, A + iB.

If $\quad z \equiv x + iy, \ \exp(iz) = \exp(ix - y) = \exp(ix)\exp(-y)$

$$= e^{-y} \operatorname{cis} x.$$

Similarly, $\exp(-iz) = e^{y} \operatorname{cis}(-x)$; and so it is easy, by using the definitions (13) and (14), to express $\cos z$ and $\sin z$ in the form $A + iB$. But it is more convenient to proceed as follows:

$$\cos(x+iy) = \cos x \cos iy - \sin x \sin iy$$

$$= \cos x \operatorname{ch} y - i \sin x \operatorname{sh} y, \quad \text{.................(27)}$$

$$\sin(x+iy) = \sin x \cos iy + \cos x \sin iy$$

$$= \sin x \operatorname{ch} y + i \cos x \operatorname{sh} y. \quad \text{.................(28)}$$

Also, $\tan(x+iy)$

$$=\frac{\sin(x+iy)\cos(x-iy)}{\cos(x+iy)\cos(x-iy)}=\frac{\sin 2x+\sin 2iy}{\cos 2x+\cos 2iy}=\frac{\sin 2x+i\,\text{sh}\,2y}{\cos 2x+\text{ch}\,2y}. \quad \ldots(29)$$

Example 3. If $\sin(x+iy)=\text{cis}\,\alpha$, prove that $\tan x\tan\alpha=\text{th}\,y$.

We have also $\quad \sin(x-iy)=\text{cis}(-\alpha)$;

$\therefore$ by subtraction, $\quad 2i\sin\alpha=\sin(x+iy)-\sin(x-iy)$

$$=2\cos x\sin iy=2i\cos x\,\text{sh}\,y;$$

$$\therefore\ \sin\alpha=\cos x\,\text{sh}\,y.$$

Similarly, by addition, $\cos\alpha=\sin x\,\text{ch}\,y$.

Dividing, we have $\quad \tan\alpha=\cot x\,\text{th}\,y$.

Note. The expressions for $\sin\alpha$ and $\cos\alpha$ also follow at once from the expansion of $\sin(x+iy)$ in (28), viz. $\cos\alpha+i\sin\alpha=\sin(x+iy)$ $=\sin x\,\text{ch}\,y+i\cos x\,\text{sh}\,y$.

Example 4. Find the sum to infinity of $\dfrac{\sin\theta}{1!}+\dfrac{\sin 3\theta}{3!}+\dfrac{\sin 5\theta}{5!}+\ldots$.

Put $\text{cis}\,\theta$ for z in equation (23).

Then

$$\text{sh}\,[\text{cis}\,\theta]=\frac{\text{cis}\,\theta}{1!}+\frac{\text{cis}\,3\theta}{3!}+\frac{\text{cis}\,5\theta}{5!}+\ldots.$$

Similarly $\text{sh}\,[\text{cis}(-\theta)]=\dfrac{\text{cis}(-\theta)}{1!}+\dfrac{\text{cis}(-3\theta)}{3!}+\dfrac{\text{cis}(-5\theta)}{5!}+\ldots$.

Now $\text{cis}(n\theta)-\text{cis}(-n\theta)=2i\sin n\theta$; $\therefore$ by subtraction we get

$$2i\left\{\frac{\sin\theta}{1!}+\frac{\sin 3\theta}{3!}+\frac{\sin 5\theta}{5!}+\ldots\right\}=\text{sh}\,[\text{cis}\,\theta]-\text{sh}\,[\text{cis}(-\theta)]$$

$$=2\,\text{ch}\left\{\frac{\text{cis}\,\theta+\text{cis}(-\theta)}{2}\right\}\text{sh}\left\{\frac{\text{cis}\,\theta-\text{cis}(-\theta)}{2}\right\}$$

$$=2\,\text{ch}(\cos\theta)\,\text{sh}(i\sin\theta)=2i\,\text{ch}(\cos\theta)\sin(\sin\theta);$$

$\therefore$ the sum to infinity $=\text{ch}(\cos\theta)\sin(\sin\theta)$.

EXERCISE X. b.

1. Deduce the values of $\cos\dfrac{\pi i}{2}$ and $i\sin\dfrac{\pi i}{2}$ from equations (13), (14).

2. Verify $\sin 2z=2\sin z\cos z$ and $\cos^2 z+\sin^2 z=1$ for the generalised functions, directly from the definitions.

3. Prove that $\text{ch}\,zi=\cos z$ and $\text{sh}\,zi=i\sin z$.

4. Verify $\operatorname{ch}(A+B) = \operatorname{ch} A \operatorname{ch} B + \operatorname{sh} A \operatorname{sh} B$

and $$\operatorname{ch} C - \operatorname{ch} D = 2 \operatorname{sh}\frac{C+D}{2} \operatorname{sh}\frac{C-D}{2},$$

by means of equations (24), (25).

Express the following in the form $a+ib$:

5. $\sin(x-iy)$.

6. $\cos^2(x+iy)$.

7. $\cot(x+iy)$.

8. $\operatorname{ch}(x+iy)$.

9. $\operatorname{th}(x-iy)$.

10. $\operatorname{cosec}(x+iy)$.

11. $\exp\{\sin(x+iy)\}$.

12. $\exp\{\operatorname{sh}(x-iy)\}$.

13. $\operatorname{sh}(x-iy)\cos(y+ix)$.

14. Simplify:

(i) $\operatorname{ch}(x+\frac{1}{2}\pi i)$; (ii) $\operatorname{sh}(x+\frac{1}{2}\pi i)$; (iii) $\operatorname{th}(x+\frac{1}{2}\pi i)$;
(iv) $\operatorname{ch}(x+\pi i)$; (v) $\operatorname{sh}(x+\pi i)$; (vi) $\operatorname{th}(x+\pi i)$.

15. Simplify $\dfrac{\cos(x+yi)}{\cos(x-yi)} + \dfrac{\cos(x-yi)}{\cos(x+yi)}$.

16. If $\sin(x+iy) = u+iv$, prove that

(i) $u^2 \operatorname{cosec}^2 x - v^2 \sec^2 x = 1$; (ii) $u^2 \operatorname{sech}^2 y + v^2 \operatorname{cosech}^2 y = 1$.

17. If $\cos(x+iy) = \cos\theta + i\sin\theta$, prove that $\cos 2x + \operatorname{ch} 2y = 2$.

18. If $\tan(x+iy) = u+iv$, prove that

(i) $(u^2+v^2-1)\tan 2x + 2u = 0$; (ii) $(u^2+v^2+1)\operatorname{th} 2y - 2v = 0$.

19. If $\operatorname{ch}(x+yi)\cos(u+iv) = 1$, and if $\cos y \neq 0$ and $\cos u \neq 0$, prove that $\tan u \operatorname{th} v = \operatorname{th} x \tan y$.

20. If $\sin(x+iy) = \tan(u+iv)$, prove that $\dfrac{\tan x}{\operatorname{th} y} = \dfrac{\sin 2u}{\operatorname{sh} 2v}$.

21. If $x+yi = c\operatorname{ch}(\alpha+i\beta)$, find the locus of the point (x, y), (i) when α is constant, and (ii) when β is constant. Prove that the loci cut orthogonally.

22. If $x+yi = (\alpha+i\beta)^2$, find the loci of (x, y) for α constant and for β constant.

23. If $x+yi = f(\alpha+i\beta)$ show that the loci of (x, y) for α constant and β constant cut orthogonally. [Assume that f denotes a function with differential coefficients, and that $\dfrac{\partial f}{\partial \beta} = i\dfrac{\partial f}{\partial \alpha}$.]

Sum to infinity the following series:

24. $\dfrac{\sin\theta}{1!} - \dfrac{\sin 3\theta}{3!} + \dfrac{\sin 5\theta}{5!} - \dots$.

25. $\dfrac{\cos\theta}{2!} - \dfrac{\cos 2\theta}{4!} + \dfrac{\cos 3\theta}{6!} - \dots$

26. $\dfrac{\sin 2\theta}{2!} + \dfrac{\sin 4\theta}{4!} + \dfrac{\sin 6\theta}{6!} + \dots$.

27. $\sin\theta\sin\theta+\dfrac{\sin^3\theta}{3!}\sin 3\theta+\dfrac{\sin^5\theta}{5!}\sin 5\theta+\dots$.

28. $1+\dfrac{\cos 4\theta}{4!}+\dfrac{\cos 8\theta}{8!}+\dots$.

29. If z is any complex number, prove that $\exp(\cos z)\sin(\sin z)$ is the sum to infinity of $\dfrac{\sin z}{1!}+\dfrac{\sin 2z}{2!}+\dfrac{\sin 3z}{3!}+\dots$.

30. If z is any complex number, express in series of powers of z:

(i) $\cos z \operatorname{ch} z$; (ii) $\sin z \operatorname{ch} z$.

MISCELLANEOUS EXAMPLES

EXERCISE X. c.

Express the following in the form $a+ib$:

1. $\exp\left\{(2n+1)\dfrac{\pi i}{2}\right\}$. 2. $\exp\{(x+iy)^2\}$. 3. $\tan\frac{1}{2}(x+iy)$.

4. $\sec(x+iy)$. 5. $\operatorname{cosec}(x-iy)$. 6. $\operatorname{cosech}(x-iy)$.

7. If $\sin(\alpha+i\beta)=\cos\theta+i\sin\theta$, prove that

$$\sin\theta=\pm\cos^2\alpha=\pm\operatorname{sh}^2\beta.$$

8. If $\tan\frac{1}{2}(x+iy)=u+iv$, prove that

(i) $\dfrac{u}{v}=\dfrac{\sin x}{\operatorname{sh} y}$; (ii) $(1-u^2-v^2)\operatorname{ch} y=(1+u^2+v^2)\cos x$.

9. If $\operatorname{th} x=\sin a \operatorname{sech} b$ and $\tan y=\sec a \operatorname{sh} b$, express $\operatorname{ch}(x+yi)$ in terms of a and b.

10. If $a\exp(\theta i)+b\exp(-3\theta i)=c$, where a, b, c are real numbers, prove that either $a+b=\pm c$ or $(a-b)(a^2-b^2)=bc^2$.

Find the sums to infinity of the following series:

11. $x\operatorname{sh}\alpha+x^2\operatorname{sh}2\alpha+x^3\operatorname{sh}3\alpha+\dots$.

12. $1-x\cos\alpha+x^2\cos(\alpha+\beta)-\dots$.

13. $\sin\alpha+\dfrac{x}{1!}\sin(\alpha+\beta)+\dfrac{x^2}{2!}\sin(\alpha+2\beta)+\dots$.

14. $x\sin\theta+\dfrac{x^2\sin 2\theta}{2!}+\dfrac{x^3\sin 3\theta}{3!}+\dots$.

15. $1+\dfrac{\cos 2\theta}{2!}+\dfrac{\cos 4\theta}{4!}+\dots$. 16. $\dfrac{\sin\theta}{1!}+\dfrac{\sin 3\theta}{3!}+\dfrac{\sin 5\theta}{5!}+\dots$.

17. Prove that $1+\dfrac{x^3}{3!}+\dfrac{x^6}{6!}+\dfrac{x^9}{9!}+\dots=\frac{1}{3}\left\{e^x+2e^{-\frac{1}{2}x}\cos\dfrac{x\sqrt{3}}{2}\right\}$.

18. Prove that

$$\frac{x^3}{3!}-\frac{x^7}{7!}+\frac{x^{11}}{11!}-\dots=\frac{1}{\sqrt{2}}\left\{\operatorname{ch}\frac{x}{\sqrt{2}}\sin\frac{x}{\sqrt{2}}-\operatorname{sh}\frac{x}{\sqrt{2}}\cos\frac{x}{\sqrt{2}}\right\}.$$

19. Expand $e^{ax}\cos bx$ in a series of powers of x.

20. If $Z=X+iY$, $z=x+iy$, $Z=\dfrac{\exp z-1}{\exp z+1}$, and $-\dfrac{\pi}{2}<y<\dfrac{\pi}{2}$, prove that $X^2+Y^2<1$.

---

21. Find the value of $\sin\alpha\sin\beta$ where α and β are the roots of $2x^2-2x\pi+\pi^2=0$.

22. Simplify $\exp\{\exp(\theta i)\}-\exp\{-\exp(-\theta i)\}$.

23. Expand $e^{x\cos\beta}\sin(\alpha+x\sin\beta)$ in a series of powers of x.

Find the sums to infinity of

24. $2\sin\theta+3\sin\theta\sin 2\theta+4\sin^2\theta\sin 3\theta$

$$+5\sin^3\theta\sin 4\theta+\dots,\ \theta\neq k\pi+\frac{\pi}{2}.$$

25. $\cos\alpha+\dfrac{1}{3!}\cos(\alpha+2\beta)+\dfrac{1}{5!}\cos(\alpha+4\beta)+\dots$.

26. $1+\dfrac{e^{\cos\theta}}{1!}\cos(\sin\theta)+\dfrac{e^{2\cos\theta}}{2!}\cos(2\sin\theta)+\dots$.

27. $\sin\theta+\dfrac{\sin 5\theta}{5!}+\dfrac{\sin 9\theta}{9!}+\dots$.

28. $\dfrac{\cos 3\theta}{3!}+\dfrac{\cos 7\theta}{7!}+\dfrac{\cos 11\theta}{11!}+\dots$.

29. Prove that the sum to infinity of

$$x+\frac{x^4}{4!}+\frac{x^7}{7!}+\dots \text{ is } \tfrac{1}{3}e^x+\tfrac{2}{3}e^{-\frac{1}{2}x}\cos\left(\frac{x\sqrt{3}}{2}-\frac{2\pi}{3}\right).$$

30 Prove that the sum to infinity of

$$\frac{x^2}{2!}+\frac{x^5}{5!}+\frac{x^8}{8!}+\dots \text{ is } \tfrac{1}{3}e^x+\tfrac{2}{3}e^{-\frac{1}{2}x}\cos\left(\frac{x\sqrt{3}}{2}+\frac{2\pi}{3}\right).$$

CHAPTER XI.

ROOTS OF EQUATIONS

Equations with Assigned Roots. Many trigonometrical results can be derived from the algebraic properties of symmetrical functions of the roots of an equation. Examples 1-3 illustrate the construction of an equation with assigned roots.

Example 1. Form the equation whose roots are

$$\cos\frac{2\pi}{7},\ \cos\frac{4\pi}{7},\ \cos\frac{6\pi}{7}$$

The equation $\cos 4\theta = \cos 3\theta$ is satisfied by $4\theta = 2m\pi \pm 3\theta$, that is by $\theta = \frac{2n\pi}{7}$, where n is any integer or zero.

Writing $\cos\theta \equiv c$, since $\cos 4\theta = 2\cos^2 2\theta - 1 = 2(2c^2 - 1)^2 - 1$, we have

$$8c^4 - 8c^2 + 1 = 4c^3 - 3c.$$

The roots of this equation are $c = \cos 0,\ \cos\frac{2\pi}{7},\ \cos\frac{4\pi}{7},\ \cos\frac{6\pi}{7}$.

But $\quad 8c^4 - 4c^3 - 8c^2 + 3c + 1 \equiv (c-1)(8c^3 + 4c^2 - 4c - 1)$;

$\therefore\ \cos\frac{2\pi}{7},\ \cos\frac{4\pi}{7},\ \cos\frac{6\pi}{7}$ are the roots of $8c^3 + 4c^2 - 4c - 1 = 0$.

Example 2. Form the equation whose roots are

$$\tan^2\frac{\pi}{7},\ \tan^2\frac{2\pi}{7},\ \tan^2\frac{3\pi}{7}.$$

First Method. Since $\tan^2\frac{\pi}{7} = \dfrac{\sin^2\frac{\pi}{7}}{\cos^2\frac{\pi}{7}} = \dfrac{1-\cos\frac{2\pi}{7}}{1+\cos\frac{2\pi}{7}}$, it follows from Example 1 that the values of x given by $x = \frac{1-c}{1+c}$, where c satisfies $8c^3 + 4c^2 - 4c - 1 = 0$, are $\tan^2\frac{\pi}{7},\ \tan^2\frac{2\pi}{7},\ \tan^2\frac{3\pi}{7}$.

Since $x + xc = 1 - c$; $\therefore\ c(1+x) = 1 - x$; $\therefore\ c = \frac{1-x}{1+x}$.

Substituting for c we have

$$8(1-x)^3+4(1-x)^2(1+x)-4(1-x)(1+x)^2-(1+x)^3=0;$$

$$\therefore\ 8(1-3x+3x^2-x^3)+4(1-x^2)(1-x-1-x)-(1+3x+3x^2+x^3)=0;$$

$$\therefore\ 7-35x+21x^2-x^3=0; \quad \therefore\ x^3-21x^2+35x-7=0.$$

Second Method. The equation $\tan 7\theta=0$ is satisfied by $\theta=\frac{n\pi}{7}$, where n is any integer or zero.

Writing $\tan\theta\equiv t$, and using equation (6), p. 172, we have

$$7t-35t^3+21t^5-t^7=0.$$

The factor t corresponds to $\theta=0$; it follows that $\pm\tan\frac{\pi}{7}$, $\pm\tan\frac{2\pi}{7}$, $\pm\tan\frac{3\pi}{7}$ are the roots of $t^6-21t^4+35t^2-7=0$. Put $x=t^2$; then $\tan^2\frac{\pi}{7}$, $\tan^2\frac{2\pi}{7}$, $\tan^2\frac{3\pi}{7}$ are the roots of $x^3-21x^2+35x-7=0$.

Example 3. Form the equation whose roots are

$$2\sin\frac{2\pi}{7},\ 2\sin\frac{4\pi}{7},\ 2\sin\frac{8\pi}{7}.$$

If, using Example 1, we eliminate c between $x=2\sqrt{(1-c^2)}$ and $8c^3+4c^2-4c-1=0$, we shall obtain an equation whose roots are

$$\pm 2\sin\frac{2\pi}{7},\ \pm 2\sin\frac{4\pi}{7},\ \pm 2\sin\frac{6\pi}{7}.$$

Thus, since $4c^2=4-x^2$, $2c(4-x^2)+4-x^2-4c-1=0$;

$$\therefore\ c(4-2x^2)=x^2-3; \quad \therefore\ (4-x^2)(4-2x^2)^2=4(x^2-3)^2,$$

which reduces to $x^6-7x^4+14x^2-7=0$.

[*Or*, using Ex. IX. e, No. 25,

$$2\sin 7\theta=7(2s)-14(2s)^3+7(2s)^5-(2s)^7,$$

$\therefore$ we see that 0, $\pm 2\sin\frac{2\pi}{7}$, $\pm 2\sin\frac{4\pi}{7}$, $\pm 2\sin\frac{6\pi}{7}$ are the roots of the equation, $7x-14x^3+7x^5-x^7=0$. This leads to the same result as before.]

Now the equation whose roots are

$$\pm 2\sin\frac{2\pi}{7},\ \pm 2\sin\frac{4\pi}{7},\ \pm 2\sin\frac{6\pi}{7}$$

may be written $x^6=7(x^2-1)^2$ or $x^3=\pm\sqrt{7}(x^2-1)$.

But $2\sin\frac{2\pi}{7}>1$ and $2\sin\frac{4\pi}{7}>1$, also $-2\sin\frac{6\pi}{7}$ lies between 0 and -1, so that these three values of x give x^3 and x^2-1 the same sign.

$\therefore\ 2\sin\frac{2\pi}{7},\ 2\sin\frac{4\pi}{7},\ -2\sin\frac{6\pi}{7}\left(\equiv 2\sin\frac{8\pi}{7}\right)$, are the roots of $x^3=+\sqrt{7}\,(x^2-1)$; it should be noted that $-2\sin\frac{2\pi}{7}$, $-2\sin\frac{4\pi}{7}$, $+2\sin\frac{6\pi}{7}$ are the roots of $x^3=-\sqrt{7}\,(x^2-1)$.

Example 4. Evaluate (i) $\sec\frac{2\pi}{7}+\sec\frac{4\pi}{7}+\sec\frac{6\pi}{7}$;

(ii) $\sec^2\frac{2\pi}{7}+\sec^2\frac{4\pi}{7}+\sec^2\frac{6\pi}{7}$.

(i) From Example 1, $\sec\frac{2\pi}{7}+\sec\frac{4\pi}{7}+\sec\frac{6\pi}{7}$ is the sum of the reciprocals of the roots of $8c^3+4c^2-4c-1=0$, that is, the sum of the roots of $y^3+4y^2-4y-8=0$;

$$\therefore\ \sec\frac{2\pi}{7}+\sec\frac{4\pi}{7}+\sec\frac{6\pi}{7}=-4.$$

(ii) Similarly, $\sec^2\frac{2\pi}{7}+\sec^2\frac{4\pi}{7}+\sec^2\frac{6\pi}{7}$ is the sum of the squares of the roots of $y^3+4y^2-4y-8=0$ and is therefore equal to

$$(-4)^2-2(-4)=16+8=24.$$

We can obtain this result also from Example 2.

$$\sec^2\frac{2\pi}{7}+\sec^2\frac{4\pi}{7}+\sec^2\frac{6\pi}{7}=3+\tan^2\frac{2\pi}{7}+\tan^2\frac{4\pi}{7}+\tan^2\frac{6\pi}{7}$$

$$=3+\tan^2\frac{2\pi}{7}+\tan^2\frac{3\pi}{7}+\tan^2\frac{\pi}{7}.$$

But $\tan^2\frac{\pi}{7}+\tan^2\frac{2\pi}{7}+\tan^2\frac{3\pi}{7}$ equals the sum of the roots of $x^3-21x^2+35x-7=0$, namely 21.

Example 5. Evaluate $\sec^4\frac{\pi}{9}+\sec^4\frac{3\pi}{9}+\sec^4\frac{5\pi}{9}+\sec^4\frac{7\pi}{9}$.

Since $\cos 3\theta=\frac{1}{2}$ is satisfied by $\theta=\frac{\pi}{9},\frac{5\pi}{9},\frac{7\pi}{9}$, it follows that $4c^3-3c=\frac{1}{2}$ is satisfied by $c=\cos\frac{\pi}{9}$, $\cos\frac{5\pi}{9}$, $\cos\frac{7\pi}{9}$; and, as these three values are all different, they must be the roots of the cubic in c.

Put $x=\dfrac{1}{c^2}$; then $\sec^2\dfrac{\pi}{9}$, $\sec^2\dfrac{5\pi}{9}$, $\sec^2\dfrac{7\pi}{9}$ are roots of $\dfrac{4}{x\sqrt{x}}-\dfrac{3}{\sqrt{x}}=\tfrac{1}{2}$,

or $2(4-3x)=x\sqrt{x}$, or $4(16-24x+9x^2)=x^3$, or $x^3-36x^2+96x-64=0$;

$$\therefore\ \sec^4\frac{\pi}{9}+\sec^4\frac{5\pi}{9}+\sec^4\frac{7\pi}{9}=36^2-2\,.\,96=1104.$$

But $\sec^4\dfrac{3\pi}{9}=16$; $\therefore$ given expression $=1104+16=1120$.

EXERCISE XI. a.

1. Use the equation $\cos 3\theta=\cos 2\theta$ to prove that

$$\cos 72^\circ=\tfrac{1}{4}(\sqrt{5}-1) \quad \text{and} \quad \cos 36^\circ=\tfrac{1}{4}(\sqrt{5}+1).$$

2. Form the equation whose roots are

$$\cos\frac{\pi}{7},\ \cos\frac{3\pi}{7},\ \text{and}\ \cos\frac{5\pi}{7}.$$

3. Prove that $\sec\dfrac{\pi}{7}\sec\dfrac{3\pi}{7}+\sec\dfrac{3\pi}{7}\sec\dfrac{5\pi}{7}+\sec\dfrac{5\pi}{7}\sec\dfrac{\pi}{7}=-4$.

4. Form the equation whose roots are

$$\sin^2\frac{\pi}{7},\ \sin^2\frac{2\pi}{7},\ \text{and}\ \sin^2\frac{3\pi}{7}.$$

5. Prove that $\tan^2\dfrac{\pi}{14}$, $\tan^2\dfrac{3\pi}{14}$, and $\tan^2\dfrac{5\pi}{14}$ are the roots of

$$7x^3-35x^2+21x-1=0.$$

6. Prove that $\tan\dfrac{\pi}{7}$, $\tan\dfrac{2\pi}{7}$, and $\tan\dfrac{4\pi}{7}$ are the roots of

$$x^3+x^2\sqrt{7}-7x+\sqrt{7}=0.$$

7. Evaluate

(i) $\operatorname{cosec}^2\dfrac{\pi}{7}+\operatorname{cosec}^2\dfrac{2\pi}{7}+\operatorname{cosec}^2\dfrac{4\pi}{7}$ and (ii) $\sec^4\dfrac{\pi}{7}+\sec^4\dfrac{2\pi}{7}+\sec^4\dfrac{4\pi}{7}$.

8. Prove that $\quad\sin\dfrac{3\pi}{7}+\sin\dfrac{5\pi}{7}+\sin\dfrac{8\pi}{7}=\tfrac{1}{2}\sqrt{7}$.

9. Prove that $\quad\sin^2\dfrac{\pi}{14}+\sin^2\dfrac{3\pi}{14}+\sin^2\dfrac{5\pi}{14}=\dfrac{5}{4}$.

10. If $k=\operatorname{cis}\dfrac{2\pi}{7}$, form the quadratic whose roots are $k+k^2+k^4$ and $k^3+k^5+k^6$, and deduce that $\sin\dfrac{2\pi}{7}+\sin\dfrac{4\pi}{7}+\sin\dfrac{8\pi}{7}=\tfrac{1}{2}\sqrt{7}$.

11. Prove that $\cos\frac{2\pi}{9}$, $\cos\frac{4\pi}{9}$, and $\cos\frac{8\pi}{9}$ are the roots of

$$8x^3-6x+1=0.$$

12. Prove that $\quad \sec\frac{\pi}{9}+\sec\frac{5\pi}{9}+\sec\frac{7\pi}{9}=-6.$

13. Show that $x=2\cos\frac{\pi}{9}$ satisfies $x^6-6x^4+9x^2-1=0.$

14. Prove that $\quad \tan^2\frac{\pi}{9}+\tan^2\frac{2\pi}{9}+\tan^2\frac{4\pi}{9}=33.$

15. Prove that $\tan\frac{\pi}{9}$, $\tan\frac{4\pi}{9}$, and $\tan\frac{7\pi}{9}$ are the roots of

$$x(3-x^2)=+\sqrt{3}(1-3x^2).$$

16. Prove that the roots of $x^5+x^4-4x^3-3x^2+3x+1=0$ are the values of $2\cos\frac{2r\pi}{11}$ for $r=1$ to 5.

17. Deduce from No. 16 the value of $\sum_1^5 \cos^2\frac{2r\pi}{11}$.

18. Find the equation whose roots are cos 55°, cos 65°, and cos 175°.

19. Prove that $\quad \sum_1^{12}\operatorname{cosec}^2\frac{r\pi}{13}=56 \quad$ and $\quad \sum_1^{12}\sec^2\frac{r\pi}{13}=168.$

20. Prove that $\quad \cot^2\frac{\pi}{14}+\cot^2\frac{3\pi}{14}+\cot^2\frac{5\pi}{14}=21.$

21. Prove that the roots of $x^4-x^3-4x^2+4x+1=0$ are

$$2\cos\frac{2r\pi}{15}, \text{ where } r=1,\ 2,\ 4,\ 7.$$

22. Prove that the roots of $x^4+4x^3-6x^2-4x+1=0$ are

$$\tan\frac{r\pi}{16}, \text{ where } r=1,\ 5,\ 9,\ 13.$$

23. Prove that $$\sum_1^8 \sec^2\frac{2r\pi}{17}=144.$$

24. Prove that tan 10°, tan 70°, and tan 130° are the roots of

$$x^3\sqrt{3}-3x^2-3x\sqrt{3}+1=0.$$

Equations of Degree n.

Example 6. Prove that $\quad \sum_{r=0}^{n-1}\cot\left(a+\frac{r\pi}{n}\right)=n\cot na.$

We form the equation whose roots are the values of $\cot\left(a+\frac{r\pi}{n}\right)$ for $r=0,\ 1,\ 2,\ \ldots,\ (n-1)$.

Consider the equation $\tan n\theta = \tan na$. This is satisfied by $\theta = a + \frac{r\pi}{n}$, where r is any integer.

But $$\tan n\theta = \frac{\binom{n}{1} t - \binom{n}{3} t^3 + \ldots}{1 - \binom{n}{2} t^2 + \ldots}, \text{ where } t \equiv \tan\theta$$

$$= \frac{\binom{n}{1} x^{n-1} - \binom{n}{3} x^{n-3} + \ldots}{x^n - \binom{n}{2} x^{n-2} + \ldots}, \text{ where } x \equiv \frac{1}{t} \equiv \cot\theta;$$

$$\therefore \quad \frac{\binom{n}{1} x^{n-1} - \binom{n}{3} x^{n-3} + \ldots}{x^n - \binom{n}{2} x^{n-2} + \ldots} = \tan na$$

is satisfied by $\theta = a + \frac{r\pi}{n}$, and therefore, regarded as an equation in x, is satisfied by $x = \cot\left(a + \frac{r\pi}{n}\right)$. But the values of x given by $r = 0, 1, 2, \ldots, (n-1)$ are all different and are therefore the n roots of this equation in x, of degree n.

The equation may be written,

$$x^n \tan na - nx^{n-1} - \ldots = 0;$$

$$\therefore \quad \sum_{r=0}^{n-1} \cot\left(a + \frac{r\pi}{n}\right) = \text{sum of roots} = n \cot na.$$

Example 7. (i) Prove that, if n is odd,

$$\operatorname{cosec}^2 \frac{\pi}{n} + \operatorname{cosec}^2 \frac{2\pi}{n} + \operatorname{cosec}^2 \frac{3\pi}{n} + \ldots + \operatorname{cosec}^2 \frac{\overline{n-1}\,\pi}{2n} = \frac{n^2-1}{6}.$$

(ii) Deduce that the sum to infinity, σ, of $\frac{1}{1^2} + \frac{1}{2^2} + \frac{1}{3^2} + \frac{1}{4^2} + \ldots$ is equal to $\frac{\pi^2}{6}$.

(i) Since n is odd, by Ex. IX. e, No. 25, writing $\sin\theta \equiv s$, we have

$$\sin n\theta = ns - \frac{n(n^2-1)}{6} s^3 + \ldots + (-1)^{\frac{1}{2}(n-1)} 2^{n-1} s^n;$$

$\therefore \quad \pm\sin \frac{r\pi}{n}$, for $r = 0, 1, 2, \ldots, \frac{1}{2}(n-1)$, are the roots of

$$s - \frac{n^2-1}{6} s^3 + \ldots + (-1)^{\frac{1}{2}(n-1)} \frac{1}{n} 2^{n-1} s^n = 0.$$

Removing the factor s which corresponds to $r=0$, and putting $x=\frac{1}{s^2}$, we see that the values of $\operatorname{cosec}^2\frac{r\pi}{n}$, for $r=1$ to $\frac{1}{2}(n-1)$, are the roots of

$$x^{\frac{1}{2}(n-1)}-\frac{n^2-1}{6}x^{\frac{1}{2}(n-3)}+\ldots=0;$$

$$\therefore \sum_{r=1}^{\frac{1}{2}(n-1)}\operatorname{cosec}^2\frac{r\pi}{n}=\text{sum of roots}=\frac{n^2-1}{6}.$$

(ii) $s_r\equiv\frac{1}{1^2}+\frac{1}{2^2}+\frac{1}{3^2}+\ldots+\frac{1}{r^2}<1+\frac{1}{1\,.\,2}+\frac{1}{2\,.\,3}+\ldots+\frac{1}{(r-1)r}$;

$$\therefore s_r<1+\left(1-\frac{1}{2}\right)+\left(\frac{1}{2}-\frac{1}{3}\right)+\ldots+\left(\frac{1}{r-1}-\frac{1}{r}\right)=2-\frac{1}{r}<2.$$

But s_r increases steadily with r. Therefore, since s_r is always less than 2, it follows that, when $r\to\infty$, s_r tends to a definite limit, say σ, and that $\sigma\leqslant 2$.

If $0<\phi<\frac{\pi}{2}$, then $\sin\phi<\phi<\tan\phi$. (*E.T.*, p. 162.)

$$\therefore \frac{1}{\phi^2}<\operatorname{cosec}^2\phi=1+\cot^2\phi<1+\frac{1}{\phi^2};$$

$$\therefore \left\{\frac{n^2}{\pi^2}+\frac{n^2}{2^2\pi^2}+\ldots+\frac{4n^2}{(n-1)^2\pi^2}\right\}<\sum_{1}^{\frac{1}{2}(n-1)}\operatorname{cosec}^2\frac{r\pi}{n}$$

$$<\frac{n-1}{2}+\left\{\frac{n^2}{\pi^2}+\frac{n^2}{2^2\pi^2}+\ldots\frac{4n^2}{(n-1)^2\pi^2}\right\};$$

$$\therefore s_{\frac{1}{2}(n-1)}<\frac{\pi^2}{n^2}\cdot\frac{n^2-1}{6}<\frac{\pi^2}{n^2}\cdot\frac{n-1}{2}+s_{\frac{1}{2}(n-1)};$$

$\therefore$ making $n\to\infty$, $\sigma\leqslant\frac{\pi^2}{6}\leqslant\sigma$; $\therefore\ \sigma=\frac{\pi^2}{6}$. See p. 228.

EXERCISE XI. b.

1. Use the equation $\cos 3x=\cos 3a$ to show that

$$\cos 3a=4\cos a\cos\left(a+\frac{2\pi}{3}\right)\cos\left(a+\frac{4\pi}{3}\right).$$

2. Prove that $\sin a\sin\left(a+\frac{\pi}{3}\right)\sin\left(a+\frac{2\pi}{3}\right)=\frac{1}{4}\sin 3a.$

3. Use the equation $\sin 3x=\sin 3a$ to show that

$$\operatorname{cosec} a+\operatorname{cosec}\left(a+\frac{2\pi}{3}\right)+\operatorname{cosec}\left(a+\frac{4\pi}{3}\right)=3\operatorname{cosec}3a.$$

4. Express $\tan\theta+\tan\left(\theta+\frac{\pi}{3}\right)+\tan\left(\theta+\frac{2\pi}{3}\right)$ in terms of $\tan 3\theta$.

5. Prove that $\sec^2\alpha+\sec^2\left(\alpha+\frac{2\pi}{3}\right)+\sec^2\left(\alpha+\frac{4\pi}{3}\right)=9\sec^2 3\alpha$.

6. Prove that $\sum_0^{n-1}\cot^2\left(\alpha+\frac{r\pi}{n}\right)=n(n\operatorname{cosec}^2 n\alpha-1)$.

7. Prove that, if n is odd, $\sum_0^{n-1}\sec^2\left(\alpha+\frac{r\pi}{n}\right)=n^2\sec^2 n\alpha$.

8. Prove that $\sum_0^{2n-1}\tan\left(\alpha+\frac{r\pi}{2n}\right)=-2n\cot 2n\alpha$.

9. Prove that $\sum_0^{n-1}\tan^2\left(\alpha+\frac{r\pi}{n}\right)=n^2\operatorname{cosec}^2\left(n\alpha+\frac{n\pi}{2}\right)-n$.

10. If n is even and >2, prove that $\sum_1^{\frac{1}{2}n-1}\sec^2\frac{r\pi}{n}=\frac{n^2-4}{6}$.

11. If n is odd and >1, prove that $\sum_1^{\frac{1}{2}(n-1)}\sec^2\frac{r\pi}{n}=\frac{n^2-1}{2}$.

12. If n is odd and >1, prove that the sum of the products two together of $\tan\frac{r\pi}{n}$ for $r=1, 2, \ldots, (n-1)$, is $\frac{1}{2}n(1-n)$.

13. Prove that, if n is odd, $\sum_0^{n-1}\sec\left(\alpha+\frac{2r\pi}{n}\right)=(-1)^{\frac{n-1}{2}}n\sec n\alpha$.

14. Prove that, if n is odd, $\sum_0^{n-1}(-1)^r\operatorname{cosec}\left(\theta+\frac{r\pi}{n}\right)=n\operatorname{cosec} n\theta$.

15. Prove that, if n is even, $\sum_0^{n-1}(-1)^r\cot\left(\theta+\frac{r\pi}{n}\right)=n\operatorname{cosec} n\theta$.

Use the result proved in Example 7 (ii) for Nos. 16-20.

16. Prove that (i) $\frac{1}{2^2}+\frac{1}{4^2}+\frac{1}{6^2}+\ldots=\frac{\pi^2}{24}$;

(ii) $\frac{1}{1^2}+\frac{1}{3^2}+\frac{1}{5^2}+\ldots=\frac{\pi^2}{8}$.

17. Sum to infinity: $\frac{1}{1^2}-\frac{1}{2^2}+\frac{1}{3^2}-\frac{1}{4^2}+\ldots$.

18. Sum to infinity: $\frac{1}{1^2}+\frac{1}{2^2}+\frac{1}{4^2}+\frac{1}{5^2}+\frac{1}{7^2}+\frac{1}{8^2}+\frac{1}{10^2}+\ldots$.

19. Prove that $\frac{1}{1^2 . 2^2}+\frac{1}{2^2 . 3^2}+\frac{1}{3^2 . 4^2}+\ldots=\frac{\pi^2}{3}-3$.

20. Prove that $\frac{1}{1^3 \cdot 2^3}+\frac{1}{2^3 \cdot 3^3}+\frac{1}{3^3 \cdot 4^3}+\ldots=10-\pi^2$.

21. If n is odd, prove that

$$\Sigma \operatorname{cosec}^4 \frac{r\pi}{n}, \text{ for } r=1 \text{ to } \frac{n-1}{2}, \text{ equals } \frac{(n^2-1)(n^2+11)}{90}.$$

Deduce that $$\frac{1}{1^4}+\frac{1}{2^4}+\frac{1}{3^4}+\ldots=\frac{\pi^4}{90}.$$

22. Prove that $\sum_{r=1}^{\infty} \sum_{s=1}^{\infty} \frac{1}{r^2 s^2}=\frac{\pi^4}{120}$, if $r=s$ is excluded.

23. Prove that, if n is odd, $\sum_0^{n-1} \sec^2\left(\theta+\frac{2r\pi}{n}\right)=n^2 \sec^2 n\theta$.

24. Prove that, if n is even, $\sum_0^{n-1} \sec^2\left(\theta+\frac{2r\pi}{n}\right)=\frac{n^2}{1-\cos n\left(\frac{\pi}{2}+\theta\right)}$.

Equations involving more than one Trigonometric Function. It is often convenient to use the phrase "*essentially distinct* roots of a trigonometrical equation" to denote angles, satisfying the equation, which do not differ from one another by a multiple of π. Thus, the equation $\sin \theta=\frac{1}{2}$ has two, and only two, essentially distinct roots, $\frac{\pi}{6}$ and $\frac{5\pi}{6}$; the equation $\tan \theta=\sqrt{3}$ has no root essentially distinct from $\frac{\pi}{3}$.

Example 8. If α, β are two essentially distinct roots of $\sin(\theta+\lambda)=m \sin 2\lambda$, prove that $m=\pm\cos\frac{1}{2}(\alpha-\beta)\operatorname{cosec}(\alpha+\beta)$.

$\sin(\alpha+\lambda)=m\sin 2\lambda=\sin(\beta+\lambda)$.

But $\alpha+\lambda=2r\pi+(\beta+\lambda)$ is excluded by the data;

$$\therefore\ \alpha+\lambda=(2r+1)\pi-(\beta+\lambda) \quad \text{or} \quad \lambda=(2r+1)\frac{\pi}{2}-\tfrac{1}{2}(\alpha+\beta);$$

$$\therefore\ \sin\left[\alpha+(2r+1)\frac{\pi}{2}-\tfrac{1}{2}(\alpha+\beta)\right]=m\sin[(2r+1)\pi-(\alpha+\beta)];$$

$$\therefore\ \pm\cos\tfrac{1}{2}(\alpha-\beta)=m\sin(\alpha+\beta);$$

$$\therefore\ m=\pm\cos\tfrac{1}{2}(\alpha-\beta)\operatorname{cosec}(\alpha+\beta).$$

Example 9. If $\alpha, \beta, \gamma, \delta$ are essentially distinct values of θ which satisfy $a\cos 2\theta+b\sin 2\theta-c\cos\theta-d\sin\theta+e=0$, prove that

$$\text{(i) } \tan\frac{\alpha+\beta+\gamma+\delta}{2}=\frac{b}{a}; \quad \text{(ii) } \Sigma\sin\alpha=\frac{bc-ad}{a^2+b^2}.$$

(i) Put $t \equiv \tan\frac{\theta}{2}$; then $\sin\theta = \frac{2t}{1+t^2}$ and $\cos\theta = \frac{1-t^2}{1+t^2}$;

$$\therefore \sin 2\theta = \frac{4t(1-t^2)}{(1+t^2)^2} \quad \text{and} \quad \cos 2\theta = \frac{1-6t^2+t^4}{(1+t^2)^2};$$

$\therefore$ $\tan\frac{\alpha}{2}, \tan\frac{\beta}{2}, \tan\frac{\gamma}{2}, \tan\frac{\delta}{2}$ are the 4 roots of the equation,

$$a(t^4-6t^2+1)+b(4t-4t^3)-c(1-t^4)-d(2t+2t^3)+e(1+2t^2+t^4)=0$$

or
$$t^4(a+c+e)-t^3(4b+2d)+t^2(2e-6a)-t(2d-4b)+a-c+e=0$$

$$\therefore \Sigma\tan\frac{\alpha}{2} = \frac{4b+2d}{a+c+e}; \text{ etc.};$$

$$\therefore \tan\frac{\alpha+\beta+\gamma+\delta}{2} = \frac{(4b+2d)-(2d-4b)}{(a+c+e)-(2e-6a)+(a-c+e)} = \frac{8b}{8a} = \frac{b}{a}.$$

(ii) The given equation may be written

$$a(1-2\sin^2\theta)-d\sin\theta+e = -2b\sin\theta\cos\theta+c\cos\theta$$

or
$$2a\sin^2\theta+d\sin\theta-(a+e) = \cos\theta(2b\sin\theta-c);$$

$$\therefore [2a\sin^2\theta+d\sin\theta-(a+e)]^2 = (1-\sin^2\theta)(2b\sin\theta-c)^2;$$

$$\therefore \sin^4\theta(4a^2+4b^2)+\sin^3\theta(4ad-4bc)+\ldots=0.$$

This equation is satisfied by $\theta=\alpha, \beta, \gamma, \delta$;

$$\therefore \Sigma\sin\alpha = -\frac{4ad-4bc}{4a^2+4b^2} = \frac{bc-ad}{a^2+b^2}.$$

Example 10. If $\frac{\cos(\alpha+\theta)}{\sin^3\alpha} = \frac{\cos(\beta+\theta)}{\sin^3\beta} = \frac{\cos(\gamma+\theta)}{\sin^3\gamma}$, where no two of the angles α, β, γ differ by a multiple of π, prove that

(i) $\alpha+\beta+\gamma=n\pi$; (ii) $\tan\theta = \cot\alpha+\cot\beta+\cot\gamma$.

Put $\frac{\cos(\alpha+\theta)}{\sin^3\alpha}=k$; then α, β, γ are values of x which satisfy the equation, $\frac{\cos(x+\theta)}{\sin^3 x}=k$.

Now
$$\frac{\cos(x+\theta)}{\sin^3 x} = \frac{\cos x\cos\theta-\sin x\sin\theta}{\sin^3 x} = \frac{\cot x\cos\theta-\sin\theta}{\sin^2 x}$$
$$= (\cot x\cos\theta-\sin\theta)(1+\cot^2 x)$$
$$= \cot^3 x\cos\theta-\cot^2 x\sin\theta+\cot x\cos\theta-\sin\theta;$$

$\therefore$ $\cot \alpha$, $\cot \beta$, $\cot \gamma$ are the roots of the equation

$$y^3 \cos \theta - y^2 \sin \theta + y \cos \theta - (k + \sin \theta) = 0;$$

$$\therefore \Sigma \cot \alpha = \tan \theta \quad \text{and} \quad \Sigma(\cot \alpha \cot \beta) = 1.$$

The latter is equivalent to $\Sigma \tan \alpha = \tan \alpha \tan \beta \tan \gamma$, from which we have $\tan(\alpha + \beta + \gamma) = 0$, so that $\alpha + \beta + \gamma = n\pi$.

EXERCISE XI. c.

1. If $\tan \alpha$, $\tan \beta$, $\tan \gamma$ are the roots of $ax^3 + x^2 + bx + 1 = 0$, prove that $\alpha + \beta + \gamma = n\pi$.

2. If $\tan \alpha$, $\tan \beta$, $\tan \gamma$ are the roots of $x^3 + px^2 + qx + r = 0$, find the condition that $\alpha + \beta + \gamma$ is an odd multiple of $\frac{\pi}{2}$.

3. If $a \cos \alpha + b \sin \alpha = c = a \cos \beta + b \sin \beta$, prove that either

$$\sin \alpha = \sin \beta \quad \text{or} \quad \sin \alpha + \sin \beta = \frac{2bc}{a^2 + b^2}.$$

4. If x and y are essentially distinct values of θ which satisfy the equation $\cos \theta = \cos \alpha \cos \beta \pm \sin \alpha \sin \beta \sqrt{(1 - k^2 \sin^2 \theta)}$,

prove that
$$\cos x + \cos y = \frac{2 \cos \alpha \cos \beta}{1 - k^2 \sin^2 \alpha \sin^2 \beta}.$$

5. If $\tan \alpha$, $\tan \beta$, $\tan \gamma$ are unequal and such that

$$\tan 3\alpha = \tan 3\beta = \tan 3\gamma,$$

prove that
$$\Sigma \cot \alpha \,.\, \Sigma \tan \alpha = 9.$$

6. If α, β, γ are three essentially distinct values of θ which give the same value to $\tan 3\theta - \tan \theta$, prove that their sum is an odd multiple of $\frac{\pi}{2}$, and that the common value of $\tan 3\theta - \tan \theta$ is equal to $2 \tan \alpha \tan \beta \tan \gamma$.

7. If α, β, γ are essentially distinct values of x which satisfy

$$\tan(x - \theta) + \tan(x - \phi) + \tan(x - \psi) = 0,$$

prove that
$$\tan(\alpha + \beta + \gamma) = \tan(\theta + \phi + \psi).$$

8. If α_1, α_2, α_3, α_4 have unequal sines and satisfy the equation

$$a + b \sin 2\theta = \cos \theta + \sin \theta, \text{ prove that } b\Sigma \sin \alpha = 1.$$

9. If α, β, γ, δ are values of x, not differing from one another or from θ by $2n\pi$, which satisfy the equation

$$\sin 2\theta (a \sin x + b \cos x) = \sin 2x (a \sin \theta + b \cos \theta),$$

prove that
$$\tan \frac{\alpha}{2} \tan \frac{\beta}{2} \tan \frac{\gamma}{2} \tan \frac{\delta}{2} = -1.$$

10. If $\theta_1, \theta_2, \theta_3, \theta_4$ are essentially distinct roots of

$$a \sin 4\theta + b \cos 4\theta = c,$$

prove that (i) $\tan\theta_1 \tan\theta_2 \tan\theta_3 \tan\theta_4 = 1$; (ii) $\Sigma \operatorname{cosec} 2\theta = 0$.

11. If $\theta_1, \theta_2, \theta_3, \theta_4$ are essentially distinct roots of

$$a \cos 2(\theta - \alpha) + b \cos(\theta - \beta) + c = 0,$$

prove that $4\alpha - \Sigma\theta$ is a multiple of 2π.

12. If $\alpha_1, \alpha_2, \alpha_3, \alpha_4$ are essentially distinct roots of

$$\sin 2\theta - m \cos\theta - n \sin\theta + r = 0,$$

prove that (i) $\Sigma\alpha = (2n+1)\pi$; (ii) $\Sigma \cos\alpha = n$; (iii) $\Sigma \sin\alpha = m$.

13. Prove that $\cot n\theta = k \cot(\theta + \alpha)$ has $n+1$ solutions for θ, no two of which differ by $r\pi$, and that the sum of their cotangents is $(kn-1)\cot\alpha$.

14. If $\theta_1, \theta_2, \dots \theta_5$ are five essentially distinct roots of

$$a \tan 3\theta + b \tan 2\theta + c \tan\theta + d = 0,$$

prove that

$$(a+b+c)\tan(\Sigma\theta) + d = 0.$$

15. If $\theta_1, \theta_2, \theta_3, \theta_4$ are essentially distinct roots of

$$a \sec\theta + b \operatorname{cosec}\theta = c,$$

prove that (i) $\Sigma \cos\theta = \dfrac{2a}{c}$; (ii) $\Sigma \sin\theta = \dfrac{2b}{c}$; (iii) $\Sigma\theta = (2n+1)\pi$.

Interpret the last result in terms of the eccentric angles of points on an ellipse.

16. If $(a + \cos\theta)\cos(\theta - \gamma) = b$ is satisfied by four values of θ between 0 and 2π, prove that

(i) $\Sigma \cos\theta = -2a$; (ii) $\Sigma \sin\theta = 0$; (iii) $\Sigma\theta = 2\gamma + 2n\pi$.

17. If $a \cos x \cos y + b \sin x \sin y = c$, $a \cos y \cos z + b \sin y \sin z = c$, $a \cos z \cos x + b \sin z \sin x = c$, and no two of x, y, z differ by a multiple of 2π, prove that $bc + ca + ab = 0$.

18. If α, β, γ are essentially distinct angles such that

$$a \cos\beta \cos\gamma + b(\sin\beta + \sin\gamma) + c = 0,$$

$$a \cos\gamma \cos\alpha + b(\sin\gamma + \sin\alpha) + c = 0,$$

$$a \cos\alpha \cos\beta + b(\sin\alpha + \sin\beta) + c = 0,$$

prove that

(i) $b^2 = ac$; (ii) $\Sigma \cos\alpha = \cos(\Sigma\alpha)$; (iii) $\sin(\Sigma\alpha) - \Sigma \sin\alpha = \dfrac{2b}{a}$.

19. If x, y, z are essentially distinct angles such that

$$\frac{\cos x \cos y}{\cos^2\alpha} + \frac{\sin x \sin y}{\sin^2\alpha} = -1 = \frac{\cos y \cos z}{\cos^2\alpha} + \frac{\sin y \sin z}{\sin^2\alpha},$$

prove that

$$\frac{\cos z \cos x}{\cos^2\alpha} + \frac{\sin z \sin x}{\sin^2\alpha} = -1.$$

20. If θ, ϕ are distinct angles such that

$$\sin\theta + \sin\phi = a \quad \text{and} \quad \cos\theta + \cos\phi = b,$$

find the equation whose roots are $\tan\frac{1}{2}\theta$ and $\tan\frac{1}{2}\phi$, and the values of $\cot^3\frac{1}{2}\theta + \cot^3\frac{1}{2}\phi$ and $\sin(\theta+\phi)$ in terms of a and b.

MISCELLANEOUS EXAMPLES

EXERCISE XI. d.

1. Deduce an equation in x from $\sin 5\theta = \sin\frac{\pi}{4}$, where $x = \sin\theta$. What are its roots?

2. Prove that $\sum_{1}^{4} \sin^2\frac{2n\pi}{5} = \frac{5}{2}$.

3. Prove that $2\cos\frac{2\pi}{7}$, $2\cos\frac{4\pi}{7}$, $2\cos\frac{8\pi}{7}$ are the roots of

$$x^3 + x^2 - 2x - 1 = 0.$$

4. Prove that $\cos\frac{\pi}{7} - \cos\frac{2\pi}{7} + \cos\frac{3\pi}{7} = \frac{1}{2}$.

5. Evaluate $\cot^2\frac{\pi}{7} + \cot^2\frac{2\pi}{7} + \cot^2\frac{3\pi}{7}$.

6. Evaluate $\cot^2\frac{\pi}{9} + \cot^2\frac{2\pi}{9} + \cot^2\frac{4\pi}{9}$.

7. Prove that the values of $4\cos^2\frac{r\pi}{9}$ for $r = 1, 2, 3, 4$ are the roots of $x^4 - 7x^3 + 15x^2 - 10x + 1 = 0$.

8. Prove that, for values of r from 1 to 5,

(i) $\Sigma\tan^2\frac{r\pi}{11} = 55$; (ii) $\Sigma\tan^4\frac{r\pi}{11} = 2365$

(iii) $\Sigma\operatorname{cosec}^2\frac{r\pi}{11} = 20$; (iv) $\Sigma\cos^2\frac{r\pi}{11} = \frac{9}{4}$.

9. Prove that $8\cos\frac{\pi}{14}\cos\frac{3\pi}{14}\cos\frac{5\pi}{14} = \sqrt{7}$.

10. Prove that $\prod_{1}^{6} \tan\frac{r\pi}{13} = \sqrt{13}$.

11. Form the quadratic whose roots are

$$\cos\frac{2\pi}{13} + \cos\frac{6\pi}{13} + \cos\frac{8\pi}{13} \quad \text{and} \quad \cos\frac{4\pi}{13} + \cos\frac{10\pi}{13} + \cos\frac{12\pi}{13}.$$

12. If $\alpha = \frac{2\pi}{13}$, prove that

$$8(\cos\alpha + \cos 5\alpha)(\cos 2\alpha + \cos 3\alpha)(\cos 4\alpha + \cos 6\alpha) = -1.$$

13. Prove that, if $\alpha=\dfrac{2\pi}{17}$;

(i) $\cos\alpha+\cos 9\alpha+\cos 13\alpha+\cos 15\alpha=\frac{1}{4}(\sqrt{17}-1)$;

(ii) $\cos 3\alpha+\cos 5\alpha+\cos 7\alpha+\cos 11\alpha=-\frac{1}{4}(\sqrt{17}+1)$.

14. If $k=\operatorname{cis}\dfrac{2\pi}{11}$, form the quadratic whose roots are

$$k+k^3+k^4+k^5+k^9 \quad\text{and}\quad k^2+k^6+k^7+k^8+k^{10}.$$

Deduce that $\sum_1^5 \cos\dfrac{2r\pi}{11}=-\frac{1}{2}$.

15. Prove that

$$\tan^2\alpha+\tan^2\left(\alpha+\frac{\pi}{3}\right)+\tan^2\left(\alpha+\frac{2\pi}{3}\right)=3(3\tan^2 3\alpha+2).$$

16. Prove that $\sum_1^{2n}\sec^2\left(\theta+\dfrac{(r-1)\pi}{2n}\right)=4n^2\operatorname{cosec}^2 2n\theta$.

17. Evaluate $\sum_1^{2n}\tan^4\left(\theta+\dfrac{(r-1)\pi}{2n}\right)$.

18. Show that the product of the different values of $\cos(\frac{1}{2}\sin^{-1}x)$ is $\dfrac{x^2-1}{16}$.

19. If n is odd and $\geqslant 3$, prove that $\sum_1^{n-1}\operatorname{cosec}^2\dfrac{r\pi}{n}=\dfrac{n^2-1}{3}$.

20. Prove that $\sum_1^{n}\sin^4\dfrac{r\pi}{2n}=\dfrac{3n+4}{8}$.

21. Prove that $\sum_1^{n}\operatorname{cosec}^2\dfrac{(2r-1)\pi}{4n}=2n^2$.

22. If n is odd, show that $\Sigma\cot^2\dfrac{r\pi}{n}$, for values of r from 1 to $\frac{1}{2}(n-1)$, is $\frac{1}{6}(n-1)(n-2)$.

23. Prove that $\sum_1^{\frac{1}{2}(n-1)}\cot^4\dfrac{r\pi}{n}=\frac{1}{90}(n-1)(n-2)(n^2+3n-13)$ where n is odd.

24. If s is an integer less than $2n$, prove that $\sum_{r=1}^{2n}\cos\dfrac{rs\pi}{n}=0$.

Deduce that $\sum_{r=1}^{2n}\cos^{2n}\dfrac{r\pi}{n}=\dfrac{n}{2^{2n-2}}\left\{1+\dfrac{(2n)!}{2(n!)^2}\right\}$.

25 Find the sums to infinity of the series whose nth terms are

$$\text{(i)}\ \frac{1}{n^4(n+1)^4};\qquad \text{(ii)}\ \frac{1}{n^5(n+1)^5}.$$

26. Find the sums to infinity of the series whose nth terms are

$$\text{(i) } \frac{1}{n^6}; \qquad \text{(ii) } \frac{1}{n^6(n+1)^6}; \qquad \text{(iii) } \frac{1}{n^7(n+1)^7}.$$

27. If α, β are essentially distinct roots of $a \cos\theta + b \sin\theta = c$, find the value of $\tan 2\alpha + \tan 2\beta$ in terms of a, b, c.

28. If the cosecants of $\theta_1, \theta_2, \dots \theta_6$ are unequal and such that $a \cos 3\theta + b \sin 3\theta = c$, find their sum.

29. If $\theta_1, \theta_2, \theta_3, \theta_4$ have unequal tangents and satisfy

$$\tan(\theta - \alpha) + \sec(\theta - \beta) = \cot(\alpha + \beta),$$

prove that $$\theta_1 + \theta_2 + \theta_3 + \theta_4 = 2n\pi.$$

30. If
$$a \cos(\beta - \gamma) + b(\cos\beta + \cos\gamma) + c = 0,$$
$$a \cos(\gamma - \alpha) + b(\cos\gamma + \cos\alpha) + c = 0,$$
$$a \cos(\alpha - \beta) + b(\cos\alpha + \cos\beta) + c = 0,$$

are satisfied by values of α, β, γ, which do not differ from one another by a multiple of 2π, prove that

$$\text{(i) } a^2 + b^2 = 2ac; \quad \text{(ii) } \Sigma \sin\alpha = 0; \quad \text{(iii) } \Sigma \cos\alpha = -\frac{a}{b}.$$

CHAPTER XII

FACTORS

It is known that, if a polynomial $p_0x^n + p_1x^{n-1} + \dots + p_n$ of degree n is zero for $x = c$, then $x - c$ is a factor of it; also that, if the polynomial is zero for n *different* values $x = c_1$, $x = c_2$, ..., $x = c_n$, it must be identical with $p_0(x - c_1)(x - c_2)\dots(x - c_n)$.

This is also true when $c_1, c_2, \dots, p_0, p_1, p_2, \dots$ are complex numbers, see p. 142, Note 2 (iii).

We proceed to apply this result to factorize various trigonometrical expressions.

Notation for Products. Corresponding to the use of the symbol Σ to represent sums, it is convenient to use Π to represent products. Thus we denote $f(1).f(2).f(3).\dots.f(n)$ by $\prod_{r=1}^{n} f(r)$, or $\prod_{1}^{n} f(r)$, or $\Pi f(r)$ for $r = 1$ to n.

FACTORS OF ALGEBRAIC FUNCTIONS

Factors of $x^n - 1$. $x^n - 1 = 0$ if x is a value of $1^{\frac{1}{n}}$, and the n different values are given by $\cos\dfrac{2r\pi}{n} + i\sin\dfrac{2r\pi}{n}$, for $r = 1, 2, \dots, n$.

Corresponding to each of these values we have a factor

$$x - \left(\cos\frac{2r\pi}{n} + i\sin\frac{2r\pi}{n}\right).$$

Since $r = -k$ and $r = n - k$ give the same value to $\operatorname{cis}\dfrac{2r\pi}{n}$ we may use $0, -1, -2, \dots$ as values of r instead of $n, n-1, n-2, \dots$.

(i) *n even.* We take $r = 0, \pm 1, \pm 2, \dots \pm(\frac{1}{2}n - 1)$, and $\frac{1}{2}n$, thus getting $(\frac{1}{2}n - 1)$ pairs of factors, and the two single factors given by 0 and $\frac{1}{2}n$, i.e. $2(\frac{1}{2}n - 1) + 2, = n$, in all.

The factors corresponding to $r = \pm k$ are

$$x - \left(\cos\frac{2k\pi}{n} + i\sin\frac{2k\pi}{n}\right) \quad \text{and} \quad x - \left(\cos\frac{2k\pi}{n} - i\sin\frac{2k\pi}{n}\right)$$

and their product

$$=\left(x-\cos\frac{2k\pi}{n}\right)^2-\left(i\sin\frac{2k\pi}{n}\right)^2=x^2-2x\cos\frac{2k\pi}{n}+1;$$

the factors corresponding to $r=0$, $r=\frac{n}{2}$ are $x-1$, $x+1$. Thus,

$$\text{if } n \text{ is even, } x^n-1=(x-1)(x+1)\prod_1^{\frac{1}{2}n-1}\left(x^2-2x\cos\frac{2k\pi}{n}+1\right). \quad \ldots\ldots(1)$$

(ii) *n odd.* Here we take $r=0, \pm 1, \pm 2, \ldots, \pm\frac{n-1}{2}$, which gives $1+2\left(\frac{n-1}{2}\right)$, $=n$, factors. Thus,

$$\text{if } n \text{ is odd, } x^n-1=(x-1)\prod_1^{\frac{1}{2}(n-1)}\left(x^2-2x\cos\frac{2k\pi}{n}+1\right). \quad \ldots\ldots\ldots\ldots(2)$$

Factors of x^n+1. $x^n+1=0$ if

$$x=(-1)^{\frac{1}{n}}=\cos\frac{(2r-1)\pi}{n}+i\sin\frac{(2r-1)\pi}{n}.$$

If n is even, we take $2r-1=\pm 1, \pm 3, \ldots, \pm(n-1)$; and if n is odd, we take $2r-1=\pm 1, \pm 3, \ldots, \pm(n-2)$, and n; this last value gives the factor $x-\cos\pi-i\sin\pi=x+1$. Thus,

$$\text{if } n \text{ is even, } x^n+1=\prod_1^{\frac{1}{2}n}\left(x^2-2x\cos\frac{(2r-1)\pi}{n}+1\right), \quad \ldots\ldots\ldots\ldots\ldots\ldots(3)$$

$$\text{if } n \text{ is odd, } x^n+1=(x+1)\prod_1^{\frac{1}{2}(n-1)}\left(x^2-2x\cos\frac{(2r-1)\pi}{n}+1\right). \quad \ldots\ldots(4)$$

It should be remarked that, although the work of this chapter is in complex algebra, the formulae (1) (2) (3) (4) *are true results of real algebra.* If the products on the right were multiplied out we know that they would come to $x^n \pm 1$ because we have proved the equality in complex algebra. The results could be obtained, although not so shortly, by using only the methods of real algebra, see Ex. XII. a, Nos. 13, 14.

EXERCISE XII. a.

1. In complex algebra what are the factors of x^3-1?

2. In real algebra what are the factors of x^3-1?

3. Obtain from first principles the factors of x^4+1, and deduce quadratic factors not involving i. Show that these can also be found by writing x^4+1 as the difference between two squares.

4. Find the complex factors of $x^5 - 1$; deduce the real quadratic factors of $x^4 + x^3 + x^2 + x + 1$. Verify the result by writing $x^4 + x^3 + x^2 + x + 1 = x^2\left(x^2 + x + 1 + \frac{1}{x} + \frac{1}{x^2}\right)$, and putting $x + \frac{1}{x} = y$.

5. Obtain from first principles the quadratic factors of

(i) $x^6 + 1$; (ii) $x^8 + y^8$;

(iii) $x^6 - a^6$; (iv) $x^8 - 256$.

6. Express $x^{10} - x^5 + 1$ in quadratic factors.

7. Find the values of $\cos\theta$ for which

(i) $\cos n\theta = 0$; (ii) $\cos n\theta = 1$; (iii) $\cos n\theta = -1$;

(iv) $\cos n\theta = \cos n\alpha \neq \pm 1$; (v) $\dfrac{\sin n\theta}{\sin\theta} = 0$.

8. If n is even, find the values of $\sin\theta$ for which

(i) $\cos n\theta = 0$; (ii) $\dfrac{\sin n\theta}{\cos\theta\sin\theta} = 0$.

9. If n is odd, find the values of $\sin\theta$ for which

(i) $\sin n\theta = 0$; (ii) $\dfrac{\cos n\theta}{\cos\theta} = 0$.

10. Solve $x^{2n} - 2x^n \cos n\alpha + 1 = 0$.

11. Write down the factors of $x^{2n} + 1$, and deduce those of $(1 + x)^{2n} + (1 - x)^{2n}$.

12. Show that the solutions of $(1 + x)^{2n} + (1 - x)^{2n} = 0$ are

$$x = \pm i \tan\frac{(2r-1)\pi}{4n}, \text{ for } r = 1, 2, \dots, n.$$

Deduce that $(1 + x)^{2n} + (1 - x)^{2n} = 2\prod_1^n \left(x^2 + \tan^2\frac{(2r-1)\pi}{4n}\right)$.

Hence prove that $\sum_1^n \sec^2\frac{(2r-1)\pi}{4n} = 2n^2$.

13. If $u_n \equiv x^{2n} - 2x^n a^n \cos n\theta + a^{2n}$, prove that

$$u_{n+1} \equiv 2xa\cos\theta\, u_n - a^2x^2 u_{n-1} + (x^{2n} + a^{2n})\, u_1.$$

Hence prove by induction that $x^2 - 2ax\cos\theta + a^2$ is a factor of $x^{2n} - 2x^n a^n \cos n\theta + a^{2n}$, and deduce that $x^2 - 2xa\cos\left(\theta + \frac{2r\pi}{n}\right) + a^2$ is also a factor, where r is any integer.

14. Use No. 13 to factorize, by the methods of real algebra, $x^n - 1$, when n is even. [Put $a = 1$, $\theta = 0$.]

FACTORS OF TRIGONOMETRIC FUNCTIONS

We have shown, in Chapter IX., that certain functions, like $\cos n\theta$, $\sin n\theta$, are polynomials in $\cos\theta$ or $\sin\theta$. The results are given in Ex. IX. e, Nos. 12-25. Corresponding to each of these results it is possible, by the method stated at the beginning of the present chapter, to obtain an expression in factors for each of these functions. An essential step in the process is to find the values of $\cos\theta$ or $\sin\theta$ for which the function vanishes, and the reader who has worked Ex. XII. a, Nos. 7, 8, 9 will already have found the values.

We will give the reasoning in full for one example, and the reader will then be able to supply it for the others. The results are given in Ex. XII. b, Nos. 3-13.

Factors of $\sin n\theta$ when n is odd. It is known from Chapter IX., that, *when n is odd*, $\sin n\theta$ is a polynomial,

$$n\sin\theta - \dots + (-1)^{\frac{n-1}{2}} 2^{n-1}\sin^n\theta,$$

of degree n in $\sin\theta$. To find the values of $\sin\theta$ for which the polynomial is zero we put $\sin n\theta = 0$; this gives $\theta = \frac{r\pi}{n}$, and the n different values of $\sin\theta$ are $\sin\frac{r\pi}{n}$ for $r = 0, \pm 1, \pm 2, \dots \pm\frac{n-1}{2}$.

The factors corresponding to $r=0$, $r=\pm k$ are $\sin\theta$, $\sin\theta - \sin\frac{k\pi}{n}$, and $\sin\theta + \sin\frac{k\pi}{n}$, and the product of the last two is $\sin^2\theta - \sin^2\frac{k\pi}{n}$; thus,

$$\sin n\theta = \mathsf{A}\sin\theta \prod_{1}^{\frac{1}{2}(n-1)} \left(\sin^2\theta - \sin^2\frac{r\pi}{n}\right), \qquad \dots\dots\dots\dots(5)$$

where A = the coefficient of $\sin^n\theta$ in the polynomial $= (-1)^{\frac{n-1}{2}} 2^{n-1}$.

It is convenient, however, to divide each factor by a term $-\sin^2\frac{r\pi}{n}$; thus,

$$\sin n\theta = \mathsf{B}\sin\theta \prod_{1}^{\frac{1}{2}(n-1)} \left(1 - \frac{\sin^2\theta}{\sin^2\frac{r\pi}{n}}\right), \qquad \dots\dots\dots\dots(6)$$

where $\mathsf{B} = \mathsf{A}\Pi\left(-\sin^2\frac{r\pi}{n}\right) = 2^{n-1}\Pi\sin^2\frac{r\pi}{n}$. It is not, however, necessary to use the value of A; dividing each side of (6) by $\sin\theta$ and making $\theta\to 0$, we get $\mathsf{B} = \lim_{\theta\to 0}\frac{\sin n\theta}{\sin\theta} = n$; thus,

$$\sin n\theta = n\sin\theta \prod_{1}^{\frac{1}{2}(n-1)} \left(1 - \frac{\sin^2\theta}{\sin^2\frac{r\pi}{n}}\right), \qquad \dots\dots\dots\dots(7)$$

and it has been proved incidentally that

$$n = 2^{n-1}\prod_{1}^{\frac{1}{2}(n-1)} \sin^2\frac{r\pi}{n}. \qquad \dots\dots\dots\dots(8)$$

From formula (8), $\sqrt{n} = \pm 2^{\frac{1}{2}(n-1)} \prod_{1}^{\frac{1}{2}(n-1)} \sin \frac{r\pi}{n}$, but all the angles have positive sines;

$$\therefore \text{ if } n \text{ is odd,} \quad \sqrt{n} = 2^{\frac{1}{2}(n-1)} \prod_{1}^{\frac{1}{2}(n-1)} \sin \frac{r\pi}{n}. \quad \text{.........(9)}$$

Other results of this kind can be deduced from the factors of $\frac{\sin n\theta}{\cos \theta}$, when n is even, and from the factors of $\cos n\theta$. See Ex. XII. b, Nos. 18-25.

Another deduction is often made from equation (7) by putting $n\theta = \phi$ and making $n \to \infty$, but, as explained in the Preface, the consideration of Infinite Products is held over for the companion volume. It is found that (see Ex. XII. f, Nos. 23, 24)

$$\frac{\sin \varphi}{\varphi} = \lim_{n\to\infty} \prod_{k=1}^{k=n} \left(1 - \frac{\varphi^2}{k^2\pi^2}\right) \equiv \prod_{1}^{\infty} \left(1 - \frac{\varphi^2}{k^2\pi^2}\right),$$

$$\cos \varphi = \prod_{1}^{\infty} \left(1 - \frac{4\varphi^2}{(2k-1)^2\pi^2}\right),$$

and, by putting $\phi = \frac{\pi}{2}$ in the first,

$$\sqrt{(\tfrac{1}{2}\pi)} = \lim_{n\to\infty} \frac{2.4.6 \ldots 2n}{1.3.5 \ldots (2n-1)} \cdot \frac{1}{\sqrt{2n+1}}.$$

EXERCISE XII. b.

[For convenience of reference some results proved in the text are included.]

1. Obtain from first principles the factors of

$$\text{(i) } \sin 5\theta; \quad \text{(ii) } \frac{\sin 6\theta}{\cos \theta}; \quad \text{(iii) } \sin 5\theta - \sin 5\alpha;$$

regarded as functions of $\sin \theta$.

2. Obtain from first principles the factors of

$$\text{(i) } \cos 5\theta; \quad \text{(ii) } \cos 6\theta; \quad \text{(iii) } \cos 6\theta - \cos 6\alpha;$$

regarded as functions of $\cos \theta$.

Verify some of the following results (Nos. 3-13):

3. $\cos n\theta - \cos n\alpha = 2^{n-1} \prod_{1}^{n} \left(\cos \theta - \cos \frac{n\alpha + 2r\pi}{n}\right)$.

4. $\cos n\theta = 2^{n-1} \prod_{1}^{n} \left(\cos \theta - \cos \frac{(2r-1)\pi}{2n}\right)$.

5. $\cos(2n+1)\theta = 2^{2n}\cos\theta \prod_1^n \left(\cos^2\theta - \cos^2\frac{(2r-1)\pi}{2(2n+1)}\right)$

$$= 2^{2n}\cos\theta \prod_1^n \left(\sin^2\frac{(2r-1)\pi}{2(2n+1)} - \sin^2\theta\right).$$

6. $\cos 2n\theta = 2^{2n-1} \prod_1^n \left(\cos^2\theta - \cos^2\frac{(2r-1)\pi}{4n}\right)$

$$= 2^{2n-1} \prod_1^n \left(\sin^2\frac{(2r-1)\pi}{4n} - \sin^2\theta\right).$$

7. $\sin n\theta = 2^{n-1}\sin\theta \prod_1^{n-1} \left(\cos\theta - \cos\frac{r\pi}{n}\right).$

8. $\sin(2n+1)\theta = 2^{2n}\sin\theta \prod_1^n \left(\cos^2\theta - \cos^2\frac{r\pi}{2n+1}\right)$

$$= 2^{2n}\sin\theta \prod_1^n \left(\sin^2\frac{r\pi}{2n+1} - \sin^2\theta\right).$$

9. $\sin 2n\theta = 2^{2n-2}\sin 2\theta \prod_1^{n-1} \left(\cos^2\theta - \cos^2\frac{r\pi}{2n}\right)$

$$= 2^{2n-2}\sin 2\theta \prod_1^{n-1} \left(\sin^2\frac{r\pi}{2n} - \sin^2\theta\right).$$

10. If n is odd, $\dfrac{\cos n\theta}{\cos\theta} = \displaystyle\prod_1^{\frac{1}{2}(n-1)} \left(1 - \frac{\sin^2\theta}{\sin^2\frac{(2r-1)\pi}{2n}}\right).$

11. If n is even, $\cos n\theta = \displaystyle\prod_1^{\frac{1}{2}n} \left(1 - \frac{\sin^2\theta}{\sin^2\frac{(2r-1)\pi}{2n}}\right).$

12. If n is odd, $\dfrac{\sin n\theta}{n\sin\theta} = \displaystyle\prod_1^{\frac{1}{2}(n-1)} \left(1 - \frac{\sin^2\theta}{\sin^2\frac{r\pi}{n}}\right).$

13. If n is even, $\dfrac{\sin n\theta}{n\sin\theta\cos\theta} = \displaystyle\prod_1^{\frac{1}{2}n-1} \left(1 - \frac{\sin^2\theta}{\sin^2\frac{r\pi}{n}}\right).$

14. Express $\dfrac{\cos 3\theta - \sin 4\theta}{\cos\theta}$ as a product of three factors.

15. Find the factors, if any, of $\sin n\theta - \sin n\alpha$, regarded as a function of $\sin\theta$.

16. Show how to deduce No. 4 from No. 3.

17. By writing $\frac{\pi}{2}-\theta$ for θ in Nos. 10, 11, 12, 13, find the values of

$$\text{(i)}\ \prod_{1}^{[\frac{1}{2}n]}\left(1-\frac{\cos^2\theta}{\sin^2\frac{(2r-1)\pi}{2n}}\right);\quad \text{(ii)}\ \prod_{1}^{[\frac{1}{2}(n-1)]}\left(1-\frac{\cos^2\theta}{\sin^2\frac{r\pi}{n}}\right),$$

where n may be odd or even. See footnote on p. 46.

18. Prove that $\sqrt{n}=2^{\frac{1}{2}(n-1)}\prod\limits_{1}^{[\frac{1}{2}(n-1)]}\sin\frac{r\pi}{n}$.

19. Prove that $1=2^{\frac{1}{2}(n-1)}\prod\limits_{1}^{[\frac{1}{2}n]}\sin\frac{(2r-1)\pi}{2n}$.

20. Prove that $1=2^{n-\frac{1}{2}}\prod\limits_{1}^{n}\sin\frac{(2r-1)\pi}{4n}$.

21. Prove that $\sqrt{n}=2^{n-1}\prod\limits_{1}^{n-1}\sin\frac{r\pi}{2n}$.

22. If n is even, prove that $\prod\limits_{1}^{\frac{1}{2}n-1}\sin\frac{r\pi}{n}=\prod\limits_{1}^{\frac{1}{2}n-1}\cos\frac{r\pi}{n}$, and that

$$\prod_{1}^{\frac{1}{2}n}\sin\frac{(2r-1)\pi}{2n}=\prod_{1}^{\frac{1}{2}n}\cos\frac{(2r-1)\pi}{2n}.$$

23. If n is odd, prove that $\prod\limits_{1}^{\frac{1}{2}(n-1)}\sin\frac{r\pi}{n}=\prod\limits_{1}^{\frac{1}{2}(n-1)}\cos\frac{(2r-1)\pi}{2n}$,

and that
$$\prod_{1}^{\frac{1}{2}(n-1)}\sin\frac{(2r-1)\pi}{2n}=\prod_{1}^{\frac{1}{2}(n-1)}\cos\frac{r\pi}{n}.$$

24. Evaluate (i) $\prod\limits_{1}^{[\frac{1}{2}(n-1)]}\cos\frac{r\pi}{n}$; (ii) $\prod\limits_{1}^{[\frac{1}{2}n]}\cos\frac{(2r-1)\pi}{2n}$;

each for n odd, and for n even.

25. Prove that the results of Nos. 20 and 21 hold when the sines are replaced by cosines.

26. Prove that $8\sin\frac{\pi}{14}\sin\frac{3\pi}{14}\sin\frac{5\pi}{14}=1$.

27. Prove that $32\cos\frac{\pi}{11}\cos\frac{2\pi}{11}\cos\frac{3\pi}{11}\cos\frac{4\pi}{11}\cos\frac{5\pi}{11}=1$.

28. Prove that $8\sin\frac{\pi}{7}\sin\frac{2\pi}{7}\sin\frac{3\pi}{7}=\sqrt{7}$.

29. Prove that $\prod\limits_{1}^{n}\cos\frac{(2r-1)\pi}{2n}=2^{1-n}\cos\frac{n\pi}{2}$.

30. Evaluate $\prod\limits_{1}^{n}\tan\frac{(2r-1)\pi}{4n}$.

31. Prove that $\prod_1^n \cos\frac{(4r-1)\pi}{4n}=(-1)^{[\frac{1}{2}(n+1)]}\,2^{\frac{1}{2}-n}$.

32. Prove that $1+\cos\theta$ is a factor of $1+\cos 5\theta$ and find the other factors.

33. Prove that $2\cos\theta+1$ is a factor of $2\cos 5\theta+1$, and find the other factors. Deduce that

$$\sec^2\frac{\pi}{15}+\sec^2\frac{2\pi}{15}+\sec^2\frac{4\pi}{15}+\sec^2\frac{8\pi}{15}=96.$$

34. Factorize $x^{2n}-2x^n\cos n\theta+1$, using Ex. XII. a, No. 10.

35. Prove that $\operatorname{ch} nx-\cos n\alpha=2^{n-1}\prod_0^{n-1}\left\{\operatorname{ch} x-\cos\frac{n\alpha+2r\pi}{n}\right\}$.

36. Prove that $\dfrac{\cos n\theta-\cos n\alpha}{1-\cos n\alpha}=\prod_1^n\left\{1-\dfrac{\sin^2\frac{\theta}{2}}{\sin^2\left(\frac{r\pi}{n}+\frac{\alpha}{2}\right)}\right\}$.

Factors of $x^{2n}-2x^n\cos n\alpha+1$. The equation

$$x^{2n}-2x^n\cos n\alpha+1=0$$

is a quadratic in x^n, with roots $\cos n\alpha\pm i\sin n\alpha$; thus the $2n$ values of x are $\cos\left(\alpha+\frac{2r\pi}{n}\right)\pm i\sin\left(\alpha+\frac{2r\pi}{n}\right)$, for $r=0, 1, 2, \dots, (n-1)$. The product of the factors corresponding to $r=k$ is

$$\left\{x-\cos\left(\alpha+\frac{2k\pi}{n}\right)-i\sin\left(\alpha+\frac{2k\pi}{n}\right)\right\}\times$$

$$\left\{x-\cos\left(\alpha+\frac{2k\pi}{n}\right)+i\sin\left(\alpha+\frac{2k\pi}{n}\right)\right\}=x^2-2x\cos\left(\alpha+\frac{2k\pi}{n}\right)+1,\text{ thus}$$

$$\mathbf{x^{2n}-2x^n\cos n\alpha+1=\prod_0^{n-1}\left\{x^2-2x\cos\left(\alpha+\frac{2k\pi}{n}\right)+1\right\}}. \quad \dots\text{(10)}$$

Many results can be deduced from formula (10).

(i) Putting $\alpha=0$, $\alpha=\frac{\pi}{n}$ and taking the square root, we get equations (1), (2), (3), (4). The reader should verify this.

(ii) Dividing by x^n,

$$x^n+\frac{1}{x^n}-2\cos n\alpha=\prod_0^{n-1}\left\{x+\frac{1}{x}-2\cos\left(\alpha+\frac{2r\pi}{n}\right)\right\},$$

and, putting $x = \cos\theta + i\sin\theta$, we have

$$2\cos n\theta - 2\cos n\alpha = \prod_0^{n-1}\left\{2\cos\theta - 2\cos\left(\alpha + \frac{2r\pi}{n}\right)\right\},$$

i.e. $$\cos n\theta - \cos n\alpha = 2^{n-1}\prod_0^{n-1}\left\{\cos\theta - \cos\left(\alpha + \frac{2r\pi}{n}\right)\right\}. \quad \text{....(11)}$$

This is the same as Ex. XII. b, No. 3, and can, of course, be proved directly in the usual manner.

(iii) Putting $x = 1$, $\alpha = 2\beta$ in (10),

$$2(1 - \cos 2n\beta) = \prod_0^{n-1} 2\left\{1 - \cos\left(2\beta + \frac{2r\pi}{n}\right)\right\};$$

$$\therefore\ \sin^2 n\beta = 2^{2n-2}\prod_0^{n-1}\sin^2\left(\beta + \frac{r\pi}{n}\right);$$

$$\therefore\ \sin n\beta = \pm 2^{n-1}\prod_0^{n-1}\sin\left(\beta + \frac{r\pi}{n}\right).$$

Now, if $0 < \beta < \frac{\pi}{n}$, each factor on the right is positive, and so is $\sin n\beta$. Also, as β increases, $\sin n\beta$ changes sign whenever β passes through a value $\frac{r\pi}{n}$, and at the same time *one* factor on the right changes sign. Thus the ambiguous sign is always a $+$, and

$$\sin n\beta = +2^{n-1}\prod_0^{n-1}\sin\left(\beta + \frac{r\pi}{n}\right). \quad \text{...............(12)}$$

Similar results to (12) can be found by the substitutions indicated in Ex. XII. c, Nos. 1-5.

(iv) From (12), by taking logarithms

$$\log\sin n\beta = (n-1)\log 2 + \sum_0^{n-1}\log\sin\left(\beta + \frac{r\pi}{n}\right)$$

and, differentiating with respect to β,

$$n\cot n\beta = \sum_0^{n-1}\cot\left(\beta + \frac{r\pi}{n}\right). \quad \text{..................(13)}$$

This may also be proved by the methods of Chapter XI.

(v) **De Moivre's and Cotes' Properties of a Circle.** If $A_0A_1A_2 \ldots A_{n-1}$ is a regular polygon inscribed in a circle centre O, radius a, and P is a point such that $OP = x$, $\angle(OA_0, OP) = \theta$, then

$$\angle(OA_r, OP) = \theta + \frac{2r\pi}{n} \quad \text{and} \quad PA_r^2 = x^2 + a^2 - 2xa\cos\left(\theta + \frac{2r\pi}{n}\right);$$

$\therefore$ by formula (10),

$$PA_0^2 \,.\, PA_1^2 \ldots PA_{n-1}^2 = \prod_0^{n-1} \left\{ x^2 - 2xa \cos\left(\theta + \frac{2r\pi}{n}\right) + a^2 \right\},$$

or $$\mathbf{PA_0 \,.\, PA_1 \ldots PA_{n-1} = (x^{2n} - 2x^n a^n \cos n\theta + a^{2n})^{\frac{1}{2}}.}$$

This is called *de Moivre's* property.

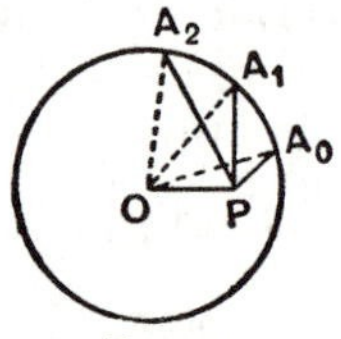

FIG. 78.

If P lies on OA_0 so that $\theta = 0$, $\mathbf{PA_0 \,.\, PA_1 \ldots PA_{n-1} = x^n \sim a^n}$.

If OP bisects $\angle A_{n-1}OA_0$, so that $\theta = \frac{\pi}{n}$, $\mathbf{PA_0 \,.\, PA_1 \ldots PA_{n-1} = x^n + a^n}$.

These special results are called *Cotes'* properties.

Comparison of Series and Products. A number of identities can be obtained by comparing values obtained for the same function as a sum and as a product.

For example, if n is odd, $\sin n\theta = n \sin\theta \prod_1^{\frac{1}{2}(n-1)} \left(1 - \dfrac{\sin^2\theta}{\sin^2 \frac{r\pi}{n}}\right)$ and also, from Ex. IX. e, No. 25,

$\sin n\theta$

$$= ns - \frac{n(n^2 - 1^2)}{3!} s^3 + \frac{n(n^2 - 1^2)(n^2 - 3^2)}{5!} s^5 - \ldots + (-1)^{\frac{n-1}{2}} 2^{n-1} s^n ;$$

$$\therefore\ ns \prod_1^{\frac{1}{2}(n-1)} \left(1 - \frac{s^2}{\sin^2 \frac{r\pi}{n}}\right) = ns - \frac{n(n^2-1)}{3!} s^3 + \frac{n(n^2-1)(n^2-9)}{5!} s^5 - \ldots$$

is an identity.

Equating coefficients of s^3 on the two sides,

$$-n \sum_1^{\frac{1}{2}(n-1)} \operatorname{cosec}^2 \frac{r\pi}{n} = -\frac{n(n^2-1)}{3!}, \quad \text{i.e.} \quad \sum_1^{\frac{1}{2}(n-1)} \operatorname{cosec}^2 \frac{r\pi}{n} = \frac{n^2-1}{6},$$

a result which has been proved (p. 209) in another way.

Other results can be obtained by equating other powers of s, and by using the formulae for $\sin n\theta$ (n even) and $\cos n\theta$.

The formula for $\Sigma \operatorname{cosec}^2(r\pi/n)$ was used on p. 210 to prove that $\Sigma 1/n^2 = \pi^2/6$. In some of the older text-books this result is obtained by equating coefficients of θ^3 in the series and product expressions for $\sin\theta$, (see pp. 80 and 223). Such a process requires careful justification, and the product expression is itself obtained, not without difficulty, from equation (7). It seemed more satisfactory to deduce the sum $\Sigma 1/n^2$ directly from $\Sigma \operatorname{cosec}^2(r\pi/n)$, the infinite product being left for the companion volume.

EXERCISE XII. c.

By substituting ± 1 for x, and 2β or $2\beta + \frac{\pi}{n}$ for α, in formula (10), prove the following results (Nos. 1-5).

1. $\cos n\beta = 2^{n-1} \prod_0^{n-1} \sin\left(\beta + \frac{(2r+1)\pi}{2n}\right)$.

2. $\cos n\beta = (-1)^{\frac{1}{2}(n-1)}\, 2^{n-1} \prod_0^{n-1} \cos\left(\beta + \frac{r\pi}{n}\right)$, if n is odd.

3. $\sin n\beta = (-1)^{\frac{1}{2}n}\, 2^{n-1} \prod_0^{n-1} \cos\left(\beta + \frac{r\pi}{n}\right)$, if n is even.

4. $\sin n\beta = 2^{n-1}(-1)^{\frac{1}{2}(n+1)} \prod_0^{n-1} \cos\left(\beta + \frac{(2r+1)\pi}{2n}\right)$, if n is odd.

5. $\cos n\beta = 2^{n-1}(-1)^{\frac{1}{2}n} \prod_0^{n-1} \cos\left(\beta + \frac{(2r+1)\pi}{2n}\right)$, if n is even.

6. Show how to deduce Nos. 4 and 5 from No. 1, and Nos. 2 and 3 from formula (12).

7. What result can be deduced from No. 1, by taking logarithms of each side and then differentiating w.r.t. β?

8. Simplify (i) $4 \sin\theta \sin\left(\theta + \frac{\pi}{3}\right) \sin\left(\theta + \frac{2\pi}{3}\right)$;

(ii) $\sin\theta \sin\left(\theta + \frac{\pi}{4}\right) \sin\left(\theta + \frac{\pi}{2}\right) \sin\left(\theta + \frac{3\pi}{4}\right)$.

9. Simplify $(\cos\theta - \cos\alpha)\left(\cos\theta - \cos\overline{\alpha + \frac{2\pi}{3}}\right)\left(\cos\theta - \cos\overline{\alpha + \frac{4\pi}{3}}\right)$.

10. Prove that $\prod_0^{2n-1} \tan\left(\phi + \frac{r\pi}{2n}\right) = \cos n\pi$.

11. Prove that $\prod_0^{n-1} \sin 2\left(\phi + \frac{r\pi}{n}\right) = 2^{2-n} \sin n\phi \sin n\left(\frac{\pi}{2} + \phi\right)$.

12. Prove that $\sum_0^{n-1} \cot\left(\phi + \frac{(2r+1)\pi}{2n}\right) = -n \tan n\phi$.

13. Prove that $\sum_0^{n-1} \operatorname{cosec}^2\left(\phi + \frac{r\pi}{n}\right) = n^2 \operatorname{cosec}^2 n\phi$.

14. $A_1A_2 \dots A_n$ is a regular polygon inscribed in a circle, centre O, and radius a. P is a point on the circumference such that $\angle POA_1 = \theta$; prove that $PA_1 . PA_2 \dots PA_n = 2a^n \sin \frac{n\theta}{2}$.

15. A square $A_1A_2A_3A_4$ and a regular pentagon $B_1B_2B_3B_4B_5$ are inscribed in a circle of radius a; prove that the continued product of the chords A_rB_s is numerically equal to $2a^{20} \sin 20\theta$, where 2θ is the angle subtended at the centre by any one of the chords.

16. $A_1A_2 \dots A_{2n}$ is a regular polygon inscribed in a circle centre O, radius a, and P is a point on the circumference such that $\angle POA_1 = \theta$; prove that the continued product of the perpendiculars from P to OA_1, OA_2, ... OA_n is $2^{1-n}a^n \sin n\theta$.

17. $A_0A_1A_2 \dots A_{2n}$ is a regular polygon inscribed in a circle of radius a of which A_0C is a diameter; prove that $CA_1 . CA_2 \dots CA_n = a^n$.

18. With the data of No. 14, show that the continued product of the chords A_rA_s is $n^{\frac{1}{2}n}a^{\frac{1}{2}n(n-1)}$.

19. With the data of No. 16, if OB bisects A_1A_2, prove that the product of the perpendiculars from A_1, A_2, ... , A_{2n} to OB is $a^{2n}2^{2-2n}$.

20. If n is even, prove that $\sum_{1}^{\frac{1}{2}n-1} \operatorname{cosec}^2 \frac{r\pi}{n} = \frac{n^2-4}{6}$.

21. If n is even, prove that $\sum_{1}^{\frac{1}{2}n} \operatorname{cosec}^2 \frac{(2r-1)\pi}{2n} = \frac{n^2}{2}$.

22. If n is odd, prove that $\sum_{1}^{\frac{1}{2}(n-1)} \operatorname{cosec}^2 \frac{(2r-1)\pi}{2n} = \frac{n^2-1}{2}$.

23. If n is odd, prove that the sum of the squares of the products two at a time of the cosecants of

$$\frac{\pi}{n}, \frac{2\pi}{n}, \frac{3\pi}{n}, \dots, \frac{(n-1)\pi}{2n} \quad \text{is} \quad \frac{(n^2-1)(n^2-9)}{120},$$

and deduce that $\sum_{1}^{\frac{1}{2}(n-1)} \operatorname{cosec}^4 \frac{r\pi}{n} = \frac{(n^2-1)(n^2+11)}{90}$.

24. Evaluate $\sum_{1}^{\frac{1}{2}n-1} \operatorname{cosec}^4 \frac{r\pi}{n}$, when n is even.

25. Evaluate $\Sigma \operatorname{cosec}^4 \frac{(2r-1)\pi}{2n}$, for $r=1$ to $\frac{1}{2}n$, when n is even, and for $r=1$ to $\frac{1}{2}(n-1)$, when n is odd.

PARTIAL FRACTIONS

If $f(x)$ and $\mathsf{F}(x)$ are polynomials in x of which $f(x)$ has the smaller degree, and if $\mathsf{F}(x) \equiv (x-a)g(x)$, where $g(a) \neq 0$, then it is known from Algebra that

$$\frac{f(x)}{\mathsf{F}(x)} \equiv \frac{f(x)}{(x-a)g(x)} \equiv \frac{\mathsf{A}}{x-a} + \frac{h(x)}{g(x)},$$

where A is independent of x, and $h(x)$ is a polynomial of smaller degree than $g(x)$.

The value of A may be proved to be $\dfrac{f(a)}{g(a)}$.

Alternatively, $\mathsf{A} = \lim\limits_{x \to a} \dfrac{(x-a)f(x)}{\mathsf{F}(x)}$. This limit can then be evaluated by algebraic methods; or, by using a theorem in the Calculus, we have

$$\mathsf{A} = \lim_{x \to a} \frac{\dfrac{d}{dx}\{(x-a)f(x)\}}{\dfrac{d}{dx}\mathsf{F}(x)} = \frac{f(a)}{\dfrac{d}{da}\mathsf{F}(a)}.$$

If $\mathsf{F}(x) \equiv c(x-a_1)(x-a_2)\ldots(x-a_n)$, where no two factors are equal, repeated applications of the above give

$$\frac{f(x)}{\mathsf{F}(x)} \equiv \sum_{1}^{n} \frac{\mathsf{A}_r}{x-a_r},$$

where A_r can be found by either of the methods described above.

These results are also true when $a_1, a_2, \ldots$ are complex.

Example 1. Express $\dfrac{1}{x^3-1}$ as the sum of three partial fractions.

Since $x^3 - 1 \equiv (x-1)(x-\omega)(x-\omega^2)$, where $\omega = \text{cis}\,\dfrac{2\pi}{3}$, we may write

$$\frac{1}{x^3-1} \equiv \frac{1}{(x-1)(x-\omega)(x-\omega^2)} \equiv \frac{\mathsf{A}}{x-1} + \frac{\mathsf{B}}{x-\omega} + \frac{\mathsf{C}}{x-\omega^2}.$$

A, B, C may be evaluated in any of the following ways :

(i) $\mathsf{A} = \dfrac{1}{(1-\omega)(1-\omega^2)}$, $\mathsf{B} = \dfrac{1}{(\omega-1)(\omega-\omega^2)}$, $\mathsf{C} = \dfrac{1}{(\omega^2-1)(\omega^2-\omega)}$.

It may be shown that these give $\mathsf{A} = \dfrac{1}{3}$, $\mathsf{B} = \dfrac{\omega}{3}$, $\mathsf{C} = \dfrac{\omega^2}{3}$.

(ii) $\mathsf{A}=\lim_{x\to 1}\frac{x-1}{x^3-1}=\lim_{x\to 1}\frac{1}{x^2+x+1}=\frac{1}{3},$

$$\mathsf{B}=\lim_{x\to\omega}\frac{x-\omega}{x^3-1}=\lim_{x\to\omega}\frac{x-\omega}{x^3-\omega^3}=\lim_{x\to\omega}\frac{1}{x^2+\omega x+\omega^2}=\frac{1}{3\omega^2}=\frac{\omega}{3},$$

$$\mathsf{C}=\lim_{x\to\omega^2}\frac{x-\omega^2}{x^3-1}=\lim_{x\to\omega^2}\frac{x-\omega^2}{x^3-\omega^6}=\lim_{x\to\omega^2}\frac{1}{x^2+\omega^2 x+\omega^4}=\frac{1}{3\omega^4}=\frac{\omega^2}{3}.$$

(iii) $\mathsf{A}=\lim_{x\to 1}\frac{x-1}{x^3-1}=\lim_{x\to 1}\frac{\frac{d}{dx}(x-1)}{\frac{d}{dx}(x^3-1)}=\lim_{x\to 1}\frac{1}{3x^2}=\frac{1}{3}.$

Similarly, $\mathsf{B}=\lim_{x\to\omega}\frac{x-\omega}{x^3-1}=\lim_{x\to\omega}\frac{1}{3x^2}=\frac{1}{3\omega^2}=\frac{\omega}{3}$

and $\mathsf{C}=\lim_{x\to\omega^2}\frac{x-\omega^2}{x^3-1}=\lim_{x\to\omega^2}\frac{1}{3x^2}=\frac{1}{3\omega^4}=\frac{\omega^2}{3};$

$$\therefore\ \frac{1}{x^3-1}\equiv\frac{\frac{1}{3}}{x-1}+\frac{\frac{1}{3}\omega}{x-\omega}+\frac{\frac{1}{3}\omega^2}{x-\omega^2}.$$

Note. The last two fractions can of course be combined, so as to give the ordinary result in real algebra,

$$\frac{1}{x^3-1}\equiv\frac{\frac{1}{3}}{x-1}-\frac{\frac{1}{3}(x+2)}{x^2+x+1}.$$

Example 2. Express $\frac{2nx}{(1+x)^{2n}-(1-x)^{2n}}$ as the sum of fractions with quadratic denominators.

The denominator is a polynomial of degree $2n-1$ in x, which is zero if $\frac{1+x}{1-x}=1^{\frac{1}{2n}}=\text{cis}\,\frac{2r\pi}{2n}=\text{cis}\,2a$, where $a=\frac{r\pi}{2n}$.

This gives $x=\frac{\text{cis}\,2a-1}{\text{cis}\,2a+1}=\frac{-1+\cos 2a+i\sin 2a}{1+\cos 2a+i\sin 2a}$

$$=\frac{-2\sin^2 a+2i\sin a\cos a}{2\cos^2 a+2i\sin a\cos a}=\frac{i\sin a\,\text{cis}\,a}{\cos a\,\text{cis}\,a}=i\tan a;$$

$$\therefore\ (1+x)^{2n}-(1-x)^{2n}=cx\prod_{1}^{n-1}\left\{\left(x-i\tan\frac{r\pi}{2n}\right)\left(x+i\tan\frac{r\pi}{2n}\right)\right\};$$

$$\therefore\ \frac{2nx}{(1+x)^{2n}-(1-x)^{2n}}=\sum_{1}^{n-1}\left\{\frac{\mathsf{A}_r}{x-i\tan a}+\frac{\mathsf{B}_r}{x+i\tan a}\right\},$$

where $$\mathsf{A}_r = \lim_{x\to i\tan\alpha} \frac{2nx(x - i\tan\alpha)}{(1+x)^{2n} - (1-x)^{2n}}$$

$$= \lim_{x\to i\tan\alpha} \frac{2nx + 2n(x - i\tan\alpha)}{2n(1+x)^{2n-1} + 2n(1-x)^{2n-1}}$$

$$= \frac{i\tan\alpha}{(1+i\tan\alpha)^{2n-1} + (1-i\tan\alpha)^{2n-1}}$$

$$= \frac{i\tan\alpha\cos^{2n-1}\alpha}{(\text{cis }\alpha)^{2n-1} + (\text{cis } -\alpha)^{2n-1}}$$

$$= \frac{(-1)^r i\sin\alpha\cos^{2n-2}\alpha}{(\text{cis }\alpha)^{-1} + (\text{cis } -\alpha)^{-1}},$$

because $(\text{cis }\alpha)^{2n} = \text{cis }(r\pi) = (-1)^r$;
hence $\mathsf{A}_r = (-1)^r \,.\, \frac{1}{2}i\sin\alpha\cos^{2n-3}\alpha$,

Writing $-i$ for i, we have $\mathsf{B}_r = -(-1)^r \,.\, \frac{1}{2}i\sin\alpha\cos^{2n-3}\alpha$;

∴ the expression

$$= \sum\left\{(-1)^r \,.\, \tfrac{1}{2}\sin\alpha\cos^{2n-3}\alpha\left(\frac{i}{x - i\tan\alpha} - \frac{i}{x + i\tan\alpha}\right)\right\}$$

$$= \sum \frac{(-1)^{r+1}\sin\alpha\cos^{2n-3}\alpha\tan\alpha}{x^2 + \tan^2\alpha}$$

$$= \sum_1^{n-1} \frac{(-1)^{r+1}\sin^2\alpha\cos^{2n-4}\alpha}{x^2 + \tan^2\alpha}, \text{ where } \alpha = \frac{r\pi}{2n}.$$

Example 3. Express $\dfrac{n\tan n\theta}{\sin\theta}$ as the sum of fractions.

$$\frac{n\tan n\theta}{\sin\theta} = \frac{\dfrac{n\sin n\theta}{\sin\theta}}{\cos n\theta} = \frac{\text{a polynomial of degree } n-1 \text{ in } \cos\theta}{\text{a polynomial of degree } n \text{ in } \cos\theta}.$$

The denominator is zero when

$$\cos\theta = \cos\frac{(2r-1)\pi}{2n}, \text{ for } r = 1, 2, 3, \ldots, n.$$

$$\therefore \frac{n\tan n\theta}{\sin\theta} = \sum_1^n \frac{\mathsf{A}_r}{\cos\theta - \cos\dfrac{(2r-1)\pi}{2n}},$$

where $$\mathsf{A}_r = \lim_{\theta\to\alpha} \frac{n\tan n\theta(\cos\theta - \cos\alpha)}{\sin\theta} \text{ and } \alpha = \frac{(2r-1)\pi}{2n}.$$

The value of A_r may be found in either of the following ways:

$$\text{(i) } A_r = \lim_{\theta\to\alpha} \frac{n \sin n\theta}{\sin\theta} \cdot \frac{\cos\theta - \cos\alpha}{\cos n\theta} = \frac{n \sin n\alpha}{\sin\alpha} \lim_{\theta\to\alpha} \frac{\cos\theta - \cos\alpha}{\cos n\theta}$$

$$= \frac{n \sin n\alpha}{\sin\alpha} \lim_{\theta\to\alpha} \frac{-\sin\theta}{-n\sin n\theta} = \frac{n\sin n\alpha}{\sin\alpha} \cdot \frac{\sin\alpha}{n \sin n\alpha} = \mathbf{1}.$$

$$\text{(ii) } A_r = \lim_{\theta\to\alpha} \frac{n \tan n\theta\,(\cos\theta - \cos\alpha)}{\sin\theta}$$

$$= \lim_{\theta\to\alpha} \frac{2n \sin\dfrac{\alpha+\theta}{2} \sin\dfrac{\alpha-\theta}{2}}{\sin\theta \tan(n\alpha - n\theta)}, \text{ since } n\alpha - n\theta = (2r-1)\frac{\pi}{2} - n\theta$$

$$= \lim_{\phi\to 0} \frac{2n \sin\dfrac{\phi}{2}}{\tan n\phi} = \mathbf{1}.$$

$$\therefore \frac{n \tan n\theta}{\sin\theta} = \sum_1^n \frac{1}{\cos\theta - \cos\dfrac{(2r-1)\pi}{2n}}.$$

Note. This result may also be obtained as follows:

From Ex. XII. b, No. 4, $\cos n\theta \equiv 2^{n-1} \prod_1^n \left[\cos\theta - \cos\dfrac{(2r-1)\pi}{2n}\right]$;

$$\therefore \log\cos n\theta \equiv (n-1)\log 2 + \sum_1^n \log\left[\cos\theta - \cos\frac{(2r-1)\pi}{2n}\right].$$

Now differentiate each side w.r.t. θ.

EXERCISE XII. d.

1. Find the real partial fractions of $\dfrac{1}{x^5 - 1}$.

2. Express $\dfrac{1}{x^8+1}$ in partial fractions with quadratic denominators.

3. Prove that $\dfrac{8\sqrt{5}}{(1+x)^5 + (1-x)^5} = \dfrac{1}{x^2 + \tan^2\dfrac{\pi}{10}} - \dfrac{1}{x^2 + \tan^2\dfrac{3\pi}{10}}$.

4. Express $\dfrac{x}{(1+x)^7 - (1-x)^7}$ in partial fractions with quadratic denominators.

5. Express $\dfrac{\cos^2\theta}{\cos n\theta}$ in partial fractions when $n > 3$.

6. If n is odd, prove that

$$\frac{n}{\sin n\theta} = \frac{1}{\sin\theta} + \sum_1^{\frac{1}{2}(n-1)} \left\{\frac{(-1)^r}{\sin\left(\theta + \dfrac{r\pi}{n}\right)} + \frac{(-1)^r}{\sin\left(\theta - \dfrac{r\pi}{n}\right)}\right\}.$$

7. Prove that $\dfrac{n\sin(n-1)\theta}{\sin\theta\cos n\theta}=\sum_{1}^{n}\dfrac{\cos\dfrac{(2r-1)\pi}{2n}}{\cos\theta-\cos\dfrac{(2r-1)\pi}{2n}}$.

8. Prove that $\dfrac{n\sin n\alpha}{\cos n\theta-\cos n\alpha}=\sum_{0}^{n-1}\dfrac{\sin\left(\alpha+\dfrac{2r\pi}{n}\right)}{\cos\theta-\cos\left(\alpha+\dfrac{2r\pi}{n}\right)}$.

9. Prove that $\dfrac{n\sin n\theta\operatorname{cosec}\theta}{\cos n\theta-\cos n\alpha}=\sum_{0}^{n-1}\dfrac{1}{\cos\theta-\cos\left(\alpha+\dfrac{2r\pi}{n}\right)}$.

10. Prove that $\sum_{1}^{n}\dfrac{1}{1-\cos\dfrac{(2r-1)\pi}{2n}}=n^2$.

11. Prove that $\sum_{1}^{n}\dfrac{1}{1-2\cos\dfrac{(2r-1)\pi}{2n}}$ equals 0 or $n(-1)^{[\frac{2}{3}n]}$,

according as n is or is not a multiple of 3.

12. Express $\dfrac{x^{n-1}}{x^{2n}-2x^n\cos n\theta+1}$ as a sum of n partial fractions.

13. Prove that $\sum_{1}^{n-1}\dfrac{x-\cos\dfrac{r\pi}{n}}{x^2-2x\cos\dfrac{r\pi}{n}+1}=\dfrac{nx^{2n-1}}{x^{2n}-1}-\dfrac{x}{x^2-1}$,

and deduce an expression for $\sum_{1}^{n-1}\dfrac{\cos\dfrac{r\pi}{n}}{\cos\dfrac{r\pi}{n}-\cos\theta}$.

14. Prove that

$$\frac{nx^{n-1}(x^n-a^n\cos n\theta)}{x^{2n}-2x^na^n\cos n\theta+a^{2n}}=\sum_{0}^{n-1}\frac{x-a\cos\left(\theta+\dfrac{2r\pi}{n}\right)}{x^2-2xa\cos\left(\theta+\dfrac{2r\pi}{n}\right)+a^2}.$$

15. If n is even and $\alpha=\dfrac{\pi}{2n}$, prove that

$$\frac{\tan n\theta}{\tan\theta}\text{ is the sum, for } r=1, 2, \dots, \tfrac{1}{2}n-1, \text{ of}$$

$$\left\{\frac{2}{n}\operatorname{cosec}(4r+2)\alpha\left(\frac{1}{\tan\theta+\tan(2r+1)\alpha}-\frac{1}{\tan\theta-\tan(2r+1)\alpha}\right)\right\}.$$

16. Prove that $\dfrac{\sin 5\theta}{\sin 7\theta}=\frac{1}{7}\sum_{1}^{6}(-1)^r\sin\dfrac{5r\pi}{7}\cot\left(\theta-\dfrac{r\pi}{7}\right)$.

17. Prove that $\dfrac{\sin 4\theta}{\sin 7\theta}=\frac{1}{7}\sum_{1}^{6}(-1)^r\sin\dfrac{4r\pi}{7}\operatorname{cosec}\left(\theta-\dfrac{r\pi}{7}\right)$.

18. Prove that $\dfrac{\sin 5\theta}{\sin 8\theta}=\frac{1}{8}\sum_{1}^{7}(-1)^r\sin\dfrac{5r\pi}{8}\operatorname{cosec}\left(\theta-\dfrac{r\pi}{8}\right)$.

19. Prove that

$$\frac{\sin x}{\sin(x-a)\sin(x-b)}=\frac{\sin a}{\sin(x-a)\sin(a-b)}+\frac{\sin b}{\sin(x-b)\sin(b-a)}.$$

and express $\dfrac{\sin^{n-1}x}{\sin(x-a_1)\sin(x-a_2)\ldots\sin(x-a_n)}$ in a similar form.

20. Prove that

$$\frac{\sin x\cos x}{\sin(x-a)\sin(x-b)\sin(x-c)}=\Sigma\frac{\sin a\cos a}{\sin(x-a)\sin(a-b)\sin(a-c)}.$$

21. Prove that

$$\frac{\sin x}{\sin(x-a)\sin(x-b)\sin(x-c)}=\Sigma\frac{\sin a\cos(x-a)}{\sin(x-a)\sin(a-b)\sin(a-c)}.$$

22. $A_1A_2\ldots A_n$ is a regular polygon inscribed in a circle, centre O, radius a; P is a point of its plane such that $OP=x$, and $\angle POA_1=\theta$ prove that

$$\sum_{1}^{n}\frac{1}{PA_r^2}=\frac{n(x^{2n}-a^{2n})}{(x^2-a^2)(x^{2n}-2x^na^n\cos n\theta+a^{2n})}.$$

EASY MISCELLANEOUS EXAMPLES

EXERCISE XII. e.

1. Express $x^{2n}-a^{2n}$ as the product of n quadratic factors.

2. Express x^6-x^3+1 in the form $\Pi(x-a)$ and prove that

$$\Sigma(x-a)^2=6x^2.$$

3. Prove that

$$x^3-3x+1=\left(x-2\cos\frac{2\pi}{9}\right)\left(x-2\cos\frac{4\pi}{9}\right)\left(x-2\cos\frac{8\pi}{9}\right).$$

4. Prove that

$$(1+x)^{2n+1}-(1-x)^{2n+1}=2x\prod_{1}^{n}\left(x^2+\tan^2\frac{r\pi}{2n+1}\right)$$

and deduce that $\displaystyle\prod_{1}^{n}\tan^2\frac{r\pi}{2n+1}=2n+1.$

5. Express $(x+1)^{2n}-(x-1)^{2n}$ in the form

$$4nx\prod_{1}^{n-1}\left(x^2+\cot^2\frac{r\pi}{2n}\right)$$

and deduce the value of $\displaystyle\prod_{1}^{n-1}\left(4+\cot^2\frac{r\pi}{2n}\right)$.

6. Express $\frac{\cos 5\theta - \sin 2\theta}{\cos\theta}$ as a product of factors linear in $\sin\theta$.

7. Prove that

$$\cos 7x - 8\cos^7 x = 7\cos x\cos 2x\left(\cos 2x - 2\cos\frac{\pi}{5}\right)\left(\cos 2x - 2\cos\frac{3\pi}{5}\right).$$

Prove the following (Nos. 8-16):

8. $\cos\theta - \cos\phi$ is a factor of $\frac{\sin n\theta}{\sin\theta} - \frac{\sin n\phi}{\sin\phi}$.

9. $\prod_{1}^{2n-1}\cos\frac{r\pi}{n} = 2^{1-2n}\{(-1)^n - 1\}$.

10. $\sum_{0}^{n-1}\frac{1}{1-\cos\left(\phi + \frac{2r\pi}{n}\right)} = \frac{1}{2}n^2\operatorname{cosec}^2\frac{n\phi}{2}$.

11. $\frac{n}{1+x^{2n}} = \sum_{1}^{n}\frac{1 - x\cos\frac{(2r-1)\pi}{2n}}{1 - 2x\cos\frac{(2r-1)\pi}{2n} + x^2}$.

12. $\frac{(2n+1)\sin(2n+1)\theta}{\cos(2n+1)\theta - \cos\alpha} = \sum_{-n}^{+n}\frac{\sin\theta}{\cos\theta - \cos\frac{\alpha + 2r\pi}{2n+1}}$.

13. $\frac{\tan n\theta}{\tan\theta} = \frac{1}{n} + \frac{2}{n}\sum_{0}^{\frac{1}{2}(n-3)}\frac{\sec^2\alpha_r}{\tan^2\alpha_r - \tan^2\theta}$,

where $\alpha_r = \frac{(2r+1)\pi}{2n}$, and n is odd.

14. $(2n+1)\operatorname{cosec}(2n+1)\theta - \operatorname{cosec}\theta$

$$= \sum_{1}^{n}(-1)^r\left\{\operatorname{cosec}\left(\theta + \frac{r\pi}{2n+1}\right) + \operatorname{cosec}\left(\theta - \frac{r\pi}{2n+1}\right)\right\}.$$

15. $\frac{\sin m\theta}{\sin n\theta} = \frac{1}{n}\sum_{1}^{n-1}(-1)^r\sin\frac{mr\pi}{n}\cot\left(\theta - \frac{r\pi}{n}\right)$,

where m and n are odd, and $0 < m < n$.

16. $\frac{\sin m\theta}{\sin n\theta} = \frac{1}{n}\sum_{1}^{n-1}(-1)^r\sin\frac{mr\pi}{n}\operatorname{cosec}\left(\theta - \frac{r\pi}{n}\right)$,

where $m+n$ is odd, and $0 < m < n$.

17. Express $\operatorname{cosec}(x-a)\operatorname{cosec}(x-b)\operatorname{cosec}(x-c)$ in the form $A\operatorname{cosec}(x-a) + B\operatorname{cosec}(x-b) + C\operatorname{cosec}(x-c)$, where A, B, C are independent of x; also extend to the case of $2n+1$ factors.

18. Express $\operatorname{cosec}(x-a)\operatorname{cosec}(x-b)$ in the form

$$A\cot(x-a) + B\cot(x-b),$$

where A, B are independent of x; also extend to the case of $2n$ factors.

19. Express $\dfrac{\cos^3 x}{\sin(x-a)\sin(x-b)\sin(x-c)}$ in the form

$$A\cot(x-a)+B\cot(x-b)+C\cot(x-c)+D,$$

where A, B, C, D are independent of x.

20. $A_1A_2 \ldots A_n$ is a regular polygon inscribed in a circle, centre O, radius a. PQ, QR are equal chords, such that $\angle POQ=\alpha$ and $\angle A_1OQ=\beta$. Prove that the product of the perpendiculars from $A_1, A_2, \ldots, A_n$ to the chord PR is $2^{1-n}a^n(\cos n\alpha-\cos n\beta)$.

HARDER MISCELLANEOUS EXAMPLES

EXERCISE XII. f.

1. Prove that $\cos 7\theta-\cos 8\theta$ is divisible by $2\cos 5\theta+1$ and find the other factor.

2. Prove that

$$\frac{(1+\sin 15\theta)(1-\sin\theta)}{(1-\sin 5\theta)(1+\sin 3\theta)}=(16\sin^4\theta-8\sin^3\theta-16\sin^2\theta+8\sin\theta+1)^2,$$

and factorize the expression on the right-hand side.

3. If m and n are odd co-prime integers, prove that $\dfrac{\sin mn\theta\sin\theta}{\sin m\theta\sin n\theta}$ is a polynomial in $\cos\theta$ of degree $mn-m-n+1$, with factors of the form $\cos\theta-\cos\dfrac{r\pi}{mn}$, for values of r from 1 to $mn-1$ which are not multiples of m or n.

4. Prove that $$\text{ch}^2 nx=\text{ch}^{2n}x\prod_1^n\left\{1+\text{th}^2x\cot^2(2r-1)\frac{\pi}{2n}\right\}.$$

5. Write down the results obtained by taking the square root of each side in No. 4, (i) if n is even, (ii) if n is odd.

6. If n is even, prove that

$$\text{sh}\,nx=n\,\text{sh}\,x\,\text{ch}^{n-1}x\prod_1^{\frac{1}{2}n-1}\left(1+\text{th}^2x\tan^2\frac{r\pi}{n}\right),$$

and find a corresponding result when n is odd.

Prove the following (Nos. 7-14):

7. $1-2\cos n\theta$

$$=(1-2\cos\theta)\left(1-2\cos\frac{n\theta+2\pi}{n}\right)\ldots\left(1-2\cos\frac{n\theta+2n\pi-2\pi}{n}\right),$$

if $n^2\equiv 1 \pmod 6$.

8. $$\sin\frac{n}{2}(\pi+2\phi)\sin n\phi=2^{n-2}\prod_0^{n-1}\sin 2\left(\phi+\frac{r\pi}{n}\right).$$

9. $\cos n\phi + \sin n\phi = 2^{n-\frac{1}{2}} \prod_0^{n-1} \sin\left\{\phi + (4r+1)\frac{\pi}{4n}\right\}.$

10. $\cos \alpha = 2^{n-1} \prod_0^{n-1} \sin \dfrac{2\alpha + (2r+1)\pi}{2n}.$

11. $\sum_0^{n-1} \cot \dfrac{2\alpha + (2r+1)\pi}{2n} = -n \tan \alpha.$

12. $\prod_0^{n-1} \sin^2\left(\theta + \dfrac{r\pi}{n}\right) + \prod_0^{n-1} \cos^2\left(\theta + \dfrac{r\pi}{n}\right) = 4^{1-n}$, if n is odd.

13. $\dfrac{(\cos\theta + \sin\theta)^{2n+1} - (\cos\theta - \sin\theta)^{2n+1}}{2^{n+1}\sin\theta} = \prod_1^n \left(1 - \cos\dfrac{2r\pi}{2n+1}\cos 2\theta\right).$

14. $\prod_1^n \dfrac{\cos 2\phi - \cos\dfrac{2r\pi}{n}}{\cos 2\phi - \cos(2r+1)\dfrac{\pi}{n}} = -\tan^2 n\phi.$

15. Prove that the product of the perpendiculars drawn from the vertices of a regular n-gon inscribed in the circle $x^2 + y^2 = 9a^2$ to a tangent to the circle $x^2 + y^2 = 25a^2$ is

$$\left(\frac{a}{2}\right)^n \{3^{2n} + 1 + 2\,.\,3^n \cos n\alpha\},$$

where α is one of the angles between the tangent and a side of the polygon.

16. A regular polygon $A_1A_2 \ldots A_n$ is inscribed in a circle, centre O; P is a point in space such that the projection of OP on the plane of the circle makes an angle α with OA_1; r_1 and r_2 are the greatest and least distances of P from the circumference, and $2s = r_1 + r_2$, $2d = r_1 - r_2$. Prove that $PA_1 \,.\, PA_2 \ldots PA_n = \sqrt{(s^{2n} - 2s^n d^n \cos n\alpha + d^{2n})}$.

17. Prove that $n \tan n\theta = \sum_1^n \dfrac{\sin\theta}{\cos\theta - \cos\dfrac{(2r-1)\pi}{2n}}.$

18. Prove that $\sum_1^n \cot^2 \dfrac{r\pi}{2n+1} = \frac{1}{3}n(2n-1).$

19. Prove that $2^{n-1} \sin\dfrac{\pi}{n} \sin\dfrac{2\pi}{n} \ldots \sin\dfrac{(n-1)\pi}{n} = n$, and deduce that

$$\int_0^\pi \log \sin x\, dx = -\pi \log 2.$$

20. If $N_1, N_2, \ldots, N_m$ are the integers less than N and prime to it and $\alpha = \dfrac{\pi}{N}$, prove that $2^m \sin N_1\alpha \sin N_2\alpha \ldots \sin N_m\alpha = 1$, unless N is itself prime, in which case the product equals N.

21. If $n_1, n_2, \ldots n_t$ are the integers less than 2^n which are not powers of 2 (2^0 being reckoned as a power of 2), and if $\alpha = \frac{\pi}{2^{n+1}+1}$, prove that $\prod_{r=1}^{t} \sec n_r\alpha = 2^{2^n-n-1}$.

22. Prove that

$$\prod_{m=1}^{n} \left\{\prod_{s=1}^{n-1} \sin\left[\frac{(m-1)\pi}{n} - \frac{(s-1)\pi+\theta}{n-1}\right]\right\} = \pm\frac{\sin n\theta}{2^{n^2-n-1}},$$

and determine the ambiguous sign.

23. Express $\sin x$ as an infinite product as follows:

(i) Differentiate $\theta \operatorname{cosec}\theta$ and $\theta\cot\theta$; hence prove, for $0<\theta<\phi<\frac{1}{2}\pi$, that $\theta\operatorname{cosec}\theta<\phi\operatorname{cosec}\phi$, and $\theta\cot\theta>\phi\cot\phi$. Deduce that

$$1-\frac{\sin^2\theta}{\sin^2\phi}<1-\frac{\theta^2}{\phi^2}<\sec^2\theta\left(1-\frac{\sin^2\theta}{\sin^2\phi}\right);$$

(ii) If n is an odd integer, and r takes the values $1, 2, \ldots, \frac{1}{2}(n-1)$, prove that, for $0<x<\pi$,

$$\prod\left(1-\frac{\sin^2\frac{x}{n}}{\sin^2\frac{r\pi}{n}}\right)<\prod\left(1-\frac{x^2}{r^2\pi^2}\right)<\sec^{n-1}\frac{x}{n}\prod\left(1-\frac{\sin^2\frac{x}{n}}{\sin^2\frac{r\pi}{n}}\right).$$

(iii) Use equation (7), p. 222, and Example 7, p. 70, to deduce that $\prod_{1}^{\frac{1}{2}(n-1)}\left(1-\frac{x^2}{r^2\pi^2}\right)$ lies between two expressions which tend to $\sin x/x$ when $n\to\infty$.

This proves that, for $0<x<\pi$,

$$\sin x = x\prod_{1}^{\infty}\left(1-\frac{x^2}{r^2\pi^2}\right).$$

(iv) If $0<(\theta, \phi)<\frac{1}{2}\pi$, prove that

$$\left|1-\frac{\sin^2\theta}{\sin^2\phi}\right| \leqslant \left|1-\frac{\theta^2}{\phi^2}\right| \leqslant \sec^2\theta\left|1-\frac{\sin^2\theta}{\sin^2\phi}\right|$$

and hence establish the result of (iii) for all values of x.

24. Use Ex. XII. b, No. 10, and the method of No. 23 above to express $\cos x$ as an infinite product.

CHAPTER XIII

MANY-VALUED FUNCTIONS OF A COMPLEX VARIABLE

Logarithms of Complex Numbers. Definition. If $w = \exp(z)$, then z is called **a natural logarithm** of w, and we write $z = \text{Log}\, w$.

Thus, if $\mathbf{u + iv = \exp(x + iy)}$, then $\mathbf{x + iy = Log\,(u + iv)}$. ....(1)

We have seen that to any value of z, $\equiv x + iy$, there corresponds one and only one value of w, $= \exp(z) \equiv \exp(x + iy)$, viz. $e^x \operatorname{cis} y$; see p. 194. We shall now show that to a given value of w there correspond an unlimited number of values of z, $= \text{Log}\, w$.

Values of Log w, where $\mathbf{w = \rho(\cos\varphi + i \sin\varphi)}$.
Let z, $\equiv x + iy$, be a natural logarithm of w.

By definition, $\rho(\cos\phi + i \sin\phi) = \exp(z) = \exp(x + iy)$

$$= e^x(\cos y + i \sin y);$$

$$\therefore \rho = e^x \quad \text{and} \quad \phi = 2k\pi + y;$$

$$\therefore x = \log\rho \quad \text{and} \quad y = \phi + 2n\pi,$$

where $\log\rho$ is the unique logarithm of ρ as defined in Ch. IV;

$$\therefore \ \mathbf{Log}\,\{\rho(\cos\varphi + i \sin\varphi)\} \equiv \text{Log}\, w = x + iy = \log\rho + i(\varphi + 2n\pi). \quad \text{...(2)}$$

Thus the natural logarithm of a complex number, as defined above, is an **infinitely many-valued** function.

If ϕ_1 is the principal value of the amplitude of w, so that

$$-\pi < \phi_1 \leqslant \pi,$$

we define the **principal value**, $\log w$, of $\text{Log}\, w$, by the relation

$$\mathbf{\log w = \log\rho + i\varphi_1}. \quad \text{..............................(3)}$$

By using the results of Ex. VIII. f, No. 8, we may express the relation (3) as follows:

$$\text{If } u > 0, \qquad \log(u + iv) = \tfrac{1}{2}\log(u^2 + v^2) + i \tan^{-1}\frac{v}{u}.$$

$$\text{If } u < 0 < v, \qquad \log(u + iv) = \tfrac{1}{2}\log(u^2 + v^2) + i\left\{\tan^{-1}\frac{v}{u} + \pi\right\}, \quad \text{...(4)}$$

$$\text{If } u < 0,\ v < 0, \ \log(u + iv) = \tfrac{1}{2}\log(u^2 + v^2) + i\left\{\tan^{-1}\frac{v}{u} - \pi\right\}.$$

Functional Law for Log w.

Let $w_1=\rho_1(\cos\phi_1+i\sin\phi_1)$ and $w_2=\rho_2(\cos\phi_2+i\sin\phi_2)$, then, by equation (2),

$$\begin{aligned}\text{Log}\, w_1+\text{Log}\, w_2&=\{\log\rho_1+i(\phi_1+2p\pi)\}+\{\log\rho_2+i(\phi_2+2q\pi)\}\\&=\log\rho_1+\log\rho_2+i(\phi_1+\phi_2+2n\pi)\\&=\log(\rho_1\rho_2)+i(\phi_1+\phi_2+2n\pi).\end{aligned}$$

$$\begin{aligned}\text{Also Log}\,(w_1w_2)&=\text{Log}\,\{\rho_1\rho_2\,\text{cis}\,(\phi_1+\phi_2)\}\\&=\log(\rho_1\rho_2)+i(\phi_1+\phi_2+2k\pi);\end{aligned}$$

∴ *every* value of $\text{Log}\, w_1+\text{Log}\, w_2$ is equal to *some* value of $\text{Log}\,(w_1w_2)$, and conversely.

We therefore write

$$\textbf{Log w}_1+\textbf{Log w}_2=\textbf{Log (w}_1\textbf{w}_2\textbf{)}. \quad\text{.....................(5)}$$

In the same way it may be proved that

$$\textbf{Log w}_1-\textbf{Log w}_2=\textbf{Log}\,\frac{\textbf{w}_1}{\textbf{w}_2}, \quad\text{.........................(6)}$$

and in particular $$\text{Log}\,\frac{1}{w}=-\text{Log}\, w, \quad\text{.....................(7)}$$

where *every* value of either side is equal to *some* value of the other side.

Further, if n is an integer,

every value of n Log w is equal to some value of Log wⁿ. ...(8)

But, in contrast with (5) and (6) it is *not* true to say that every value of $\text{Log}\, w^n$ is one of the values of $n\,\text{Log}\, w$; for example,

$$\text{Log}\, i^2=\text{Log}\,(\cos\pi+i\sin\pi)=i(\pi+2n\pi);$$

but $$2\,\text{Log}\, i=2\,\text{Log}\left(\cos\frac{\pi}{2}+i\sin\frac{\pi}{2}\right)=2i\left(\frac{\pi}{2}+2m\pi\right)=i(\pi+4m\pi);$$

thus only some of the values of $\text{Log}\, i^2$ are values of $2\,\text{Log}\, i$; the rest are values of $2\,\text{Log}\,(-i)$.

It should be noted that (5) and (6) are not necessarily true for principal values, i.e. if Log is replaced by log. Thus, in (5), $\text{am}\, w_1+\text{am}\, w_2$ may be outside the limits, $-\pi$ to $+\pi$; for example,

$$\log\left(\text{cis}\,\frac{3\pi}{4}\right)+\log\left(\text{cis}\,\frac{3\pi}{4}\right)=\frac{3\pi i}{4}+\frac{3\pi i}{4}=\frac{3\pi i}{2};$$

but $$\log\left[\text{cis}\left(\frac{3\pi}{4}+\frac{3\pi}{4}\right)\right]=\log\left(\text{cis}\,\frac{3\pi}{2}\right)=\log\left(\text{cis}\,\frac{-\pi}{2}\right)=-\frac{\pi i}{2}.$$

Example 1. Find the results corresponding to those in equation (4), p. 241, for $\log(u+iv)$, when (i) $u=0$, (ii) $u<0$, $v=0$.

(i) If $v>0$, $\log(iv)=\log\left\{v\left(\cos\frac{\pi}{2}+i\sin\frac{\pi}{2}\right)\right\}=\frac{1}{2}\log v^2+\frac{\pi}{2}i.$

If $v<0$, $\log(iv)=\log\left\{(-v)\left(\cos\frac{\pi}{2}-i\sin\frac{\pi}{2}\right)\right\}=\frac{1}{2}\log v^2-\frac{\pi}{2}i.$

(ii) $\log u=\log\{(-u)(\cos\pi+i\sin\pi)\}=\frac{1}{2}\log u^2+\pi i.$

It should be noticed that, in (i), $\frac{1}{2}\log v^2$ may be replaced by $\log|v|$; it cannot be replaced by $\log v$ unless $v>0$.

Example 2. Are the functions (i) $\exp(\text{Log}\, w)$, (ii) $\exp(i\,\text{Log}\, w)$, one-valued or many-valued ?

(i) If z denotes $\text{Log}\, w$, by definition $w=\exp(z)$;

$$\therefore\ \exp(\text{Log}\, w)=\exp(z)=w;$$

$$\therefore\ \exp(\text{Log}\, w) \text{ is one valued and equal to } w.$$

(ii) If $w=\rho(\cos\phi+i\sin\phi)$, $\text{Log}\, w=\log\rho+(\phi+2n\pi)i$, by (2);

$$\therefore\ i\,\text{Log}\, w=i\log\rho-(\phi+2n\pi);$$

$$\therefore\ \exp(i\,\text{Log}\, w)=\exp\{-(\phi+2n\pi)\}\exp(i\log\rho), \text{ see p. 194,}$$

$$=e^{-\phi-2n\pi}\operatorname{cis}(\log\rho);$$

$$\therefore\ \exp(i\,\text{Log}\, w) \text{ is infinitely many-valued.}$$

Example 3. If x is real, prove that

$$\text{Tan}^{-1}x=\frac{1}{2i}\text{Log}\frac{1+ix}{1-ix}.$$

If $x=\tan\theta$, $\frac{1}{2i}\text{Log}\frac{1+ix}{1-ix}=\frac{1}{2i}\text{Log}\frac{\cos\theta+i\sin\theta}{\cos\theta-i\sin\theta}=\frac{1}{2i}\text{Log}(\operatorname{cis}2\theta)$

$$=\frac{1}{2i}i(2\theta+2n\pi), \text{ by (2)}, =\theta+n\pi=\text{Tan}^{-1}x.$$

This result holds in the sense that *any* value of either function is equal to *some* value of the other.

EXERCISE XIII. a.

1. Find (i) the general value, (ii) the principal value, of

(*a*) Log 1 ; (*b*) Log i ; (*c*) Log (-2) ; (*d*) Log $(-2i)$.

2. Find (i) the general value, (ii) the principal value, of

(*a*) Log $(1+i)$; (*b*) Log $(\sqrt{i})$; (*c*) Log $(-3+4i)$; (*d*) Log $(-3-4i)$.

3. What is the general value of Log $(-e)$?

4. Find (i) the general value, (ii) the principal value, of

$$\text{Log}(\cos a + i \sin a), \text{ where } (2n-1)\pi < a \leqslant (2n+1)\pi.$$

5. What are the values of a and b if $\log(a+ib) = 2 - \dfrac{3\pi i}{4}$?

6. Express $\log(1+\cos 2\theta + i \sin 2\theta)$ in the form $a+ib$, given that $\dfrac{\pi}{2} < \theta < \dfrac{3\pi}{2}$.

7. Find the general values of (i) i Log i ; (ii) Log $(1+i\sqrt{3})$.

8. Express $\log(1+i\tan a)$ in the form $a+ib$, when $\dfrac{\pi}{2} < a < \dfrac{3\pi}{2}$.

9. What are the values of a and b if $\log(a+ib) = \dfrac{\pi}{6}(1+i)^3$?

10. If $a < 0 < b$ and $\tan a = \dfrac{b}{a}$, where $-\dfrac{\pi}{2} < a < \dfrac{\pi}{2}$, find the values of

$$\text{(i) } \log(a+ib) - \log(a-ib)\text{ ; (ii) } \log\frac{a+ib}{a-ib}.$$

11. Express Log $(1-\cos\theta - i\sin\theta)$ in the form $a+ib$, giving rules for determining its principal value.

12. If $(a+bi)\log(c+di) = f+gi$, express c and d in terms of a, b, f, g.

13. If $\omega = \text{cis}\,\dfrac{2\pi}{3}$, in what sense is Log $1 = 3$ Log ω ? Express some other values of Log 1 in a similar form.

14. If $\log z = a+bi$ and $z \log z = p+qi$, prove that

$$\text{Tan}^{-1}\frac{p}{q} = \text{Tan}^{-1}\frac{a}{b} - b.$$

15. What is the general value of $\exp\{\tfrac{1}{2}\text{Log}(a+bi)^2\}$ when $b=0$?

16. What are the values of

$$\text{(i) } \exp\{(1+i)\text{ Log } i\}\text{ ; (ii) } \exp\{(1+i)\text{ Log }(1+i)\}\text{ ?}$$

17. What is the value of $\exp\{i\text{ Log}(\cos a + i\sin a)\}$?

18. Express Log $\{\text{Log}(\cos\theta + i\sin\theta)\}$ in the form $a+ib$.

19. If $z \equiv x+iy$ and $x \neq 1$, prove that

$$\log(z-1) = \tfrac{1}{2}\log\{(x-1)^2+y^2\} + i\tan^{-1}\frac{y}{x-1} + k\pi i,$$

and give the values of k for different positions of the point P which represents z in the Argand Diagram.

20. Give an expression corresponding to that in No. 19 for $\log(z+1)$, and answer the same question about k.

21. Answer the same question as in No. 19 for $\log \frac{z-1}{z+1}$, considering the cases when P is (i) inside, (ii) outside, the circle $|z|=1$.

22. Find the value of $\log \frac{z-1}{z+1}$, when $z=\cos\theta+i\sin\theta$.

23. If $\log \frac{x-a+yi}{x+a+yi}=p+iq$, and if p is constant, find the locus of the point P representing $x+iy$. Also find the locus when q is constant.

24. If $\log \frac{z-a}{z+a}=p+iq$, and if the point P representing z moves round the circle $|z|=a$, discuss the behaviour of q. Also state what happens when P moves on an arbitrary circle through $(a, 0)$ and $(-a, 0)$.

25. If $\log \frac{z+i}{z-i}=\log(z+i)-\log(z-i)$, where $z\equiv x+iy$, prove that x is positive or zero, or else $x<0$ and $y\geqslant 1$ or $y<-1$.

The Logarithmic Series. It has been proved on pp. 84, 85 that, if $-1<x\leqslant 1$,

$$\log(1+x)=x-\frac{x^2}{2}+\frac{x^3}{3}-\frac{x^4}{4}+\dots.$$

We now proceed to consider the corresponding series

$$w-\frac{w^2}{2}+\frac{w^3}{3}-\frac{w^4}{4}+\dots, \qquad \text{.........................(9)}$$

where w is a complex number equal to $\rho(\cos\phi+i\sin\phi)$.

Since the series
$$\rho+\frac{\rho^2}{2}+\frac{\rho^3}{3}+\frac{\rho^4}{4}+\dots$$

is convergent for $0\leqslant\rho<1$, it follows that the series (9) is absolutely convergent, and therefore convergent, for $|w|<1$. The series (9) may be written,

$$(\rho\cos\phi-\tfrac{1}{2}\rho^2\cos 2\phi+\tfrac{1}{3}\rho^3\cos 3\phi-\dots)$$
$$+i(\rho\sin\phi-\tfrac{1}{2}\rho^2\sin 2\phi+\tfrac{1}{3}\rho^3\sin 3\phi-\dots) \qquad \text{...............(10)}$$

We proceed to evaluate separately the two parts of this function.

(i) From Ch. IX., equation (8), p. 175, we have, writing $-x$ for x,

$$1-x\cos\phi+x^2\cos 2\phi-\dots+(-1)^{n-1}x^{n-1}\cos(n-1)\phi$$
$$=\frac{1+x\cos\phi-(-x)^n\{\cos n\phi+x\cos(n-1)\phi\}}{1+2x\cos\phi+x^2}.$$

Subtract each side from unity and divide by x, then

$$\cos\phi - x\cos 2\phi + \dots + (-1)^n x^{n-2}\cos(n-1)\phi$$
$$= \frac{\cos\phi + x + (-1)^n x^{n-1}\{\cos n\phi + x\cos(n-1)\phi\}}{1+2x\cos\phi + x^2}.$$

Integrate w.r.t. x both sides of this identity from $x=0$ to $x=\rho$, then

$$\rho\cos\phi - \tfrac{1}{2}\rho^2\cos 2\phi + \dots + (-1)^n\frac{\rho^{n-1}}{n-1}\cos(n-1)\phi$$
$$= \tfrac{1}{2}\log(1+2\rho\cos\phi+\rho^2) + (-1)^n K,$$

where $$K \equiv \int_0^\rho \frac{x^{n-1}\{\cos n\phi + x\cos(n-1)\phi\}\,dx}{1+2x\cos\phi+x^2}.$$

We shall suppose that $0 \leqslant \rho < 1$. It follows that, within the limits of integration, $$|\cos n\phi + x\cos(n-1)\phi| < 2$$

and $$|1+2x\cos\phi+x^2| \geqslant (1-2x+x^2) = (1-x)^2 > (1-\rho)^2.$$

$$\therefore\ |K| < \frac{2}{(1-\rho)^2}\int_0^\rho x^{n-1}dx = \frac{2}{(1-\rho)^2}\cdot\frac{\rho^n}{n} < \frac{2}{(1-\rho)^2}\cdot\frac{1}{n};$$

$$\therefore\ |K| \to 0, \text{ when } n\to\infty;\ \therefore \text{ also } K\to 0, \text{ when } n\to\infty.$$

Therefore for $0 \leqslant \rho < 1$, the first series of (10) is convergent and

$$\rho\cos\phi - \tfrac{1}{2}\rho^2\cos 2\phi + \tfrac{1}{3}\rho^3\cos 3\phi - \dots = \tfrac{1}{2}\log(1+2\rho\cos\phi+\rho^2). \quad (11)$$

(ii) Again from Ch. IX, equation (9), we have, writing $-x$ for x,

$$-x\sin\phi + x^2\sin 2\phi - \dots + (-1)^{n-1}x^{n-1}\sin(n-1)\phi$$
$$= \frac{-x\sin\phi - (-x)^n\sin n\phi + (-x)^{n+1}\sin(n-1)\phi}{1+2x\cos\phi+x^2};$$

$$\therefore\ \sin\phi - x\sin 2\phi + \dots + (-1)^n x^{n-2}\sin(n-1)\phi$$
$$= \frac{\sin\phi + (-1)^n x^{n-1}\{\sin n\phi + x\sin(n-1)\phi\}}{1+2x\cos\phi+x^2}.$$

Integrate w.r.t. x from $x=0$ to $x=\rho$, as before, then

$$\rho\sin\phi - \tfrac{1}{2}\rho^2\sin 2\phi + \dots + (-1)^n\frac{\rho^{n-1}}{n-1}\sin(n-1)\phi$$

$$= \int_0^\rho \frac{\sin\phi\,dx}{1+2x\cos\phi+x^2} + (-1)^n H,$$

where $$H \equiv \int_0^\rho \frac{x^{n-1}\{\sin n\phi + x\sin(n-1)\phi\}\,dx}{1+2x\cos\phi+x^2}.$$

In $\int_0^\rho \frac{\sin\phi\, dx}{1+2x\cos\phi+x^2}$, put $\frac{x\sin\phi}{1+x\cos\phi}=z$, then

$$\sin\phi(1+x\cos\phi)-x\sin\phi\cos\phi$$
$$=(1+x\cos\phi)^2\frac{dz}{dx}; \quad \therefore \sin\phi=(1+x\cos\phi)^2\frac{dz}{dx};$$

also $1+2x\cos\phi+x^2=(1+x\cos\phi)^2+x^2\sin^2\phi=(1+x\cos\phi)^2(1+z^2)$;

$$\therefore \text{ the integral becomes } \int_0^{\frac{\rho\sin\phi}{1+\rho\cos\phi}} \frac{1}{1+z^2}\,dz,$$

and this, see Ex. IV. a, No. 17, p. 57, is the *principal value* of

$$\tan^{-1}\frac{\rho\sin\phi}{1+\rho\cos\phi}.$$

Also, by the same argument as before, it follows that, for $0 \leqslant \rho < 1$, $H \to 0$ when $n \to \infty$.

Therefore, for $0 \leqslant \rho < 1$, the second series of (10) is convergent and

$$\rho\sin\phi-\tfrac{1}{2}\rho^2\sin 2\phi+\tfrac{1}{3}\rho^3\sin 3\phi-\ldots=\tan^{-1}\frac{\rho\sin\phi}{1+\rho\cos\phi}. \quad \ldots(12)$$

From (11) and (12) it follows that, if $|w|<1$,

$$w-\frac{w^2}{2}+\frac{w^3}{3}-\frac{w^4}{4}+\ldots$$
$$=\tfrac{1}{2}\log(1+2\rho\cos\phi+\rho^2)+i\tan^{-1}\frac{\rho\sin\phi}{1+\rho\cos\phi}. \quad \ldots\ldots\ldots(13)$$

Now $\qquad 1+w=1+\rho\cos\phi+i\rho\sin\phi$;

$\therefore |1+w|=+\sqrt{\{(1+\rho\cos\phi)^2+\rho^2\sin^2\phi\}}=+\sqrt{(1+2\rho\cos\phi+\rho^2)}$;

also, since $1+\rho\cos\phi>0$, the principal value of the amplitude of $1+w$ is $\tan^{-1}\frac{\rho\sin\phi}{1+\rho\cos\phi}$; therefore the principal value, $\log(1+w)$, of $\operatorname{Log}(1+w)$ is equal to the sum to infinity of (9); we therefore write

$$\log(1+\mathrm{w})=\mathrm{w}-\frac{\mathrm{w}^2}{2}+\frac{\mathrm{w}^3}{3}-\frac{\mathrm{w}^4}{4}+\ldots, \text{ if } |\mathrm{w}|<1. \quad \ldots\ldots\ldots(14)$$

The Circle of Convergence. The series (9) is not convergent when $|w|>1$. If w is represented by a point P in the Argand Diagram, the series (9) is convergent if P lies *inside* the circle $|w|=1$, and is not convergent if P lies *outside* the circle $|w|=1$. We therefore call this circle the **circle of convergence** for the power series. It is usually easy to determine the circle of convergence for a given power series, but in general it is difficult to discover how the series behaves when P lies on the circle itself. Here, we know from Ch. V., p. 84, that (14) holds for $w=1$, i.e. when P is at the point (1, 0) and that the series is divergent when $w=-1$, i.e. when P is at the point $(-1, 0)$. It can be proved that (14) holds for all values of w for which $|w|=1$, except $w=-1$, i.e. for all positions of P on the circle of convergence except the point $(-1, 0)$. Although a rigorous proof of this result is beyond the scope of this book, we shall state here some of the results which follow when this fact is assumed.

In (14), put $w=\cos a+i\sin a$, where $a\neq(2n+1)\pi$, then

$$\log(1+\cos a+i\sin a)=\text{cis}\, a-\tfrac{1}{2}\,\text{cis}\, 2a+\tfrac{1}{3}\,\text{cis}\, 3a-\dots.$$

But $\log(1+\cos a+i\sin a)$

$$=\tfrac{1}{2}\log\left\{(1+\cos a)^2+\sin^2 a\right\}+i\tan^{-1}\frac{\sin a}{1+\cos a}$$

$$=\tfrac{1}{2}\log(2+2\cos a)+i\tan^{-1}\left(\frac{2\sin\frac{a}{2}\cos\frac{a}{2}}{2\cos^2\frac{a}{2}}\right)$$

$$=\tfrac{1}{2}\log\left(4\cos^2\frac{a}{2}\right)+i\tan^{-1}\left(\tan\frac{a}{2}\right);$$

$$\therefore\ \log|2\cos\tfrac{1}{2}a|=\cos a-\tfrac{1}{2}\cos 2a+\tfrac{1}{3}\cos 3a-\dots \qquad \text{.........(15)}$$

and $$\tan^{-1}(\tan\tfrac{1}{2}a)=\sin a-\tfrac{1}{2}\sin 2a+\tfrac{1}{3}\sin 3a-\dots, \qquad \text{.........}(16_1)$$

provided that $a\neq(2n+1)\pi$.

The sum to infinity of the series in (16_1) is the *principal value* of $\text{Tan}^{-1}\left(\tan\frac{a}{2}\right)$; this may be expressed as follows:

$$\text{if } -\pi<a<\pi,\quad \tan^{-1}\left(\tan\frac{a}{2}\right)=\frac{a}{2};$$

$$\text{if } \pi<a<3\pi,\quad \tan^{-1}\left(\tan\frac{a}{2}\right)=\frac{a}{2}-\pi,\ \text{etc.}$$

In general,

$$\text{if } (2n-1)\pi<a<(2n+1)\pi,\quad \tan^{-1}\left(\tan\frac{a}{2}\right)=\frac{a}{2}-n\pi;$$

hence $$\frac{a}{2}-n\pi=\sin a-\tfrac{1}{2}\sin 2a+\tfrac{1}{3}\sin 3a-\dots. \qquad \text{.........}(16_2)$$

Also, in the excluded case, $a=(2n+1)\pi$, the sum of the series (16_2) is obviously zero.

Using these results, the reader should draw the graph of

$$y = \sin x - \tfrac{1}{2}\sin 2x + \tfrac{1}{3}\sin 3x - \dots,$$

which, he will find, is discontinuous at the points $x = (2n+1)\pi$.

For example, if $x \to \pi$ from above, $y \to -\frac{\pi}{2}$, but, if $x \to \pi$ from below $y \to +\frac{\pi}{2}$, while at $x = \pi$, $y = 0$.

Example 4. Find the sum to infinity of

$$r \sin \alpha + \tfrac{1}{2}r^2 \sin 2\alpha + \tfrac{1}{3}r^3 \sin 3\alpha + \dots \quad \text{when } |r| < 1.$$

From equation (14),

$$\log(1 - r \operatorname{cis} \alpha) = -r \operatorname{cis} \alpha - \tfrac{1}{2}r^2 \operatorname{cis} 2\alpha - \tfrac{1}{3}r^3 \operatorname{cis} 3\alpha - \dots,$$

since $|-r \operatorname{cis} \alpha| = r < 1$.

But $\log(1 - r \operatorname{cis} \alpha) = \log(1 - r\cos\alpha - ir\sin\alpha)$

$$= \tfrac{1}{2}\log\{(1 - r\cos\alpha)^2 + r^2\sin^2\alpha\} + i\tan^{-1}\frac{-r\sin\alpha}{1 - r\cos\alpha},$$

since $1 - r\cos\alpha > 0$, (see p. 241);

$$\therefore\ r\sin\alpha + \tfrac{1}{2}r^2\sin 2\alpha + \dots = -\tan^{-1}\frac{-r\sin\alpha}{1 - r\cos\alpha} = \tan^{-1}\frac{r\sin\alpha}{1 - r\cos\alpha}.$$

Example 5. Find a series in powers of x, involving α, whose sum to infinity is one of the values of θ satisfying the equation,

$$\tan\theta = x + \cot\alpha, \quad \text{where } |x\sin\alpha| < 1.$$

$$\frac{\sin\theta}{\cos\theta} = x + \frac{\cos\alpha}{\sin\alpha} = \frac{x\sin\alpha + \cos\alpha}{\sin\alpha};$$

$$\therefore\ \frac{\exp(\theta i)}{\exp(-\theta i)} \equiv \frac{\cos\theta + i\sin\theta}{\cos\theta - i\sin\theta} = \frac{\sin\alpha + i(x\sin\alpha + \cos\alpha)}{\sin\alpha - i(x\sin\alpha + \cos\alpha)};$$

$$\therefore\ \exp(2\theta i) = \frac{x\sin\alpha + \cos\alpha - i\sin\alpha}{-x\sin\alpha - \cos\alpha - i\sin\alpha} = \frac{\operatorname{cis}(-\alpha)\{1 + x\sin\alpha\operatorname{cis}\alpha\}}{-\operatorname{cis}\alpha\{1 + x\sin\alpha\operatorname{cis}(-\alpha)\}}$$

$$= \operatorname{cis}(\pi - 2\alpha)\frac{1 + y\operatorname{cis}\alpha}{1 + y\operatorname{cis}(-\alpha)}, \quad \text{where } y \equiv x\sin\alpha;$$

$\therefore$ $2\theta i$ is any value of

$$\operatorname{Log}\operatorname{cis}(\pi - 2\alpha) + \operatorname{Log}\{1 + y\operatorname{cis}\alpha\} - \operatorname{Log}\{1 + y\operatorname{cis}(-\alpha)\};$$

$$\therefore\ 2\theta i = 2n\pi i + (\pi - 2\alpha)i + \sum_1^\infty\left\{(-1)^{n-1}\frac{y^n}{n}[\operatorname{cis} n\alpha - \operatorname{cis}(-n\alpha)]\right\},$$

since $|y| < 1$;

$$\therefore\ \theta = n\pi + \frac{\pi}{2} - \alpha + \frac{1}{2i}\sum_1^\infty\left\{(-1)^{n-1}\frac{x^n\sin^n\alpha}{n}2i\sin n\alpha\right\};$$

$$\therefore\ \theta = n\pi + \frac{\pi}{2} - \alpha + x\sin\alpha\sin\alpha - \frac{x^2}{2}\sin^2\alpha\sin 2\alpha + \dots.$$

Note. It follows from p. 247 that the value of the series

$$x \sin \alpha \sin \alpha - \frac{x^2}{2} \sin^2\alpha \sin 2\alpha + \dots \text{ is between } -\frac{\pi}{2} \text{ and } +\frac{\pi}{2}.$$

Thus the value of θ given by taking $n=0$ is that solution of

$$\tan \theta = x + \cot \alpha \quad \text{which lies between } -\alpha \text{ and } \pi - \alpha.$$

EXERCISE XIII. b.

1. If $|z|<1$, express as a series of powers of z:

$$\text{(i) } \log(1-z); \quad \text{(ii) } \log(1+z) - \log(1-z).$$

Is $\log \frac{1+z}{1-z}$ always equal to $\log(1+z) - \log(1-z)$, excluding $z = \pm 1$?

2. If $|r|<1$, sum to infinity: $r \cos \alpha + \frac{1}{2} r^2 \cos 2\alpha + \frac{1}{3} r^3 \cos 3\alpha + \dots$.

3. If $\alpha \neq n\pi$, sum to infinity:

$$\cos \alpha \cos \beta + \tfrac{1}{2} \cos^2\alpha \cos 2\beta + \tfrac{1}{3} \cos^3\alpha \cos 3\beta + \dots .$$

4. ABC is a triangle such that $b<c$; prove that $\frac{c\text{B}}{b}$ is the sum to infinity of the series, $\sin \text{A} + \frac{b}{2c} \sin 2\text{A} + \frac{b^2}{3c^2} \sin 3\text{A} + \dots$.

5. ABC is a triangle, prove that $\frac{1}{2}\text{C}$ is the sum to infinity of the series:

$$\frac{c \sin \text{B}}{a+b} + \frac{1}{2} \frac{c^2 \sin 2\text{B}}{(a+b)^2} + \frac{1}{3} \frac{c^3 \sin 3\text{B}}{(a+b)^3} + \dots .$$

6. Find the sum to infinity of

$$\cos \theta \sin \theta + \tfrac{1}{2} \cos^2\theta \sin 2\theta + \tfrac{1}{3} \cos^3\theta \sin 3\theta + \dots,$$

$$\text{(i) when } 0<\theta<\pi, \quad \text{(ii) when } \pi<\theta<2\pi.$$

7. If $-\frac{\pi}{2}<\theta<\frac{\pi}{2}$, prove that

$$\log(2 \cos \theta) = \cos 2\theta - \tfrac{1}{2} \cos 4\theta + \tfrac{1}{3} \cos 6\theta - \dots .$$

What is the sum to infinity of this series, if $\frac{\pi}{2}<\theta<\frac{3\pi}{2}$?

8. Find the range of values of x, if the sum to infinity of the series, $\sin 2x - \frac{1}{2} \sin 4x + \frac{1}{3} \sin 6x - \dots$, is (i) x, (ii) $x - \pi$.

9. If $n\pi<\theta<(n+1)\pi$, find the sum to infinity of the series,

$$\sin 2\theta + \tfrac{1}{2} \sin 4\theta + \tfrac{1}{3} \sin 6\theta + \dots .$$

10. ABC is a triangle; prove that $\log \frac{b}{a}$ is the sum to infinity of

$$(\cos 2\text{A} - \cos 2\text{B}) + \tfrac{1}{2}(\cos 4\text{A} - \cos 4\text{B}) + \tfrac{1}{3}(\cos 6\text{A} - \cos 6\text{B}) + \dots .$$

11. Find the sum to infinity of the series,

$$\cos a \cos \beta - \tfrac{1}{2} \cos 2a \cos 2\beta + \tfrac{1}{3} \cos 3a \cos 3\beta - \dots,$$

where neither $a+\beta$ nor $a-\beta$ is an odd multiple of π.

12. If $x \neq n\pi$, find the sum to infinity of the series,

$$\cos^2 x - \tfrac{1}{2} \sin^2 2x + \tfrac{1}{3} \cos^2 3x - \tfrac{1}{4} \sin^2 4x + \dots .$$

13. If $y = x \sin a + \frac{1}{2}x^2 \sin 2a + \frac{1}{3}x^3 \sin 3a + \dots$, where $|x| < 1$, prove that $\sin y = x \sin (y+a)$.

14. If $y = x - t \sin 2x + \frac{1}{2}t^2 \sin 4x - \frac{1}{3}t^3 \sin 6x + \dots$, where

$$t = \tan^2\phi < 1, \quad \text{prove that } \tan y = \cos 2\phi \tan x.$$

15. If $|x| < 1$, find the sum to infinity of the series,

$$x \sin \theta + \tfrac{1}{3}x^3 \sin 3\theta + \tfrac{1}{5}x^5 \sin 5\theta + \dots;$$

16. If $\frac{\pi}{4} > a > -\frac{\pi}{4}$, prove that

(i) $\tan a \sin \theta - \frac{1}{3} \tan^3 a \sin 3\theta + \frac{1}{5} \tan^5 a \sin 5\theta - \dots$

$$= \tfrac{1}{4} \log \frac{1 + \sin 2a \sin \theta}{1 - \sin 2a \sin \theta};$$

(ii) $\tan a \cos \theta - \frac{1}{3} \tan^3 a \cos 3\theta + \frac{1}{5} \tan^5 a \cos 5\theta - \dots$

$$= \tfrac{1}{2} \tan^{-1}(\tan 2a \cos \theta).$$

17. If $(1+x) \tan \theta = (1-x) \tan \phi$, and if $|x| < 1$, expand θ in ascending powers of x.

18. If $\tan a = \cos 2\omega \tan \lambda$, and $\tan^2\omega < 1$, expand λ in ascending powers of $\tan^2\omega$.

19. If $\frac{\pi}{2} < \theta < \frac{3\pi}{2}$, find the sum to infinity of the series

$$\cos \theta - \tfrac{1}{3} \cos 3\theta + \tfrac{1}{5} \cos 5\theta - \dots .$$

20. If $|x| < 1$, expand $\tan^{-1} \dfrac{2x \cos \theta}{1 - x^2}$ in ascending powers of x.

Generalised Indices. If a is positive and n is rational, it has been proved in Ch. IV, (see p. 65), that

$$a^n = e^{n \log a} = \exp (n \log a). \quad \dots\dots\dots\dots\dots\dots (17)$$

If z is complex and equal to $r(\cos \theta + i \sin \theta)$, where $r \neq 0$, and if n is rational and equal to $\frac{p}{q}$, where p, q are co-prime integers, it has been proved in Ch. IX, (see p. 165), that

$$z^n \equiv z^{\frac{p}{q}} = r^{\frac{p}{q}} \operatorname{cis} \frac{p\theta + 2k\pi}{q}, \text{ for } k = 0, 1, 2, \dots, (q-1).$$

We shall now show that this is the same as exp (n Log z);

$$\frac{p}{q}\operatorname{Log} z = \frac{p}{q}\{\log r + i(\theta + 2s\pi)\} = \frac{p}{q}\log r + i\frac{p\theta + 2sp\pi}{q};$$

$$\therefore \exp(n \operatorname{Log} z) = \exp\left(\frac{p}{q}\log r\right)\operatorname{cis}\frac{p\theta + 2sp\pi}{q}$$

$$= \exp\left(\frac{p}{q}\log r\right)\operatorname{cis}\frac{p\theta + 2k\pi}{q},$$

since $\operatorname{cis}\dfrac{p\theta + 2sp\pi}{q}$ for $s = 0, 1, 2, \ldots (q-1)$ is the same as $\operatorname{cis}\dfrac{p\theta + 2k\pi}{q}$ for $k = 0, 1, 2, \ldots (q-1)$ when p and q are co-prime.

But, by (17), $\exp\left(\dfrac{p}{q}\log r\right) = r^{\frac{p}{q}}$;

$$\therefore \exp(n \operatorname{Log} z) = r^{\frac{p}{q}}\operatorname{cis}\frac{p\theta + 2k\pi}{q}, \quad k = 0, 1, 2, \ldots, (q-1);$$

$$\therefore z^n = \exp(n \operatorname{Log} z), \quad \text{..........................(18)}$$

where n is any rational number, and $z \neq 0$.

Definition. If z is any complex number except 0, and if w is any complex number, the function z^w is defined by the relation

$$\mathbf{z^w = \exp(w \operatorname{Log} z)}. \quad \text{..........................(19)}$$

The relations (17), (18) show that the definitions of Ch. IV and Ch. IX for a^n and z^n are consistent with, and in fact suggest, the definition (19).

Modulus and Amplitude of $\mathbf{z^w}$. If $z = r(\cos\theta + i\sin\theta)$, $r \neq 0$, and $w = u + iv$, by (19),

$$z^w = \exp\{(u+iv)\operatorname{Log} z\} = \exp\{(u+iv)(\log r + i\theta + 2k\pi i)\}$$
$$= \exp[\{u\log r - v(\theta + 2k\pi)\} + i\{v\log r + u(\theta + 2k\pi)\}]$$
$$= \exp(u\log r)\exp\{-v(\theta + 2k\pi)\}\operatorname{cis}\{v\log r + u(\theta + 2k\pi)\}$$
$$= r^u e^{-v(\theta + 2k\pi)}\operatorname{cis}\{v\log r + u(\theta + 2k\pi)\}, \text{ by (17)};$$

$$\therefore |\mathbf{z^w}| = \mathbf{r^u e^{-v(\theta+2k\pi)}}; \ \mathbf{Am(z^w) = v\log r + u(\theta + 2k\pi) + 2n\pi}. \quad \text{...(20)}$$

Thus z^w is an *infinitely many-valued* function, unless $v = 0$ and u is rational.

If $v = 0$ and $w = u = \dfrac{p}{q}$, we have already seen that z^w is q-valued.

The principal value of z^w is defined to be $\mathbf{\exp(w\log z)}$; with the notation just used, this may be written

$$r^u e^{-v\theta}\operatorname{cis}(v\log r + u\theta), \text{ if } -\pi < \theta \leqslant \pi.$$

One special case of (20) should be noted:

$$e^{ia} = \exp [ia \operatorname{Log} e] = \exp [ia(1 + 2k\pi i)]$$
$$= \exp (ia - 2k\pi a) = \exp (-2k\pi a) \operatorname{cis} \alpha;$$

$\therefore$ writing n for $-k$,

$$e^{ia} = e^{2n\pi a} (\cos a + i \sin a). \quad \text{......................(21)}$$

Thus e^{ia} is an infinitely many-valued function, and $\cos a + i \sin a$ is merely its principal value. It is best, therefore, to avoid writing e^{ia} for the sum of the series $1 + (ia) + \frac{(ia)^2}{2!} + \ldots$; the sum of this series is $\exp (ia)$ or $\operatorname{cis} a$ and is, like these functions, one-valued.

The Binomial Series. The investigation of the binomial series

$$\phi(m, z) \equiv 1 + \frac{m}{1!} z + \frac{m(m-1)}{2!} z^2 + \frac{m(m-1)(m-2)}{3!} z^3 + \ldots, \quad \text{...(22)}$$

lies outside the scope of this book. We shall merely state the facts, and give some applications of them in Ex. XIII. c.

(i) If m is a positive integer or zero, the number of terms in the series is finite and the sum is the one-valued function $(1+z)^m$, for all values of z.

(ii) If m is not a positive integer or zero, suppose that $m = a + i\beta$.

(*a*) If $|z| < 1$, the series is absolutely convergent, and its sum to infinity is the principal value of $(1+z)^m$, that is to say,

$$\phi(m, z) = \exp \{m \log (1+z)\}.$$

(*b*) If $|z| = 1$, $z \neq -1$, the series is absolutely convergent if $a > 0$, and is convergent, but not absolutely, if $-1 < a \leqslant 0$, and in either case its sum to infinity is $\exp \{m \log (1+z)\}$. If $a \leqslant -1$, the series is divergent.

(*c*) If $z = -1$, the series is absolutely convergent if $a > 0$, and its sum to infinity is 0. If $a = 0$, $\beta \neq 0$, or if $a < 0$, the series is divergent.

(*d*) If $|z| > 1$, the series is divergent.

Note. Although the complete statement is necessarily elaborate, there is one simple fact that covers the vast majority of cases that occur: namely, on the circle of convergence, excluding $z = -1$, the series is convergent if $a > -1$.

Logarithms to an Arbitrary Base. Definition. If ζ is any one of the values given by $z^w = \zeta$, we say that w is a logarithm of ζ to the base z, and we write

$$\mathbf{Log_z \zeta = w}. \quad \text{..................................(23)}$$

By definition (19), $\exp\{w \operatorname{Log} z\} = z^w = \zeta$;

$\therefore$ by definition (1), p. 241, $w \operatorname{Log} z = \operatorname{Log} \zeta$;

$$\therefore \operatorname{Log}_z \zeta = w = \frac{\operatorname{Log} \zeta}{\operatorname{Log} z}. \quad \text{.......................(24)}$$

If $\zeta = \rho(\cos\psi + i \sin\psi)$ and $z = r(\cos\theta + i\sin\theta)$, then

$$\operatorname{Log}_z \zeta = \frac{\log\rho + i(\psi + 2k\pi)}{\log r + i(\theta + 2n\pi)} \quad \text{.................................(25)}$$

$$= \frac{\{\log\rho + i(\psi + 2k\pi)\}\{\log r - i(\theta + 2n\pi)\}}{(\log r)^2 + (\theta + 2n\pi)^2},$$

which shows that $\operatorname{Log}_z \zeta$ is a *doubly infinitely many-valued* function.

The principal value of $\operatorname{Log}_z \zeta$ is defined by the relation,

$$\log_z \zeta = \frac{\log \zeta}{\log z}. \quad \text{..........................(26)}$$

Note. Sometimes a more restricted definition is given for $\operatorname{Log}_z \zeta$ as follows: If ζ is the *principal value* of z^w, then w is called a logarithm of ζ to the base z.

In this case, $w = \operatorname{Log}_z \zeta$ is equivalent to $\zeta = \exp(w \log z)$;

$\therefore$ by definition (1), p. 241, $w \log z = \operatorname{Log} \zeta$;

$$\therefore \operatorname{Log}_z \zeta = w = \frac{\operatorname{Log} \zeta}{\log z}.$$

And, in particular, $\operatorname{Log}_e \zeta = \dfrac{\operatorname{Log} \zeta}{\log e} = \operatorname{Log} \zeta$.

With this definition, $\operatorname{Log}_z \zeta$ is a *singly infinitely many-valued* function and becomes identical with $\operatorname{Log} \zeta$ when $z = e$.

EXERCISE XIII. c.

Express the general values of the following in the form $a + ib$ or in the modulus-amplitude form:

1. 2^i. 2. 1^{1+i}. 3. i^i. 4. $(1-i)^i$.

5. $(1+i)^{-i}$. 6. $(-i)^{-i}$. 7. $e^{\pi i}$. 8. $e^{\frac{1}{2}\pi i}$.

9. $e^{n\pi i}$. 10. e^{x-yi}. 11. $i^{-\pi}$. 12. π^{-i}.

Write the following in a form which shows their many-valuedness:

13. $\operatorname{Log}_{10} 2$. 14. $\operatorname{Log}_1 3$. 15. $\operatorname{Log}_i i$.

16. $\operatorname{Log}_9 3$. 17. $\operatorname{Log} e^{u+iv}$. 18. $(\cos\theta - i\sin\theta)^i$.

19. What value does the definition $z^w = \exp(w \operatorname{Log} z)$ give for i^2?

20. What is the condition for the principal value of e^{u+iv} to be of the form $a+0i$?

21. Prove that the principal value of $i^{\log(1+i)}$ is $e^{-\frac{\pi^2}{8}} \operatorname{cis}\left(\frac{\pi}{4}\log 2\right)$.

22. Find the modulus and amplitude of the principal value of x^x, if x is a negative number.

23. Find the general value of $e^{\exp(\theta i)} \times e^{\exp(-\theta i)}$.

24. If $p+0i$ is the principal value of $(x+yi)^{a+bi}$, where $x>0$, prove that $\frac{1}{2}b\log(x^2+y^2)+a\tan^{-1}\frac{y}{x}$ is a multiple of π.

How is this affected if $x<0$ and $y<0$?

25. Prove that the ratio of the principal values of $(1+i)^{1-i}$ and $(1-i)^{1+i}$ is $\sin(\log 2)+i\cos(\log 2)$.

26. Find the values of z for which $i^z = \operatorname{cis} a\pi$.

27. (i) Is every value of $z^{w_1} \times z^{w_2}$ a value of $z^{w_1+w_2}$?

(ii) Is every value of $z^{w_1+w_2}$ a value of $z^{w_1} \times z^{w_2}$?

(iii) Are the principal values of $z^{w_1} \times z^{w_2}$ and $z^{w_1+w_2}$ equal ?

28. Prove that the points in the Argand Diagram which represent the values of $e^{\cos a + i\sin a}$ lie on the equiangular spiral whose polar equation is $r = ce^{-\theta\tan a}$, where $\log c = \sec a$.

29. If $|x|<1$, prove that the sum to infinity of the series,

$$\cos a + nx\cos(a+\theta) + \frac{n(n-1)}{1.2}x^2\cos(a+2\theta)+\ldots$$

is $(1+2x\cos\theta+x^2)^{\frac{n}{2}}\cos(a+n\phi)$, where $\phi = \tan^{-1}\dfrac{x\sin\theta}{1+x\cos\theta}$.

30. If, in the triangle ABC, $a<b$, prove that $\left(\frac{b}{c}\right)^n \cos n\mathrm{A}$ is the sum to infinity of $1+n\frac{a}{b}\cos \mathrm{C} + \frac{n(n+1)}{1.2}\frac{a^2}{b^2}\cos 2\mathrm{C}+\ldots$.

31. If n is real, state the conditions for convergence of the series:

$$\text{(i) } 1+n\cos 2\theta + \frac{n(n-1)}{1.2}\cos 4\theta+\ldots;$$

$$\text{(ii) } n\sin 2\theta + \frac{n(n-1)}{1.2}\sin 4\theta+\ldots;$$

and find the sums to infinity if

$$(a)\ 2k\pi - \frac{\pi}{2} < \theta < 2k\pi + \frac{\pi}{2};\quad (b)\ 2k\pi + \frac{\pi}{2} < \theta < 2k\pi + \frac{3\pi}{2}.$$

32. If $n < 1$, find the sum to infinity of the series,

$$n \sin a + \frac{n(n+1)}{1.2} \sin 2a + \frac{n(n+1)(n+2)}{1.2.3} \sin 3a + \dots,$$

if (i) $(4k-2)\pi < a < 4k\pi$; (ii) $4k\pi < a < (4k+2)\pi$; (iii) $a = 2r\pi$.

33. Prove that, if $0 < \theta < \pi$, the sum to infinity of the series,

$$1 - \tfrac{1}{2}\cos\theta + \frac{1.3}{2.4}\cos 2\theta - \frac{1.3.5}{2.4.6}\cos 3\theta + \dots$$

is $\frac{1}{2}\cos\frac{\theta}{4}\sqrt{(2\sec\frac{1}{2}\theta)}$. Find the sum to infinity if $\pi < \theta < 2\pi$.

34. If $-\frac{\pi}{4} < \theta < \frac{\pi}{4}$, prove that $\cos n\theta \cos^n\theta$ is the sum to infinity of the series,

$$1 - \frac{n(n+1)}{2!}\tan^2\theta + \frac{n(n+1)(n+2)(n+3)}{4!}\tan^4\theta - \dots.$$

Find the sum to infinity if $\frac{3\pi}{4} < \theta < \frac{5\pi}{4}$.

35. If $|x| < 1$, find the sum to infinity of the series,

$$\sin 2\theta + 2x\sin 3\theta + 3x^2\sin 4\theta + 4x^3\sin 5\theta + \dots.$$

The Generalised Inverse Circular and Hyperbolic Functions. The function $\mathrm{Tan}^{-1}\mathbf{z}$ is defined to be any value of w which satisfies the equation, $z = \tan w$. The other inverse functions are defined in a similar manner; all of them are many-valued. Thus

$$\mathrm{Sin}^{-1}z = n\pi + (-1)^n \sin^{-1}z, \qquad \mathrm{Sh}^{-1}z = n\pi i + (-1)^n \mathrm{sh}^{-1}z,$$
$$\mathrm{Cos}^{-1}z = 2n\pi \pm \cos^{-1}z, \qquad \mathrm{Ch}^{-1}z = 2n\pi i \pm \mathrm{ch}^{-1}z,$$
$$\mathrm{Tan}^{-1}z = n\pi + \tan^{-1}z, \qquad \mathrm{Th}^{-1}z = n\pi i + \mathrm{th}^{-1}z.$$

The inverse hyperbolic functions can be expressed as logarithmic functions. For example,

let $\mathrm{Th}^{-1}z = w$, then $z = \mathrm{th}\, w = \dfrac{\exp(2w) - 1}{\exp(2w) + 1}$;

$$\therefore \exp(2w) = \frac{1+z}{1-z};$$

$$\therefore 2w = \mathrm{Log}\frac{1+z}{1-z};$$

$$\therefore \mathbf{Th^{-1}z = w = \tfrac{1}{2}\,Log\,\frac{1+z}{1-z}}. \quad \dots\dots\dots(27)$$

In expressing the inverse functions in the form $a + i\beta$, care is required. The work for $\mathrm{Tan}^{-1}(x+iy)$ is given in full below, and the other results are stated in Ex. XIII. d, Nos. 15, 16.

Expression of $\mathrm{Tan}^{-1}(x+iy)$ in the form $\alpha+i\beta$. From the relation, $\mathrm{Tan}^{-1}(x+iy)=a+i\beta$, we have

$$\tan(a+i\beta)=x+iy\,; \quad \therefore \tan(a-i\beta)=x-iy\,;$$

$$\therefore 2x=\tan(a+i\beta)+\tan(a-i\beta)=\frac{\sin\{(a+i\beta)+(a-i\beta)\}}{\cos(a+i\beta)\cos(a-i\beta)}\,;$$

$$\therefore 2x=\frac{\sin 2a}{\{|\cos(a+i\beta)|\}^2}, \text{ see p. 143.}$$

Also
$$x^2+y^2=(x+iy)(x-iy)=\frac{\sin(a+i\beta)\sin(a-i\beta)}{\cos(a+i\beta)\cos(a-i\beta)}$$
$$=\frac{\mathrm{ch}\,2\beta-\cos 2a}{\cos 2a+\mathrm{ch}\,2\beta}\,;$$

$$\therefore 1-x^2-y^2=\frac{2\cos 2a}{\cos 2a+\mathrm{ch}\,2\beta}=\frac{\cos 2a}{\{|\cos(a+i\beta)|\}^2}\,;$$

$\therefore$ $\sin 2a$ has the same sign as $2x$, and $\cos 2a$ has the same sign as $1-x^2-y^2$;

$$\therefore \sin 2a : \cos 2a : 1 = 2x : 1-x^2-y^2 : +\sqrt{\{4x^2+(1-x^2-y^2)^2\}}. \quad (28)$$

Unless both x and $1-x^2-y^2$ are zero, that is unless $x=0$ *and* $y=\pm 1$, equation (28) gives *one, and only one,* value of $2a$, say $2a_0$, such that $-\pi<2a_0\leqslant\pi$, and the general value is given by

$$2a=2n\pi+2a_0, \quad \text{or} \quad a=n\pi+a_0,$$

where $-\frac{\pi}{2}<a_0\leqslant\frac{\pi}{2}$.

Also
$$\tan 2i\beta=\tan\{(a+i\beta)-(a-i\beta)\}=\frac{(x+iy)-(x-iy)}{1+(x+iy)(x-iy)}\,;$$

$$\therefore i\,\mathrm{th}2\beta=\frac{2iy}{1+x^2+y^2}\,;$$

$$\therefore \{\exp(4\beta)-1\}(1+x^2+y^2)=\{\exp(4\beta)+1\}2y\,;$$

$$\therefore \exp(4\beta)=\frac{1+x^2+y^2+2y}{1+x^2+y^2-2y},$$

which gives a unique (real) value of β, namely

$$\tfrac{1}{4}\log\frac{x^2+(1+y)^2}{x^2+(1-y)^2}\,;$$

$$\therefore \mathbf{Tan^{-1}(x+iy)=n\pi+\alpha_0+\frac{i}{4}\log\frac{x^2+(1+y)^2}{x^2+(1-y)^2}}, \quad \text{.........(29)}$$

where α_0 is the unique angle satisfying equation (28), such that $-\frac{\pi}{2} < \alpha_0 \leqslant \frac{\pi}{2}$. For the *principal value*, we have

$$\tan^{-1}(x+iy) = \alpha_0 + \tfrac{1}{4}i \log \frac{x^2+(1+y)^2}{x^2+(1-y)^2}. \quad \text{............(30)}$$

The function $\operatorname{Tan}^{-1}(x+iy)$ is not defined for either of the pairs of values, $x=0$, $y=1$, or $x=0$, $y=-1$.

Note. The reader should observe the necessity for determining the sign of $\sin 2\alpha$ and $\cos 2\alpha$. It might seem to him quicker to say

$$\tan 2\alpha = \tan\{(\alpha+i\beta)+(\alpha-i\beta)\} = \frac{(x+iy)+(x-iy)}{1-(x+iy)(x-iy)} = \frac{2x}{1-x^2-y^2}$$

and to deduce that $\alpha = \frac{1}{2}\operatorname{Tan}^{-1}\frac{2x}{1-x^2-y^2}$, but this gives *two* values of α between $-\frac{\pi}{2}$ and $+\frac{\pi}{2}$, and it is evident from the previous work that one of these is incorrect.

Also equation (28) is written in a proportion form to avoid considering separately the special cases $1-x^2-y^2=0$, $x=0$.

EXERCISE XIII. d.

1. Prove that, in equation (29), if $x^2+y^2 \neq 1$, α_0 may be replaced by $\frac{1}{2}k\pi + \frac{1}{2}\tan^{-1}\frac{2x}{1-x^2-y^2}$, where

(i) $k=0$, if $x^2+y^2<1$; (ii) $k=1$, if $x^2+y^2>1$ and $x>0$;

(iii) $k=-1$, if $x^2+y^2>1$ and $x<0$.

2. Prove that $\operatorname{Ch}^{-1}z = 2n\pi i \pm i\cos^{-1}z$; find similar expressions for $\operatorname{Sh}^{-1}z$ and $\operatorname{Th}^{-1}z$.

3. Prove that $\operatorname{Sh}^{-1}z = in\pi + (-1)^n \log\{z+\sqrt{(1+z^2)}\}$.

4. Prove that $\operatorname{Ch}^{-1}z = 2in\pi \pm \log\{z+\sqrt{(z^2-1)}\}$.

5. If $\cos^{-1}(\alpha+i\beta) = u+iv$, prove that

(i) $\alpha^2\sec^2 u - \beta^2\operatorname{cosec}^2 u = 1$; (ii) $\alpha^2\operatorname{sech}^2 v + \beta^2\operatorname{cosech}^2 v = 1$;

(iii) $\cos^2 u$ and $\operatorname{ch}^2 v$ are the roots of the equation

$$\lambda^2 - \lambda(1+\alpha^2+\beta^2) + \alpha^2 = 0.$$

6. If $\sin^{-1}(\alpha+i\beta) = u+iv$, prove that $\sin^2 u$ and $\operatorname{ch}^2 v$ are the roots of the equation $\mu^2 - \mu(1+\alpha^2+\beta^2) + \alpha^2 = 0$.

7. Prove that one value of $\operatorname{Tan}^{-1}\frac{x+iy}{x-iy}$, where $x>y>0$, is

$$\frac{\pi}{4} + \frac{i}{2}\log\frac{x+y}{x-y}.$$

8. If $\text{ch}^{-1}(x+iy)+\text{ch}^{-1}(x-iy)=\text{ch}^{-1}a$, prove that

$$2(a-1)x^2+2(a+1)y^2=a^2-1.$$

9. If $\text{Sin}^{-1}2=u+iv$, prove that $u=2n\pi+\frac{\pi}{2}$ and $v=\pm\log(2+\sqrt{3})$.

10. Solve $\cos z=1\frac{1}{4}$.

11. Solve $\sin z=1\frac{2}{3}$.

12. Evaluate $\text{Cos}^{-1}(-2\cdot6)$.

13. Evaluate $\text{Sin}^{-1}i$.

14. Express $\tan^{-1}(\cos\theta+i\sin\theta)$ in the form $u+iv$, if

$$\text{(i)}\ -\frac{\pi}{2}<\theta<\frac{\pi}{2};\quad \text{(ii)}\ \frac{\pi}{2}<\theta<\frac{3\pi}{2}.$$

15. If $\text{Cos}^{-1}(x+iy)=u+iv$, where $x>0$ and $y>0$, prove that

(i) $\cos^2 u$ and $\text{ch}^2 v$ are the roots of $\lambda^2-(1+x^2+y^2)\lambda+x^2=0$;

(ii) $\text{ch}\, v=t_1+t_2$ and $\cos u=t_1-t_2$, where $t_1=+\frac{1}{2}\sqrt{\{(x+1)^2+y^2\}}$ and $t_2=+\frac{1}{2}\sqrt{\{(x-1)^2+y^2\}}$; (iii) $\sin u$ and $\text{sh}\, v$ are of opposite sign.

Deduce that $\text{Cos}^{-1}(x+iy)=2n\pi\pm\{\cos^{-1}(t_1-t_2)-i\,\text{ch}^{-1}(t_1+t_2)\}$,

where $0<\cos^{-1}(t_1-t_2)<\frac{\pi}{2}$ and $0<\text{ch}^{-1}(t_1+t_2)$.

16. If $\text{Sin}^{-1}(x+iy)=u+iv$, where $x>0$ and $y>0$, prove that, with the notation of No. 15,

$$\text{Sin}^{-1}(x+iy)=n\pi+(-1)^n\{\sin^{-1}(t_1-t_2)+i\,\text{ch}^{-1}(t_1+t_2)\},$$

where $0<\sin^{-1}(t_1-t_2)<\frac{\pi}{2}$ and $0<\text{ch}^{-1}(t_1+t_2)$.

17. If $\tan(\alpha+i\beta)=\tan\phi+i\sec\phi$, prove that, for $0<\phi<\frac{\pi}{2}$,

$$\text{(i)}\ 2\alpha=2n\pi+\frac{\pi}{2}+\phi;\quad \text{(ii)}\ \beta=\tfrac{1}{2}\log\cot\frac{\phi}{2}.$$

Find the values of 2α and β for $-\frac{\pi}{2}<\phi<0$.

18. Prove that, for real values of θ,

$$\text{Sin}^{-1}(\text{cosec}\,\theta)=n\pi+\frac{\pi}{2}+i\log\cot\frac{n\pi+\theta}{2}.$$

19. If $\sin^{-1}(\cos x+i\sin x)=u+iv$, prove that, for principal values,

(i) if $0<x<\pi$, $u=\sin^{-1}\left(\dfrac{\cos x}{\sqrt{(1+\sin x)}}\right)$,

$$v=\log\{\sqrt{(\sin x)}+\sqrt{(1+\sin x)}\};$$

(ii) if $\pi<x<2\pi$, $u=\sin^{-1}\left(\dfrac{\cos x}{\sqrt{(1-\sin x)}}\right)$,

$$v=\log\{\sqrt{(-\sin x)}+\sqrt{(1-\sin x)}\}.$$

EASY MISCELLANEOUS EXAMPLES

EXERCISE XIII. e.

Express in the form $a+bi$:

1. $e^{(2n+1)\frac{\pi i}{2}}$, ($n$ integral). **2.** $e^{(x+yi)^2}$. **3.** $(\sin\theta+i\cos\theta)^i$.

4. $\operatorname{Log}\{\operatorname{Log}(\cos\theta+i\sin\theta)\}$. **5.** $\tan\left\{\theta-i\log\left(\tan\frac{\theta}{2}\right)\right\}$.

6. Find all the values of $(1+i)^{1+i}$.

7. Prove that, for $0<x<\pi$,

$$\log\sin(x+iy)=\tfrac{1}{2}\log(\operatorname{ch}^2 y-\cos^2 x)+i\tan^{-1}(\cot x\operatorname{th} y).$$

Find a similar expression for $\log\sin(x+iy)$, if $\pi<x<\frac{3\pi}{2}$ and $y>0$.

8. If $\log\tan\left(\frac{\pi}{4}+\frac{\theta}{2}\right)=c$, prove that $\theta=i\operatorname{Log}\cot\left(\frac{\pi}{4}+\tfrac{1}{2}ci\right)$.

9. Sum to infinity:

$$\cos\theta\frac{\cos\theta}{1}-\sin 2\theta\frac{\cos^2\theta}{2}-\cos 3\theta\frac{\cos^3\theta}{3}+\sin 4\theta\frac{\cos^4\theta}{4}+\ldots.$$

10. If $\sin\theta=\operatorname{th}\frac{\pi}{2}$ and if $\cos\theta>0$, prove that one value of

$$\{\tan^{-1}\exp(\theta i)\}^m+\{\tan^{-1}\exp(-\theta i)\}^m \text{ is } 2^{1-\frac{3m}{2}}\pi^m\cos\frac{m\pi}{4}.$$

11. If $\tan\frac{\alpha}{2}=2\tan\frac{\beta}{2}$, and if $0<\alpha<\pi$, $0<\beta<\pi$, prove that

$$\frac{1}{1.3}\sin\beta+\frac{1}{2.3^2}\sin 2\beta+\frac{1}{3.3^3}\sin 3\beta+\ldots=\tfrac{1}{2}(\alpha-\beta).$$

12. If $(1-c)\tan x=(1+c)\tan(x-y)$, and $|c|<1$, express y as a series of powers of c, involving x.

13. If $2\pi<\theta<4\pi$, find the sum to infinity of the series,

$$\sin\theta+\tfrac{1}{2}\sin 2\theta+\tfrac{1}{3}\sin 3\theta+\ldots.$$

14. If $0<\theta<2\pi$, find the sum to infinity of the series,

$$\tfrac{1}{2}\sin\theta+\frac{1.3}{2.4}\sin 2\theta+\frac{1.3.5}{2.4.6}\sin 3\theta+\ldots.$$

15. If $n<1$ and $-\pi<\theta<\pi$, prove that $\cos\frac{n\theta}{2}\left(2\cos\frac{\theta}{2}\right)^{-n}$ is the sum to infinity of

$$1-\frac{n}{1!}\cos\theta+\frac{n(n+1)}{2!}\cos 2\theta-\frac{n(n+1)(n+2)}{3!}\cos 3\theta+\ldots;$$

Find the sum if $\pi<\theta<3\pi$.

16. Find the sum to infinity of the series,

$$\cos\theta\cos\theta - \tfrac{1}{3}\cos^3\theta\cos 3\theta + \tfrac{1}{5}\cos^5\theta\cos 5\theta - \dots .$$

17. Find all real values of x, y, such that $\log\cos(x+iy) = i\tan^{-1}\tfrac{3}{4}$.

18. Find all the values of z, if $i^z = e(\cos a + i\sin a)$.

19. If $-\dfrac{\pi}{2} < \theta < \dfrac{\pi}{6}$, find the sum to infinity of

$$\frac{1}{3}\sin 3\theta + \frac{1\,.\,4}{3\,.\,6}\sin 6\theta + \frac{1\,.\,4\,.\,7}{3\,.\,6\,.\,9}\sin 9\theta + \dots ;$$

20. Prove that, if $\text{Th}^{-1}(\tan x) = a + ib$ and $\tfrac{1}{2}i\,\text{Sec}^{-1}(\cos 2x) = c + id$, then $a = \pm c$.

HARDER MISCELLANEOUS EXAMPLES

EXERCISE XIII. f.

1. If $\log\sin(x+iy) = \alpha + i\beta$, prove that $2e^{2\alpha} = \text{ch}\,2y - \cos 2x$.

2. If $|h| < 1$, expand $\log(1 + 2h\cos\theta + h^2)$ in powers of h.

3. If $\tan^{-1}(\xi + i\eta) = \sin^{-1}(x+iy)$, prove that

$$\xi^2 + \eta^2 = (x^2+y^2)/\sqrt{(x^4 + 2x^2y^2 + y^4 - 2x^2 + 2y^2 + 1)}.$$

4. Simplify $\sin(\text{Log}\, i^i)$.

5. Prove that, for $a > 0$, one value of $(a+bi)^{\alpha+\beta i}$ is of the form vi, if $\tfrac{1}{2}\beta\log(a^2+b^2) + \alpha\tan^{-1}\dfrac{b}{a}$ is an odd multiple of $\dfrac{\pi}{2}$.

6. If $\tan\log(a+ib) = x+iy$, where $x^2+y^2 \neq 1$, prove that

$$2x = (1 - x^2 - y^2)\tan\log(a^2+b^2).$$

7. If $\text{ch}(x+iy) = \sec(u+iv)$, prove that

(i) $\tan^2 u = \text{sh}^2 x\sec^2 y$; (ii) $\text{th}^2 v = \text{sech}^2 x\sin^2 y$.

8. Prove that $\log\dfrac{\cos(x-iy)}{\cos(x+iy)} = 2i\tan^{-1}(\tan x\,\text{th}\,y)$.

9. If $3\tan(\theta - \alpha) = \tan\theta$, prove that one value of α is the sum to infinity of

$$\frac{\sin 2\theta}{1\,.\,2} - \frac{\sin 4\theta}{2\,.\,2^2} + \frac{\sin 6\theta}{3\,.\,2^3} - \dots .$$

10. If $-\dfrac{\pi}{2} < \theta < \dfrac{\pi}{2}$, find the sum to infinity of the series,

$$1 + \tfrac{1}{2}\cos 2\theta - \frac{1}{2\,.\,4}\cos 4\theta + \frac{1\,.\,3}{2\,.\,4\,.\,6}\cos 6\theta - \frac{1\,.\,3\,.\,5}{2\,.\,4\,.\,6\,.\,8}\cos 8\theta + \dots .$$

11. If $0 < \theta < \pi$, find the sum to infinity of the series,

$$\cos\theta + \tfrac{1}{3}\cos 3\theta + \tfrac{1}{5}\cos 5\theta + \dots .$$

12. If the principal value of $\frac{1}{i}\log\frac{x+iy-1}{x+iy+1}$ is $u+iv$, prove that, when y tends to zero through positive values, u tends to π or zero according as x does or does not lie between -1 and $+1$.

13. If $0<x<\pi$, express $\sin^i(x+yi)$ in the modulus-amplitude form.

14. If $\tan(x+iy)=\cot\theta-i\operatorname{cosec}\theta$, find the principal values of x and y in terms of θ, (i) if $0<\theta<\frac{\pi}{2}$; (ii) if $-\frac{\pi}{2}<\theta<0$.

15. Prove that one value of $\operatorname{Sin}^{-1}\frac{5\sqrt{7}+9i}{16}$ is $\cos^{-1}\frac{3}{4}+i\log 2$.

16. If $x>0$, $y>0$, express $\operatorname{Ch}^{-1}(x+iy)$ in the form $a+ib$.

17. If $n>-1$, find the sum to infinity of the series,

$$1-\frac{n(n-1)}{2!}+\frac{n(n-1)(n-2)(n-3)}{4!}-\ldots.$$

18. Find the sum to infinity of the series,

$$\frac{\cos\theta}{1.2}-\frac{\cos 2\theta}{2.3}+\frac{\cos 3\theta}{3.4}-\ldots,$$

(i) if $-\pi<\theta<\pi$; (ii) if $\pi<\theta<3\pi$.

19. If $y=\log\tan\left(\frac{\pi}{4}+\frac{x}{2}\right)$, prove that $x=-i\operatorname{Log}\tan\left(\frac{iy}{2}+\frac{\pi}{4}\right)$.

If y is expressed as a convergent series in powers of x,

$$a_1x+a_3x^3+a_5x^5+\ldots,$$

and if x can be expressed as a convergent series in powers of y, find the expansion for x in powers of y.

20. Prove that $\sum_1^{\infty}{}'\frac{\sin 2nx}{2n}$, where Σ' denotes that terms for which n is a multiple of 3 are omitted, is

(i) $\frac{\pi}{6}$, if $0<x<\frac{\pi}{3}$; (ii) 0, if $\frac{\pi}{3}<x<\frac{2\pi}{3}$; (iii) $-\frac{\pi}{6}$, if $\frac{2\pi}{3}<x<\pi$.

21. Determine the conditions under which

(i) $\cos\frac{\pi}{3}\cos\theta+\frac{1}{2}\cos\frac{2\pi}{3}\cos 2\theta+\frac{1}{3}\cos\frac{3\pi}{3}\cos 3\theta+\ldots$
$=-\frac{1}{2}\log(1-2\cos\theta)$;

(ii) $\cos\frac{\pi}{3}\sin\theta+\frac{1}{2}\cos\frac{2\pi}{3}\sin 2\theta+\frac{1}{3}\cos\frac{3\pi}{3}\sin 3\theta+\ldots=-\frac{1}{2}\theta$.

22. If P, P_1, P_2 represent the complex numbers z, z_1, z_2, and if the half-line from P parallel to Ox either cuts P_1P_2 externally or does not cut P_1P_2 at all, prove that $\log\frac{z-z_1}{z-z_2}=\log(z-z_1)-\log(z-z_2)$.

CHAPTER XIV

MISCELLANEOUS RELATIONS

Numerical, Single-letter, and Two-letter Identities.

Example 1. Prove that $\cos^2\theta + \cos^2(\alpha+\theta) - 2\cos\alpha\cos\theta\cos(\alpha+\theta)$ is independent of θ.

First Method. The expression

$$=\cos^2\theta + \cos^2(\alpha+\theta) - \{\cos(\alpha-\theta) + \cos(\alpha+\theta)\}\cos(\alpha+\theta)$$
$$=\cos^2\theta - \cos(\alpha-\theta)\cos(\alpha+\theta)$$
$$=\tfrac{1}{2}(1+\cos 2\theta) - \tfrac{1}{2}(\cos 2\theta + \cos 2\alpha)$$
$$=\tfrac{1}{2}(1-\cos 2\alpha) = \sin^2\alpha.$$

Second Method.

If $\quad f(\theta) \equiv \cos^2\theta + \cos^2(\alpha+\theta) - 2\cos\alpha\cos\theta\cos(\alpha+\theta)$,

$$\frac{d}{d\theta}f(\theta) = -2\cos\theta\sin\theta - 2\cos(\alpha+\theta)\sin(\alpha+\theta)$$
$$+2\cos\alpha\{\sin\theta\cos(\alpha+\theta) + \cos\theta\sin(\alpha+\theta)\}$$
$$= -\{\sin 2\theta + \sin(2\alpha+2\theta)\} + 2\cos\alpha\sin(\alpha+2\theta) = 0;$$
$$\therefore\ f(\theta) \text{ is independent of } \theta.$$

Third Method. Take a circle with diameter OD of unit length, and draw chords OB, OA making angles θ, $\alpha+\theta$ with that diameter, as in Fig. 79; then $\text{OB} = \cos\theta$, $\text{OA} = \cos(\alpha+\theta)$, and $\text{AB} = \sin\alpha$. The expression

$$= \text{OB}^2 + \text{OA}^2 - 2\text{OB}\,.\,\text{OA}\,.\cos \text{AOB} = \text{AB}^2,$$

which is independent of θ.

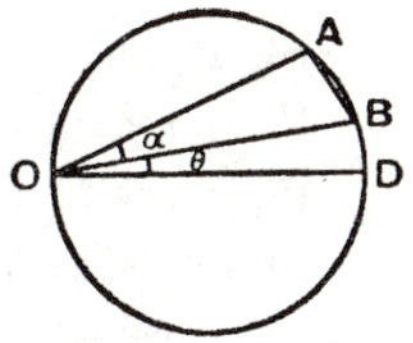

FIG. 79.

Fourth Method. Take a triangle ABC having $\text{A} = \frac{\pi}{2} + \theta$, $\text{B} = \frac{\pi}{2} - \theta - \alpha$, and hence $\text{C} = \alpha$; then

$c^2 = a^2 + b^2 - 2ab\cos\text{C}$;

$$\therefore\ \sin^2\text{C} = \sin^2\text{A} + \sin^2\text{B} - 2\sin\text{A}\sin\text{B}\cos\text{C};$$
$$\therefore\ \sin^2\alpha = \sin^2\left(\frac{\pi}{2}+\theta\right) + \sin^2\left(\frac{\pi}{2}-\theta-\alpha\right)$$
$$-2\sin\left(\frac{\pi}{2}+\theta\right)\sin\left(\frac{\pi}{2}-\theta-\alpha\right)\cos\alpha$$
$$= \cos^2\theta + \cos^2(\alpha+\theta) - 2\cos\theta\cos(\theta+\alpha)\cos\alpha.$$

It should be noticed that the geometrical methods have to be modified, or interpreted in accordance with certain conventions for some values of α and θ.

EXERCISE XIV. a.

Prove the following :

1. $(\cos A + \sin A)(\cot A + \tan A) = \operatorname{cosec} A + \sec A.$
2. $(2 - \cos^2 B)(2 + \tan^2 B) = (1 + 2\tan^2 B)(2 - \sin^2 B).$
3. $\tan^2 C + \cot^2 C = \operatorname{cosec}^2 C \sec^2 C - 2.$
4. If $\sec D + \operatorname{cosec} D = \sqrt{2}$, then $\cos^3 D + \sin^3 D = -\frac{1}{\sqrt{}}.$
5. $\cos^2 22\tfrac{1}{2}^\circ - \cos^2 67\tfrac{1}{2}^\circ = \cos 45^\circ.$
6. $2\cos 5^\circ\, 37'\, 30'' = \sqrt{[2 + \sqrt{\{2 + \sqrt{(2 + \sqrt{2})}\}}]}.$
7. $4\cos 24^\circ \cos 36^\circ \cos 84^\circ = \sin 18^\circ.$
8. $\tan 9^\circ - \tan 27^\circ - \tan 63^\circ + \tan 81^\circ = 4.$
9. $\cos 12^\circ + \cos 60^\circ + \cos 84^\circ = \cos 24^\circ + \cos 48^\circ.$
10. $\sin 40^\circ \sin 50^\circ = \sin 30^\circ \sin 80^\circ.$
11. $\tan 20^\circ \tan 40^\circ = \tan 10^\circ \tan 60^\circ.$
12. If $x : y : z : 1 = \sin 40^\circ : \sin 60^\circ : \sin 80^\circ : \sin 20^\circ$, then
$$x^2 - 1 : y^2 - 1 : z^2 - 1 = y : xz : yz.$$
13. $\cos^2 14^\circ - \cos 7^\circ \cos 21^\circ = \sin^2 7^\circ.$
14. If $\tan\frac{\theta}{2} = t$, then $\tan\theta + \sec\theta = \frac{1+t}{1-t}.$
15. If $\cot A - \tan A = x$, then $\tan 4A = \frac{4x}{x^2 - 4}.$
16. If $\sec 2A = 2 + \sec A$, then $\cos 2A + \cos 3A = 0.$
17. $\operatorname{cosec} 2\theta + \cot 4\theta = \cot\theta - \operatorname{cosec} 4\theta.$
18. $4(\cos 2\theta + \cos 6\theta)(\cos 6\theta + \cos 8\theta) = 1 + \sin 15\theta \operatorname{cosec}\theta.$
19. $\cot^3\theta + \tan^3\theta = 8\operatorname{cosec}^3 2\theta - 6\operatorname{cosec} 2\theta.$
20. $\sin^3\theta \sin 3\theta + \cos^3\theta \cos 3\theta = \cos^3 2\theta.$
21. $\sin^3(60^\circ + \theta) + \sin^3(60^\circ - \theta) = \frac{3\sqrt{3}\cos\theta}{4}.$
22. $3\tan\theta - 2\cot\theta = \operatorname{cosec} 2\theta - 5\cot 2\theta.$
23. If $\alpha + \beta = \frac{\pi}{2}$, then $\left(1 + \tan\frac{\alpha}{2}\right)\left(1 + \tan\frac{\beta}{2}\right) = 2.$
24. If $\tan^2 A = 1 + 2\tan^2 B$, then $\cos 2B = 1 + 2\cos 2A.$
25. If $\tan 2\theta = \mu \operatorname{cosec} 2\alpha - \cot 2\alpha$, then $\tan(2\theta - \alpha) = \frac{\mu - 1}{\mu + 1}\cot\alpha.$
26. If $(1 + 3\sin^2\phi)^{\frac{1}{3}} = \sin^{\frac{2}{3}}\tfrac{1}{2}\theta + \cos^{\frac{2}{3}}\tfrac{1}{2}\theta$, then
$$(1 + 3\tan^2\phi)\tan\theta = \pm 2\tan^3\phi.$$

27 If $(1+e\cos\alpha)(1-e\cos\beta)=1-e^2$, and e is not zero, then

$$(1-e)\tan^2\tfrac{1}{2}\alpha=(1+e)\tan^2\tfrac{1}{2}\beta.$$

28. If $\sin\theta:\sin\phi=\sqrt{(1-2e\cos\theta+e^2)}:\sqrt{(1-2e\cos\phi+e^2)}$, then

$$(1-2e\cos\theta+e^2)(1-2e\cos\phi+e^2)=(1-e^2)^2.$$

29. $\cos^2\alpha-\cos^2(\alpha+\beta)=\sin^2\beta+2\sin\alpha\sin\beta\cos(\alpha+\beta)$.

30. $\cos 3\alpha+\cos 3\beta$

$$=4(\cos\alpha+\cos\beta)(\cos\overline{\alpha+120^\circ}+\cos\beta)(\cos\overline{\alpha-120^\circ}+\cos\beta).$$

31. If $\dfrac{\cos^3\alpha}{\cos\beta}+\dfrac{\sin^3\alpha}{\sin\beta}=1$, then $\sin 2\alpha+2\sin(\alpha+\beta)=0$.

32. If $(2\cos\alpha-\cos 2\alpha)\sin 3\beta-(2\sin\alpha-\sin 2\alpha)\cos 3\beta=3\sin\beta$ and $\alpha\neq 2(\beta+n\pi)$, then $\dfrac{\sin\alpha}{\cos\alpha-2}=\tan\beta$.

Conditional Identities.

Example 2. If $A+B+C=\pi$, prove that

$$\Sigma\{\sin^4 A(\sin^2 B+\sin^2 C-\sin^2 A)\}$$
$$=2\sin^2 A\sin^2 B\sin^2 C(1+4\cos A\cos B\cos C).$$

We have $\Sigma x(y+z)^2\equiv(y+z)(z+x)(x+y)+4xyz$.

Putting $b^2+c^2-a^2,\ c^2+a^2-b^2,\ a^2+b^2-c^2$ for x, y, z, the identity gives

$$4\Sigma a^4(b^2+c^2-a^2)=8a^2b^2c^2+4(b^2+c^2-a^2)(c^2+a^2-b^2)(a^2+b^2-c^2)$$
$$=8a^2b^2c^2+4\,.\,2bc\cos A\,.\,2ca\cos B\,.\,2ab\cos C,$$

where A, B, C are the angles of the triangle with sides a, b, c. Since $a:b:c=\sin A:\sin B:\sin C$, it follows that

$$\Sigma\{\sin^4 A(\sin^2 B+\sin^2 C-\sin^2 A)\}$$
$$=2\sin^2 A\sin^2 B\sin^2 C(1+4\cos A\cos B\cos C).$$

Example 3. If $x+y+z=0$, prove that

$$\Big(\sum_{x,y,z}\sin 3x\Big)[\cos(2\theta+y-z)+\cos(2\theta+z-x)+\cos(2\theta+x-y)]$$
$$=-\Sigma\{\cos 2(\theta-y+z)\sin 3x\}.$$

The left-hand side may be written

$$\Sigma\{\sin 3x\cos(2\theta+y-z)\}$$
$$+\Sigma\{\sin 3y\cos(2\theta+x-y)+\sin 3z\cos(2\theta+z-x)\}.$$

The first of these terms

$$=\Sigma\{\sin(2\theta+3x+y-z)-\sin(2\theta-3x+y-z)\}$$
$$=\Sigma\{\sin(2\theta-2z+2x)-\sin(2\theta-2x+2y)\},\text{ since } x+y+z=0$$
$$=0.$$

The second term

$$=\tfrac{1}{2}\Sigma\{\sin(2\theta+x+2y)-\sin(2\theta+x-4y)+\sin(2\theta+4z-x)-\sin(2\theta-2z-x)\}$$
$$=\tfrac{1}{2}\Sigma\{\sin(2\theta+y-z)-\sin(2\theta+x-4y)+\sin(2\theta+4z-x)-\sin(2\theta+y-z)\}$$
$$=\Sigma\{\cos(2\theta-2y+2z)\sin(2y+2z-x)\}$$
$$=-\Sigma\{\cos 2(\theta-y+z)\sin 3x\},\text{ since } x+y+z=0.$$

EXERCISE XIV. b.

In this Exercise, it is to be assumed that $A+B+C=\pi$, *or that* $A+B+C+D=2\pi$.

Prove the following :

1. $\cos^2 A+\cos^2 B+\cos^2 C+2\cos A\cos B\cos C=1.$

2. $\sin^2 B+\sin^2 C-2\sin B\sin C\cos\left(\frac{\pi}{3}+A\right)$ is symmetrical.

3. $1+\cos 2A+\cos 2B+\cos 2C=-4\cos A\cos B\cos C.$

4. $\sin 3A+\sin 3B+\sin 3C=-4\cos\frac{3A}{2}\cos\frac{3B}{2}\cos\frac{3C}{2}.$

5. $\dfrac{1-\cos A+\cos B+\cos C}{1+\cos A+\cos B-\cos C}=\tan\tfrac{1}{2}A\cot\tfrac{1}{2}C.$

6. $\cos\frac{A}{2}+\cos\frac{B}{2}-\cos\frac{C}{2}=4\cos\frac{\pi+A}{4}\cos\frac{\pi+B}{4}\cos\frac{\pi-C}{4}.$

7. $\sin 2A\sin^2 A+\sin 2B\sin^2 B+\sin 2C\sin^2 C$
$=\sin 2A\sin 2B\sin 2C+2\sin A\sin B\sin C.$

8. $\sin^4 A+\sin^4 B+\sin^4 C$
$=2(\sin^2 B\sin^2 C+\sin^2 C\sin^2 A+\sin^2 A\sin^2 B)-4\sin^2 A\sin^2 B\sin^2 C.$

9. $\sin 5A+\sin 5B+\sin 5C=4\cos\frac{5A}{2}\cos\frac{5B}{2}\cos\frac{5C}{2}.$

10. $4(\cos^5 A+\cos^5 B+\cos^5 C)$
$=4+10\Pi(\sin\tfrac{1}{2}A)-5\Pi(\sin\tfrac{3}{2}A)+\Pi(\sin\tfrac{5}{2}A).$

11. $\sin 2nA+\sin 2nB+\sin 2nC=-4\cos n\pi\sin nA\sin nB\sin nC.$

12. $\cot A+\cot B+\cot C=\cot A\cot B\cot C+\operatorname{cosec} A\operatorname{cosec} B\operatorname{cosec} C.$

13. $(\tan A + \tan B + \tan C)(\cot A + \cot B + \cot C) - \sec A \sec B \sec C$ is constant.

14. If

$$\cos A = \cot\beta \cot\gamma, \quad \cos B = \cot\gamma \cot\alpha, \quad \text{and} \quad \cos C = \cot\alpha \cot\beta,$$

then $\cos^2\alpha + \cos^2\beta + \cos^2\gamma = 1$.

15. If $\dfrac{\sin A + \sin B}{\sin A + \sin C} = \dfrac{\sin \frac{1}{2}B}{\sin \frac{1}{2}C}$, then $B = C$.

Deduce that, if two angle-bisectors of a triangle are equal, then the triangle is isosceles.

16. If $(m+n)\tan A + (n+l)\tan B + (l+m)\tan C = 0$, then

$$\Sigma\{(m+n)^2 \tan A\} = \Sigma\{(m-n)^2 \cot A\}.$$

17. $\sin B \sin z \sin(B + x - y) - \sin C \sin(B - y)\sin(A + z - x)$

$$+ \sin(A - x)\sin(B - y)\sin(C - z) = \sin x \sin y \sin z.$$

18. $\sin A + \sin B + \sin C + \sin D = 4 \sin \dfrac{A+B}{2} \sin \dfrac{A+C}{2} \sin \dfrac{A+D}{2}$.

19. $\frac{1}{4}(\cos 2A + \cos 2B + \cos 2C + \cos 2D)$

$$= \cos(A+B)\cos(A+C)\cos(A+D)$$

$$= \cos A \cos B \cos C \cos D - \sin A \sin B \sin C \sin D.$$

20. $\dfrac{\tan A \tan B - \tan C \tan D}{\tan A \tan C - \tan B \tan D} = \dfrac{\tan(A+C)}{\tan(A+B)}$.

21. If $\tan\alpha = \dfrac{\tan\beta + \tan\gamma}{1 + \tan\beta \tan\gamma}$, then $\sin 2\alpha = \dfrac{\sin 2\beta + \sin 2\gamma}{1 + \sin 2\beta \sin 2\gamma}$.

22. If $\tan\phi = \dfrac{\sin\alpha \sin\theta}{\cos\theta - \cos\alpha}$, then $\tan\theta = \dfrac{\sin\alpha \sin\phi}{\cos\phi \pm \cos\alpha}$.

23. $\sin\alpha \sin(\beta - \gamma) + \sin\beta \sin(\gamma - \alpha) + \sin\gamma \sin(\alpha - \beta) = 0$,
and deduce that

$$\sin^3\alpha \sin^3(\beta - \gamma) + \sin^3\beta \sin^3(\gamma - \alpha) + \sin^3\gamma \sin^3(\alpha - \beta)$$

$$= 3 \sin\alpha \sin\beta \sin\gamma \sin(\beta - \gamma)\sin(\gamma - \alpha)\sin(\alpha - \beta).$$

24. $\Sigma \sin\beta \sin\gamma \sin(\beta - \gamma) = -\sin(\beta - \gamma)\sin(\gamma - \alpha)\sin(\alpha - \beta)$.

25. $\Sigma \sin 3\alpha \cos(\beta - \gamma)$

$$= \sin(\alpha + \beta + \gamma)\{4\cos(\beta - \gamma)\cos(\gamma - \alpha)\cos(\alpha - \beta) - 1\}.$$

26. $\dfrac{\Sigma\{\sin 3\alpha \sin(\beta - \gamma)\}}{\Sigma\{\cos 3\alpha \sin(\beta - \gamma)\}} = \tan(\alpha + \beta + \gamma)$.

27. $\Sigma \sin\beta \sin\gamma \sin(\beta + \gamma)$

$$= \sin(\beta + \gamma)\sin(\gamma + \alpha)\sin(\alpha + \beta) - 2\sin\alpha \sin\beta \sin\gamma \cos(\alpha + \beta + \gamma).$$

28. $\Sigma\{\sin 2\alpha \sin(\gamma - \beta)\} = \Sigma \sin(\beta - \gamma) \,.\, \Sigma \sin(\beta + \gamma)$.

29. If $\cos 2\alpha + 2\sin\beta \sin\gamma = 0 = \cos 2\beta + 2\sin\gamma \sin\alpha$, prove that, in general, $\cos 2\gamma + 2\sin\alpha \sin\beta = 0$.

30. If $\frac{\sin(\alpha+\theta)}{\sin(\alpha+\phi)}=\frac{\sin(\beta+\theta)}{\sin(\beta+\phi)}$, prove that $\alpha-\beta$ or $\theta-\phi$ is $n\pi$.

31. If $\sin^3\theta+\sin(A+\theta)\sin(B+\theta)\sin(C+\theta)=0$, prove that

$$\cot A+\cot B+\cot C+\cot\theta=0.$$

32. $\sin s\sin(s-a)\sin(s-b)\sin(s-c)$

$$+\cos s\cos(s-a)\cos(s-b)\cos(s-c)=\cos a\cos b\cos c,$$

where $a+b+c=2s$.

Miscellaneous Transformations.

Example 4. If $\tan\alpha=\tan^3\beta$ and $\tan 2\beta=2\tan\gamma$, prove that

$$\alpha+\beta-\gamma=n\pi.$$

$$\tan(\alpha+\beta)=\frac{\tan\alpha+\tan\beta}{1-\tan\alpha\tan\beta}=\frac{\tan^3\beta+\tan\beta}{1-\tan^4\beta}$$

$$=\frac{\tan\beta(\tan^2\beta+1)}{(1-\tan^2\beta)(1+\tan^2\beta)}=\frac{\tan\beta}{1-\tan^2\beta}$$

$$=\tfrac{1}{2}\tan 2\beta=\tan\gamma;$$

$$\therefore\ \alpha+\beta=n\pi+\gamma.$$

EXERCISE XIV. c.

1. If $\cos A=\frac{\cos a-\cos b\cos c}{\sin b\sin c}$, prove that

$$\tan\tfrac{1}{2}A=\sqrt{\frac{\sin(s-b)\sin(s-c)}{\sin s\sin(s-a)}}.$$

where $2s=a+b+c$.

2. If $\tan(B+C-A)+\tan(C+A-B)+\tan(A+B-C)=\tan(A+B+C)$, prove that A or B or C is $\frac{n\pi}{2}$.

3. If $\sin x+\sin y+\sin z=4\sin\frac{y+z}{2}\sin\frac{z+x}{2}\sin\frac{x+y}{2}$, prove that $x+y+z=n\pi$.

4. If $\cos 2A+\cos 2B+\cos 2C+\cos(2A+2B+2C)=0$, prove that

$$B+C \text{ or } C+A \text{ or } A+B=(2n+1)\frac{\pi}{2}.$$

5. If $\cos 2A+\cos 2B+\cos 2C+1+4\cos A\cos B\cos C=0$, prove

$$A\pm B\pm C=(2n+1)\pi.$$

6. If $(\sin\alpha+\sin\beta+\sin\gamma)^2+(\cos\alpha+\cos\beta+\cos\gamma)^2=1$, prove that two of the angles differ by $(2n+1)\pi$.

7. If $\dfrac{\cos\alpha+\cos\beta+\cos\gamma}{\cos(\alpha+\beta+\gamma)}=\dfrac{\sin\alpha+\sin\beta+\sin\gamma}{\sin(\alpha+\beta+\gamma)}$, prove that $\Sigma\sin(\beta+\gamma)=0$, and that each fraction $=\Sigma\cos(\beta+\gamma)$.

8. If $\sin\theta\cot\dfrac{\phi+\psi}{2}=\sin\phi\cot\dfrac{\psi+\theta}{2}$ and $\theta-\phi\neq 2n\pi$, prove that $\Sigma\sin\theta=-\sin(\Sigma\theta)$, and that $\sin\psi\cot\dfrac{\theta+\phi}{2}=\sin\theta\cot\dfrac{\phi+\psi}{2}$.

9. If
$$\frac{\sin(2\theta-\phi-\psi)}{\cos(2\theta+\phi+\psi)}\quad\text{and}\quad\frac{\sin(2\phi-\theta-\psi)}{\cos(2\phi+\theta+\psi)}$$
are equal, and if $\theta-\phi\neq n\pi$, prove that $\dfrac{\sin(2\psi-\theta-\phi)}{\cos(2\psi+\theta+\phi)}$ is equal to them.

10. If $\sin 2x=k\tan\beta$ and $\sin 2y=k\tan\alpha$, and $x-y=\alpha-\beta\neq 0$, $\alpha+\beta\neq n\pi$, prove that $\sin 2x+\sin 2y=-2\sin(\alpha+\beta)\cos(x+y)$, and that $\sin(\alpha+\beta-x-y)=k\tan(x+y)$.

11. If $\cos\alpha=\dfrac{\cos\beta\cos\gamma}{1+\sin\beta\sin\gamma}$, prove that $\cos\beta=\dfrac{\cos\gamma\cos\alpha}{1\pm\sin\gamma\sin\alpha}$.

12. If $\Sigma\cos\alpha=0=\Sigma\sin\alpha$ for three angles α, β, γ between $+\pi$ and $-\pi$, prove that these are of the form $\theta-\dfrac{2\pi}{3}$, θ, $\theta+\dfrac{2\pi}{3}$, and deduce that

$$\Sigma\cos 2\alpha=0=\Sigma\sin 2\alpha,$$
$$\Sigma\cos(\beta+\gamma)=0=\Sigma\sin(\beta+\gamma),$$
$$\Sigma\cos 3\alpha=3\cos(\alpha+\beta+\gamma),$$
$$\Sigma\sin 3\alpha=3\sin(\alpha+\beta+\gamma).$$

13. Show how to use de Moivre's Theorem to prove the last two results in No. 12.

14. If $\Sigma\cos\alpha=0=\Sigma\sin\alpha$, prove
$$\Sigma\cos(2\alpha+\beta+\gamma)=0=\Sigma\sin(2\alpha+\beta+\gamma).$$

15. If $\dfrac{\cos(\mathrm{B}+\mathrm{C})}{\cos\mathrm{A}}+\dfrac{\cos(\mathrm{C}+\mathrm{A})}{\cos\mathrm{B}}=2\dfrac{\cos(\mathrm{A}+\mathrm{B})}{\cos\mathrm{C}}$, prove that either $\mathrm{A}+\mathrm{B}+\mathrm{C}=n\pi$ or $\tan\mathrm{A}+\tan\mathrm{B}=2\tan\mathrm{C}$.

16. If $\sin^2x+\sin^2y+\sin^2z=1+2\sin x\sin y\sin z$, prove that
$$x\pm y\pm z=(2n+1)\frac{\pi}{2}.$$

17. If a triangle inscribed in an ellipse has its centroid at the centre of the ellipse, show that the eccentric angles of its vertices satisfy $\Sigma\cos\alpha=0=\Sigma\sin\alpha$, and conversely.

18. If a triangle inscribed in an ellipse has its centroid at the centre, prove that the tangents at the vertices are parallel to the opposite sides. [Use No. 12 or geometry.]

Elimination. One Variable.

Some easy examples have been given in *E.T.*, p. 266. The following illustrate other methods.

Example 5. Eliminate θ from the equations

$$a\cos\theta+b\sin\theta+c=0\,;\quad a'\cos\theta+b'\sin\theta+c'=0.$$

We have
$$\frac{\cos\theta}{bc'-b'c}=\frac{\sin\theta}{ca'-c'a}=\frac{1}{ab'-a'b}\,;$$

$$\therefore\ (bc'-b'c)^2+(ca'-c'a)^2=(ab'-a'b)^2.$$

Example 6. Eliminate θ from the equations

$$x^2+y^2=\frac{x\cos 3\theta+y\sin 3\theta}{\cos^3\theta}=\frac{y\cos 3\theta-x\sin 3\theta}{\sin^3\theta}.$$

Each expression

$$=\frac{\sin\theta(x\cos 3\theta+y\sin 3\theta)+\cos\theta(y\cos 3\theta-x\sin 3\theta)}{\sin\theta\cos^3\theta+\cos\theta\sin^3\theta}$$

$$=\frac{x(\sin\theta\cos 3\theta-\cos\theta\sin 3\theta)+y(\sin\theta\sin 3\theta+\cos\theta\cos 3\theta)}{\sin\theta\cos\theta(\cos^2\theta+\sin^2\theta)}$$

$$=\frac{-x\sin 2\theta+y\cos 2\theta}{\frac{1}{2}\sin 2\theta}\,;$$

$$\therefore\ (x^2+y^2+2x)\sin 2\theta=2y\cos 2\theta.\ \ \ldots\ldots\ldots\ldots\ldots\ldots\text{(i)}$$

Also each expression

$$=\frac{\cos\theta(x\cos 3\theta+y\sin 3\theta)-\sin\theta(y\cos 3\theta-x\sin 3\theta)}{\cos^4\theta-\sin^4\theta}$$

$$=\frac{x(\cos\theta\cos 3\theta+\sin\theta\sin 3\theta)+y(\sin 3\theta\cos\theta-\cos 3\theta\sin\theta)}{(\cos^2\theta-\sin^2\theta)(\cos^2\theta+\sin^2\theta)}$$

$$=\frac{x\cos 2\theta+y\sin 2\theta}{\cos 2\theta}\,;$$

$$\therefore\ (x^2+y^2-x)\cos 2\theta=y\sin 2\theta.\ \ \ldots\ldots\ldots\ldots\ldots\ldots\text{(ii)}$$

From (i) and (ii),

$$(x^2+y^2+2x)(x^2+y^2-x)=2y^2\,;$$

$$\therefore\ (x^2+y^2)^2+x(x^2+y^2)-2x^2-2y^2=0\,;$$

$$\therefore\ (x^2+y^2)(x^2+y^2+x-2)=0.$$

EXERCISE XIV. d.

Eliminate θ from :

1. $\tan\alpha=\cos\theta\tan\beta,\ \tan\theta\sin\alpha=\tan\gamma.$
2. $a\sin\theta=b,\ a\sin(\theta-\alpha)=c.$
3. $a\cos^2\theta+b\sin^2\theta=c,\ (b-c)\tan^2\theta+(c-a)\cot^2\theta=d.$

4. $a \sin \theta = b \sin 2\theta$, $c \cos \theta = d \cos 2\theta$.

5. $\sin \theta + \sin 2\theta = a$, $\cos \theta + \cos 2\theta = b$.

6. $\sin^3\theta = a$, $\cos^3\theta = b$; also express the result in a rational form.

7. $3 \sin \theta + 2 \cos \theta = a$, $2 \sin \theta + 3 \cos \theta = b$.

8. $3 \cos \theta + \cot \theta = a$, $4 \cos \theta - \cot \theta = b$.

9. $x + \cos \theta = \sec \theta$, $y + \sin \theta = \operatorname{cosec} \theta$.

10. $a \sin \theta(4 \cos^2\theta - 1) = x$, $b \cos \theta(4 \sin^2\theta - 1) = y$.

11. $1 + \sin^2\theta = a \sin \theta$, $1 + \cos^2\theta = b \cos \theta$.

12. $x \cos \theta + y \sin \theta = a \sin 2\theta$, $-x \sin \theta + y \cos \theta = 2a \cos 2\theta$.

13. $x \cos \theta + y \sin \theta = a \operatorname{cosec} \theta$, $x \sin \theta - y \cos \theta = a \operatorname{cosec} \theta \cot \theta$.

14. $x \cos \theta + y \sin \theta = c = x \cos (\theta + a) + y \sin (\theta + a)$.

15. $a + b \cos \theta + c \cos 2\theta = 0$, $2a \cos \theta + b \cos 2\theta + c \cos 3\theta = 0$.

16. $ax \sec \theta - by \operatorname{cosec} \theta = c^2$, $ax \sec \theta \tan \theta + by \operatorname{cosec} \theta \cot \theta = 0$.

17. $\dfrac{x}{a} \cos \theta + \dfrac{y}{b} \sin \theta = 1$, $x \sin \theta - y \cos \theta = \sqrt{a^2 \sin^2\theta + b^2 \cos^2\theta}$.

18. $a \sin \theta + b \cos \theta = c$, $a \operatorname{cosec} \theta + b \sec \theta = d$.

19. $\tan \theta + \tan 2\theta = c$, $\cot \theta + \cot 2\theta = d$.

20. $\tan (a - \theta) = \dfrac{c^2 \sin 2a}{1 + c^2 \cos 2a}$, $\tan \left(\dfrac{\pi}{4} - a\right) = \dfrac{\sin (\theta - \beta)}{\sin (\theta + \beta)}$.

21. $(a + b) \tan (\theta - a) = (a - b) \tan (\theta + a)$, $a \cos 2a + b \cos 2\theta = c$.

22. $\dfrac{\cos (a - 3\theta)}{\cos^3\theta} = \dfrac{\sin (a - 3\theta)}{\sin^3\theta} = x$.

23. $x : a : b = \cos \theta + e \cos a : \sin \theta : 1 + e \cos (\theta + a)$, where $b^2 = a^2(1 - e^2)$.

24. $x = \tan^{-1}(\theta + a) + \tan^{-1}(\theta - a)$, $y = \sin^{-1}\theta$.

Elimination. Two Variables.

Example 7. Eliminate θ and ϕ from the equations,

$$(x - a) \cos \theta + y \sin \theta = a \; ; \quad (x - a) \cos \phi + y \sin \phi = a \; ;$$

$$\tan \frac{\theta}{2} - \tan \frac{\phi}{2} = 2e$$

given that θ and ϕ are unequal and between 0 and 2π.

$$\cos \theta = \frac{1 - \tan^2 \dfrac{\theta}{2}}{1 + \tan^2 \dfrac{\theta}{2}} \quad \text{and} \quad \sin \theta = \frac{2 \tan \dfrac{\theta}{2}}{1 + \tan^2 \dfrac{\theta}{2}} \; ;$$

$\therefore \tan\frac{\theta}{2}$ and $\tan\frac{\phi}{2}$ (being unequal) are the two roots of

$$(x-a)\frac{1-t^2}{1+t^2}+y\frac{2t}{1+t^2}=a.$$

This reduces to $xt^2-2yt+2a-x=0$;

$$\therefore \tan\frac{\theta}{2}+\tan\frac{\phi}{2}=\frac{2y}{x} \quad \text{and} \quad \tan\frac{\theta}{2}\tan\frac{\phi}{2}=\frac{2a-x}{x}.$$

But $$\tan\frac{\theta}{2}-\tan\frac{\phi}{2}=2e;$$

$$\therefore \tan\frac{\theta}{2}=\frac{y}{x}+e; \quad \tan\frac{\phi}{2}=\frac{y}{x}-e;$$

$$\therefore \left(\frac{y}{x}+e\right)\left(\frac{y}{x}-e\right)=\frac{2a-x}{x};$$

$$\therefore y^2-e^2x^2=2ax-x^2;$$

$$\therefore x^2(1-e^2)+y^2=2ax.$$

EXERCISE XIV. e.

Eliminate θ and ϕ in Nos. 1 to 12:

1. $\sin\theta-\sin\phi=2a,\ \cos\theta-\cos\phi=2b,\ \theta-\phi=2\gamma.$
2. $x=a\sin(\theta-\phi),\ y=2a\cos\theta\cos\phi,\ \theta+\phi=a.$
3. $\sin\theta+\sin\phi=a,\ \cos\theta+\cos\phi=b,\ \tan\theta-\tan\phi=c\sec\theta\sec\phi.$
4. $\sin\theta+\sin\phi=x,\ \cos\theta+\cos\phi=y,\ \tan\frac{\theta}{2}\tan\frac{\phi}{2}=z.$
5. $\sin\theta+\sin\phi=s,\ \tan\theta+\tan\phi=t,\ \sec\theta+\sec\phi=k.$
6. $\cos\theta+\cos\phi=a,\ \cos 2\theta+\cos 2\phi=b,\ \cos 3\theta+\cos 3\phi=c.$
7. $y\tan\theta=x\tan^2\theta+a,\ y\tan\phi=x\tan^2\phi+a, \tan\theta\tan\phi=-1.$
8. $\frac{x}{a}\cos\phi+\frac{y}{b}\sin\phi=1=\frac{x}{a}\cos\theta+\frac{y}{b}\sin\theta,$

   $$b^2\cos\theta\cos\phi+a^2\sin\theta\sin\phi=0.$$
9. $\cos\theta=\cos\alpha\cos\gamma,\ \cos\phi=\cos\beta\cos\gamma,\ \tan\frac{\theta}{2}\tan\frac{\phi}{2}=\tan\frac{\gamma}{2}.$
10. $x\cos\theta+y\sin\theta=x\cos\phi+y\sin\phi=2a,\ 2\cos\frac{\theta}{2}\cos\frac{\phi}{2}=1.$
11. $\sin 2\theta\cos\phi=\frac{2p}{a-b},\ a\cos^2\theta+b\sin^2\theta=q,$

   $$(a\sin^2\theta+b\cos^2\theta)\cos^2\phi+c\sin^2\phi=r.$$

12. $\dfrac{a \sin^2\theta + b \sin^2\phi}{b \cos^2\theta + c \cos^2\phi} = \dfrac{b \sin^2\theta + c \sin^2\phi}{c \cos^2\theta + a \cos^2\phi} = \dfrac{c \sin^2\theta + a \sin^2\phi}{a \cos^2\theta + b \cos^2\phi}$.

13. Eliminate A and B from $a \sin \mathrm{B} = b \sin \mathrm{A}$, $c = a \cos \mathrm{B} + b \cos \mathrm{A}$, and $d = \cos(\mathrm{A} + \mathrm{B})$.

14. Eliminate θ, ϕ, and ψ from

$$\cos\theta = \cos\beta \cos\gamma, \quad \cos\phi \cos\alpha = \cos\beta, \quad \cos\psi \cos\alpha = \cos\gamma,$$
$$\theta = \phi + \psi.$$

15. Eliminate θ, ϕ, ψ from

$$\begin{cases} \cos\theta + \cos\phi + \cos\psi = a, \\ \sin\theta + \sin\phi + \sin\psi = b, \\ \cos 2\theta + \cos 2\phi + \cos 2\psi = c, \\ \sin 2\theta + \sin 2\phi + \sin 2\psi = d. \end{cases}$$

16. Eliminate α_1, α_2, α_3, β_1, β_2, β_3 from

$$\begin{cases} \cos^2\alpha_1 + \cos^2\beta_1 + \cos^2\theta_1 \\ \quad = \cos^2\alpha_2 + \cos^2\beta_2 + \cos^2\theta_2 = \cos^2\alpha_3 + \cos^2\beta_3 + \cos^2\theta_3 = 1, \\ \cos\alpha_2 \cos\alpha_3 + \cos\beta_2 \cos\beta_3 + \cos\theta_2 \cos\theta_3 \\ \quad = \cos\alpha_3 \cos\alpha_1 + \cos\beta_3 \cos\beta_1 + \cos\theta_3 \cos\theta_1 \\ \quad = \cos\alpha_1 \cos\alpha_2 + \cos\beta_1 \cos\beta_2 + \cos\theta_1 \cos\theta_2 = 0. \end{cases}$$

17. Eliminate l and m from

$$l \cos\alpha + m \cos(\beta - \gamma) = l \cos\beta + m \cos(\gamma - \alpha) = l \cos\gamma + m \cos(\alpha - \beta),$$

where no two of the angles α, β, γ differ by a multiple of 2π.

Also prove that $l = m(\cos\alpha + \cos\beta + \cos\gamma)$.

18. If $\sin(\phi + \psi) + \sin(\psi + \theta) + \sin(\theta + \phi) = 0$ and

$$\lambda = \cos\frac{\theta + \phi}{2} \sec\frac{\theta - \phi}{2} \sec\psi, \quad \mu = \sin\frac{\theta + \phi}{2} \sec\frac{\theta - \phi}{2} \operatorname{cosec}\psi,$$

where $\sin(\theta + \phi) \neq 0$, prove that $\lambda\mu + \lambda + \mu = 0$.

19. Eliminate x, y, z from

$$\begin{cases} x = y \cos\gamma + z \cos\beta, \\ y = z \cos\alpha + x \cos\gamma, \\ z = x \cos\beta + y \cos\alpha, \end{cases}$$

and express the result in factors.

20. Eliminate m and m' between

$$bm^2 + 2hm + a = 0 = bm'^2 + 2hm' + a, \quad \tan^{-1}m + \tan^{-1}m' = 2 \tan^{-1}\frac{y}{x}.$$

Inequalities.

Example 8. If $\theta_1, \theta_2, \theta_3, \theta_4, \theta_5$ are five positive acute angles such that their sum is $5a$, find the maximum value of $\sum_1^5 \sin\theta_r$.

Suppose that θ_1 is as large as any of the 5 angles and that θ_2 is as small as any of the 5 angles, so that $\theta_1 > a > \theta_2 > 0$.

Then $[\sin\theta_1 + \sin\theta_2] - [\sin a + \sin(\theta_1 + \theta_2 - a)]$

$$= 2\sin\frac{\theta_1+\theta_2}{2}\cos\frac{\theta_1-\theta_2}{2} - 2\sin\frac{\theta_1+\theta_2}{2}\cos\frac{\theta_1+\theta_2-2a}{2}$$

$$= 2\sin\frac{\theta_1+\theta_2}{2}\left(\cos\frac{\theta_1-\theta_2}{2} - \cos\frac{\theta_1+\theta_2-2a}{2}\right)$$

$$= 4\sin\frac{\theta_1+\theta_2}{2}\sin\frac{\theta_1-a}{2}\sin\frac{\theta_2-a}{2} < 0;$$

$$\therefore \sin\theta_1 + \sin\theta_2 < \sin a + \sin(\theta_1+\theta_2-a).$$

If then, in $\sum_1^5 \sin\theta_r$, we replace $\sin\theta_1, \sin\theta_2$ by $\sin a, \sin(\theta_1+\theta_2-a)$, we have not altered the sum of the angles but we have increased the sum of their sines.

This process can be repeated until each angle is a and this stage is reached after 4 steps at most; $\therefore$ the maximum value is $5\sin a$.

EXERCISE XIV. f.

Discuss the maximum and minimum values of (Nos. 1-5):

1. $4\tan x + 3\cot x$.
2. $1 - \sin x + \sin^2 x$.
3. $5 - 4\sin x + \sin^2 x$.
4. $5\sec\theta - 3\tan\theta$.
5. $10\sin^2\theta + 15\sin\theta\cos\theta + 18\cos^2\theta$.
6. Show that $\tan 3x \cot x$ is not between 3 and $\frac{1}{3}$.
7. Find the least numerical value of $\dfrac{a - b\cos\theta}{\sin\theta}$ when $a > b$.
8. Find the maximum and minimum values of $\tan 3x \cot^3 x$.
9. Show that the maximum and minimum values of
$$a\cos^2\theta + 2b\sin\theta\cos\theta + c\sin^2\theta$$
are the roots of $(x-a)(x-c) = b^2$.

Find the greatest values of the following (Nos. 10-13):

10. $a\cos\theta + b\cos\phi$, subject to $\theta + \phi = \alpha$.

11. $\tan\theta\tan\phi$, subject to $\theta+\phi=a<\frac{\pi}{2}$, and $0<(\theta,\phi)<\frac{\pi}{2}$.

12. $\sin\theta\sin\phi\sin\psi$, subject to $\theta+\phi+\psi=3a$, and $0<(\theta,\phi,\psi)<\frac{\pi}{2}$.

13. $\cos\theta\cos\phi\cos\psi$, subject to $\theta+\phi+\psi=3a$, and $0<(\theta,\phi,\psi)<\frac{\pi}{2}$.

14. Find the minimum value of $\tan^2 A+\tan^2 B+\tan^2 C$, where A, B, C are three acute angles whose sum is a right angle.

15. Find the least values of (i) $\Sigma\tan A$, (ii) $\Sigma\cot A$, (iii) $\Sigma\operatorname{cosec} A$, when A, B, C are positive acute angles with a constant sum 3D.

In a triangle ABC, prove the following results (Nos. 16-22):

16. $1<\cos A+\cos B+\cos C\leqslant\frac{3}{2}$.

17. $\cos 2A+\cos 2B+\cos 2C\geqslant-\frac{3}{2}$.

18. $8\cos A\cos B\cos C\leqslant 1$, and only $=1$ if $A=B=C$.

19. $\cos^2\frac{1}{2}(B-C)+\cos^2\frac{1}{2}(C-A)+\cos^2\frac{1}{2}(A-B)>1$.

20. $\tan A+\tan B+\tan C\geqslant 3\sqrt{3}$, if the angles are acute.

21. $8\sin\frac{1}{2}A\sin\frac{1}{2}B\sin\frac{1}{2}C\leqslant 1$.

22. $x^2+y^2+z^2-2yz\cos A-2zx\cos B-2xy\cos C>0$, unless

$$x:y:z=a:b:c.$$

23. Prove that the least value of $\cos^2\theta+\cos^2\phi+\cos^2\psi$, subject to $a\cos\theta+b\cos\phi+c\cos\psi=d$, is $\dfrac{d^2}{a^2+b^2+c^2}$.

24. If $0<(\alpha,\beta,\gamma)<\frac{1}{2}\pi$, prove that

$$\sin\alpha\sin\beta\sin\gamma\geqslant\sin(\beta+\gamma-\alpha)\sin(\gamma+\alpha-\beta)\sin(\alpha+\beta-\gamma).$$

MISCELLANEOUS EXAMPLES

EXERCISE XIV. g.

1. Prove that $\sin 16^\circ+\sin 20^\circ+\sin 92^\circ=\sin 52^\circ+\sin 56^\circ$.

2. Prove that

$$\sin(36^\circ+\theta)-\sin(36^\circ-\theta)-\sin(72^\circ+\theta)+\sin(72^\circ-\theta)=\sin\theta.$$

3. Prove that $\dfrac{3+\cos 4\theta}{1-\cos 4\theta}=\frac{1}{2}(\cot^2\theta+\tan^2\theta)$.

4. Express $\sin(\alpha+\beta)$ and $\cos(\alpha+\beta)$ in terms of $\sin\alpha+\sin\beta\,(\equiv s)$ and $\cos\alpha+\cos\beta\,(\equiv c)$, and prove that

$$\text{(i) } s\sin\alpha+c\cos\alpha=\frac{s^2+c^2}{2};\quad \text{(ii) } \tan\frac{\alpha}{2}+\tan\frac{\beta}{2}=\frac{4s}{s^2+c^2+2c}.$$

5. If $\tan\theta = \dfrac{n\sin\phi\cos\phi}{1-n\sin^2\phi}$, prove that $\tan(\theta-\phi)=(n-1)\tan\phi$.

6. If $e(\sin\phi-\sin\phi')=\sin(\phi-\phi')$ and ϕ, ϕ' do not differ by $2n\pi$, prove that $e^2\sin\phi\sin\phi'=(e^2-1)\cos^2\frac{1}{2}(\phi-\phi')$.

7. Prove that $\dfrac{\cos^3 2a-\cos 2\beta\cos^2(\beta-3a)}{\cos(2\beta-4a)+2\cos 2a}=\sin^2(\beta-a)$.

8. If $\dfrac{\cos a}{\cos\beta}+\dfrac{\sin a}{\sin\beta}=-1$, prove that $\dfrac{\cos^3\beta}{\cos a}+\dfrac{\sin^3\beta}{\sin a}=1$.

9. If $A+B+C+D=2\pi$, prove that

$$\cos 2C+\cos 2D-\cos 2A-\cos 2B$$
$$=4(\cos A\cos B\sin C\sin D-\sin A\sin B\cos C\cos D).$$

10. If c_1, c_2, c_3, c_4 are the cosines of the angles of a quadrilateral, prove that

$$(c_1^2+c_2^2+c_3^2+c_4^2-2c_1c_2c_3c_4-2)^2=4(c_1^2-1)(c_2^2-1)(c_3^2-1)(c_4^2-1).$$

11. Prove that

$$\Sigma\{\sin(\beta+\gamma-a)\}-\sin(a+\beta+\gamma)=4\sin a\sin\beta\sin\gamma.$$

12. If $\sin(\beta+\gamma)+\sin(\gamma+a)+\sin(a+\beta)$
$$=\sin(\beta+\gamma)+\sin(\gamma+\delta)+\sin(\delta+\beta),$$
prove that $a+\beta+\gamma+\delta=(2n+1)\pi$, if the angles are essentially distinct

13. If $\cos a+\cos\beta+\cos\gamma=-\cos a\cos\beta\cos\gamma$, prove that

$$\operatorname{cosec}^2 a+\operatorname{cosec}^2\beta+\operatorname{cosec}^2\gamma=1\pm 2\operatorname{cosec} a\operatorname{cosec}\beta\operatorname{cosec}\gamma.$$

14. If $\dfrac{\sin(x+a)-\sin(b+c)}{\cos(b+c)}=\dfrac{\sin(x+b)-\sin(c+a)}{\cos(c+a)}$, prove that each of them is equal to the third similar expression, unless $a-b=n\pi$.

15. If $a+b+c=2s$, prove that

$$\Sigma\{\sin 3(s-a)\sin(b-c)\}=4\sin s\sin(b-c)\sin(c-a)\sin(a-b).$$

16. Prove that

$$\frac{\Sigma\{\sin 2x\tan(y-z)\cos^2(y+z)\}}{\Sigma\{\cos 2x\tan(y-z)\sin^2(y+z)\}}=\tan 2x\tan 2y\tan 2z.$$

Eliminate θ in Nos. 17-24.

17. $a\sin\theta+b\cos\theta=c$, $a\cos\theta-b\sin\theta=d$.

18. $2\cos^2\theta+\sin\theta=a$, $2\sin^2\theta+\cos\theta=b$.

19. $x\cos\theta+y\sin\theta=\cos 3\theta$, $x\sin\theta-y\cos\theta=3\sin 3\theta$.

20. $a\sin(\theta+a)+b\sin(\theta+\beta)+c\sin(\theta+\gamma)=0$
$$=a\sec(\theta-a)+b\sec(\theta-\beta)+c\sec(\theta-\gamma).$$

21. $\cos^3\theta+a\cos\theta=b$, $\sin^3\theta+a\sin\theta=c$.

22. $16 \sin^5\theta - \sin 5\theta = 5x$, $16 \cos^5\theta - \cos 5\theta = 5y$.

23. $\sin^2\theta \tan \alpha + \sin^2\alpha \tan \theta = p$, $\cos^2\theta \cot \alpha + \cos^2\alpha \cot \theta = q$.

24. $\dfrac{\cos (3\theta - \alpha)}{\cos (\theta - \beta)} = x = \dfrac{\cos (3\theta + \alpha)}{\cos (\theta + \beta)}$.

25. Find the greatest possible value of $\dfrac{\operatorname{cosec}^2\theta - \tan^2\theta}{\cot^2\theta + \tan^2\theta - 1}$.

26. Prove that $\cot \theta - \cot 4\theta > 2$, if $0 < \theta < \frac{1}{4}\pi$.

27. Prove that

$$\frac{x^2 - 2xy \cos \theta + y^2}{x^2 - 2xy \cos \phi + y^2} \text{ lies between } \frac{1 - \cos \theta}{1 - \cos \phi} \text{ and } \frac{1 + \cos \theta}{1 + \cos \phi}.$$

28. In any triangle ABC prove that $\Sigma \tan^2 \frac{1}{2}\mathrm{A} \geqslant 1$.

29. Prove that the area of the pedal triangle DEF $\leqslant \frac{1}{4}\Delta$.

30. If $\sin \theta = \mu \sin \phi$, where $\mu > 1$, and θ, ϕ lie between 0 and $\frac{\pi}{2}$, prove that $\theta - \phi$ increases when ϕ increases.

MISCELLANEOUS EXAMPLES

On Chapters I to XIV

EXERCISE XV.

1. A man sees two objects in a straight line due E. of him. He walks a distance a due N. and then observes that they subtend an angle 2α at his eye; after walking a further distance $2a$, he observes that they subtend α. Show that the distance between them is $16a \tan\alpha / (5 + \tan^2\alpha)$.

2. In any triangle, prove that

$$a^3 - 2sa^2 + (r^2 + s^2 + 4Rr)a - 4R\Delta = 0.$$

3. Solve the equation $\sin 3\theta - 2\sin\theta = 4\cos^2\theta - 3$.

4. If x is positive, prove that

$\frac{1}{3}x^3 - x + (1 - x^2)\tan^{-1}x + x\log(1 + x^2)$ is also positive.

5. If the point of the Argand Diagram which represents the complex number Z describes concentric circles, centre the origin, and $z = \frac{1}{2}\left(Z + \frac{1}{Z}\right)$, what curves are described by the point which represents z?

6. Is it true to say that i^i pounds is 1 dollar?

7. Prove that $\cot^4\frac{\pi}{7} + \cot^4\frac{2\pi}{7} + \cot^4\frac{3\pi}{7} = 19$.

8. Factorize: $1 - \cos^2 x - \cos^2 y - \cos^2 z + 2\cos x\cos y\cos z$.

---

9. ABC is a triangle, right-angled at C, in a horizontal plane; $AC = l = CB$, and vertical posts AL, BM, CN of lengths p, q, r are erected at the vertices. Show that the angle θ between the planes ABC, LMN is given by $l^2\tan^2\theta = (p - r)^2 + (q - r)^2$.

10. If P is a point in the plane of a quadrilateral ABCD, such that the triangles ABD, PBC are directly similar, show that ABP, DBC are also similar and deduce that

$$AC^2 = AP^2 + PC^2 - 2AP \cdot PC\cos(A + C),$$

and

$$x^2y^2 = a^2c^2 + b^2d^2 - 2abcd\cos(A + C).$$

11. Find the solutions between 0 and π of

$$\cos x + \cos 2y = \cos y + \cos 2z = \cos z + \cos 2x = 0.$$

12. If $|x| < 1$, find the sum to infinity of $1 - \frac{2x}{1} + \frac{3x^2}{2} - \frac{4x^3}{3} + \dots$.

13. Prove that

$$\sin^{2n}\theta+\cos^{2n}\theta=1-n\sin^2\theta\cos^2\theta+\frac{n(n-3)}{2!}\sin^4\theta\cos^4\theta-\ldots,$$

and give the last term.

14. If α, β, and γ are real numbers such that

$$\cos(\alpha+i\beta)=\cos\gamma+i\sin\gamma,$$

prove that $\sin^2\gamma=\sin^4\alpha$ and $\text{sh}^2\beta=\sin^2\alpha$.

15. Find the values of

$$\cos\frac{2\pi}{21}+\cos\frac{8\pi}{21}+\cos\frac{10\pi}{21}\quad\text{and}\quad\cos\frac{4\pi}{21}+\cos\frac{16\pi}{21}+\cos\frac{20\pi}{21}.$$

16. Given $\sin\theta:\sin(\theta+\phi):\sin(\theta+2\phi)=a:b:c$, find the ratios

$$\cos\theta:\cos(\theta+\phi):\cos(\theta+2\phi).$$

---

17. From the top of a hill of uniform slope the angle of depression of a point in the plain below is 30°, and from a spot $\frac{3}{4}$ of the way down it is 15°; find the slope of the hill.

18. ABC is a triangle in which $B=80°=C$, and points E, F are taken in AC, AB so that $EA=EB$ and $BF=BC$; prove that $\angle FEB=30°$.

19. Find the conditions for the equation $\cos 2\theta+b^2\cos\theta+c=0$ to give two possible values of $\cos\theta$.

20. Expand $e^x\sin x$ in a series of powers of x.

21. Prove that the points of the Argand Diagram which represent the complex numbers z_1, z_2, z_3, z_4 are concyclic if real numbers a, b, c exist such that

$$(b-c)(z_1z_2+z_3z_4)+(c-a)(z_1z_3+z_2z_4)+(a-b)(z_1z_4+z_2z_3)=0.$$

22. Prove that $\text{ch}\,ax\cos bx=1+\Sigma\dfrac{x^{2n}c^{2n}\cos 2n\alpha}{(2n)!}$ where

$$\tan\alpha=\frac{b}{a}\quad\text{and}\quad c^2=a^2+b^2.$$

23. Prove that $\sum_0^{2n}\sec\left(\theta+\dfrac{2r\pi}{2n+1}\right)=(-1)^n(2n+1)\sec(2n+1)\theta.$

24. If

$$\frac{a-b}{1+ab}=\frac{c-d}{1+cd},\text{ prove that }\frac{1+ac}{\sqrt{1+a^2}\sqrt{1+c^2}}=\pm\frac{1+bd}{\sqrt{1+b^2}\sqrt{1+d^2}}.$$

---

25. The planes of two intersecting circles of radii a and b are inclined at an angle θ and the length of the common chord of the circles is $2c$; show that the radius R of the sphere which passes through them is given by

$$(R^2-c^2)\sin^2\theta=a^2-2c^2+b^2-2\cos\theta\sqrt{\{(a^2-c^2)(b^2-c^2)\}}.$$

26. In a quadrilateral, if AB and CD are parallel, prove that

$$\cos^2 \tfrac{1}{2}(\mathrm{A}+\mathrm{C})=(s-a)(s-c)(b-d)^2/\{bd(a-c)^2\}.$$

27. Find the sum, s_n, to n terms, of $\sin x+\sin 3x+\sin 5x+\ldots$ and show that $\lim\limits_{n\to\infty}\frac{1}{n}(s_1+s_2+\ldots+s_n)=\frac{1}{2}\operatorname{cosec} x$, unless $x=k\pi$.

28. Find the sum to infinity of $\frac{x^3}{3!}+\frac{x^7}{7!}+\frac{x^{11}}{11!}+\ldots$.

29. Use the identity $\Sigma\frac{a^2(x-b)(x-c)}{(a-b)(a-c)}=x^2$ to prove that

$$\Sigma\frac{\cos 2(\theta+a)\sin(\theta-\beta)\sin(\theta-\gamma)}{\sin(a-\beta)\sin(a-\gamma)}=\cos 4\theta.$$

30. If $r \operatorname{cis}\theta=\frac{2c}{\operatorname{ch}(\xi+i\eta)-1}$, prove that $r\operatorname{sh}^2\xi=2c(\cos\theta\pm\operatorname{ch}\xi)$.

31. Form the quadratic whose roots are $a+a^4$ and a^2+a^3, where $a=\operatorname{cis}\frac{2\pi}{5}$; deduce the values of $\cos\frac{2\pi}{5}$ and $\cos\frac{4\pi}{5}$.

32. If $x=\theta-\frac{\pi}{3}$, θ, $\theta+\frac{\pi}{3}$ satisfy $a\cos 2x+b\sin 2x=c\cos x+d\sin x$, prove that $a^2+b^2=c^2+d^2$.

---

33. Prove that

$\tan(\theta+h)-2\tan\theta+\tan(\theta-h)=2\tan\theta\sin^2 h\sec(\theta+h)\sec(\theta-h)$, and, taking $\theta=82°\ 3'$, compare, using four-figure tables, the value found for $\tan\theta$ by this formula and by interpolation from tan 82° and tan 82° 6′.

34. In a triangle ABC the trisectors of B and C which are the nearer to BC meet at U. V and W are similar points. Find the length of VW, and, by reducing the result to a symmetrical form, prove that UVW is equilateral.

35. Simplify:

$$\tan^{-1}\frac{a_1x-y}{a_1y+x}+\tan^{-1}\frac{a_2-a_1}{a_1a_2+1}+\tan^{-1}\frac{a_3-a_2}{a_2a_3+1}+\ldots \text{ to } n \text{ terms},$$

and deduce a series for π by taking $x=y$, $a_n=2n-1$.

36. Prove that $\lim\limits_{\theta\to 0}\frac{\operatorname{sh}\theta-\sin\theta}{\theta(\operatorname{ch}\theta-\cos\theta)}=\frac{1}{3}$.

37. Deduce a trigonometrical identity from

$$\Sigma(b-c)(1+ab)(1+ac)=(b-c)(c-a)(a-b).$$

38. If $\cos(x+iy)=\tan(\xi+i\eta)$, prove that

$$\operatorname{ch}^2 2y-2\operatorname{ch}2y\frac{\sin^2 2\xi+\operatorname{sh}^2 2\eta}{(\cos 2\xi+\operatorname{ch}2\eta)^2}+\frac{2(\sin^2 2\xi-\operatorname{sh}^2 2\eta)}{(\cos 2\xi+\operatorname{ch}2\eta)^2}=1.$$

39. Prove that $\sum_{1}^{n} \sec^2 \frac{(2r-1)\pi}{4n} = 2n^2$.

40. If $xyz = x + y + z$, prove that $\Sigma\{x(1-y^2)(1-z^2)\} = 4xyz$.

---

41. A man walking along a straight road notices two objects in line with him in a direction making an angle a with the road. After walking a distance c along the road he notices that they subtend an angle a, and again after walking a further distance d that they subtend the same angle a. Find the distance between them.

42. ABC is a triangle with points D, E in AB, AC respectively, such that BD $= a =$ CE; prove that DE is perpendicular to OI.

43. Sum the series $3 \sin^3 \frac{\theta}{3} + 3^2 \sin^3 \frac{\theta}{3^2} + 3^3 \sin^3 \frac{\theta}{3^3} + \dots$ to n terms, and deduce the sum to infinity.

44. By expanding $\log \{(1-ax)(1-\beta x)(1-\gamma x)(1-\delta x)\}$ in two ways, show that $a^5 + \beta^5 + \gamma^5 + \delta^5 + 5\Sigma a\beta \,.\, \Sigma a\beta\gamma$ is divisible by $a + \beta + \gamma + \delta$ and find the other factor.

45. If s_n is the sum to n terms of $1 + z + z^2 + z^3 + \dots$, where $z = \text{cis}\, a$ and $a \neq 2k\pi$, prove that

$$\lim_{n\to\infty} \frac{s_1 + s_2 + \dots + s_n}{n} = \frac{1}{1-z}.$$

46. Find the sums to infinity of

$$\text{(i)} \quad \frac{1}{1^2 \,.\, 3^2} + \frac{1}{3^2 \,.\, 5^2} + \frac{1}{5^2 \,.\, 7^2} + \dots ;$$

$$\text{(ii)} \quad \frac{1}{1^3 \,.\, 3^3} + \frac{1}{3^3 \,.\, 5^3} + \frac{1}{5^3 \,.\, 7^3} + \dots .$$

47. Prove that

$$\text{cosec}\, \frac{\pi}{2n} - \text{cosec}\, \frac{3\pi}{2n} + \text{cosec}\, \frac{5\pi}{2n} - \dots \text{cosec}\, \frac{(n-2)\pi}{2n},$$

where n is odd, is equal to $\frac{1}{2}\{n + (-1)^{\frac{1}{2}(n+1)}\}$.

48. If

$$\cos(\beta + \gamma + \theta) + \cos\beta + \cos\gamma = 1 = \cos(\gamma + a + \theta) + \cos\gamma + \cos a,$$

and $\quad a - \beta \neq 2n\pi \quad$ and $\quad \gamma + \theta \neq (2n+1)\pi$,

prove that $\quad \cos(a + \beta + \theta) + \cos a + \cos\beta = 1.$

---

49. AB is a horizontal line on an inclined plane OAB; OA, OB, and the plane OAB make angles a, β, γ with the vertical; prove that the cosine of $\angle$AOB satisfies the equation

$$(1 - x^2)\cos^2\gamma + 2x \cos a \cos\beta = \cos^2 a + \cos^2\beta.$$

50. In a cyclic quadrilateral, prove that $\frac{a}{c} + \frac{c}{a} > \frac{x}{y} + \frac{y}{x}$, unless $c = a$.

51. Evaluate

$$\cos\alpha+\cos\beta\cos(\alpha+\beta)+\cos^2\beta\cos(\alpha+2\beta)+\ldots+\cos^n\beta\cos(\alpha+n\beta).$$

52. If $x>1$, prove that $\dfrac{1}{x}+\dfrac{1}{2x(x-1)}>\log\dfrac{x}{x-1}>\dfrac{2}{2x-1}$.

53. If $f(n)=1-\dfrac{n^2}{2!}+\dfrac{n^2(n^2-1^2)}{4!}-\dfrac{n^2(n^2-1^2)(n^2-2^2)}{6!}+\ldots$ to $n+1$ terms, prove that $f(2n)=(-1)^n f(n)$.

54. If $|r|<1$, prove that

$$r\sin\theta+2r^2\sin 2\theta+3r^3\sin 3\theta+\ldots=\frac{r(1-r^2)\sin\theta}{(1-2r\cos\theta+r^2)^2}.$$

55. If m and n are integers such that $0<m<n$, and $t=\text{cis}\,\dfrac{2\pi}{n}$, prove that $\dfrac{nx^{m-1}}{x^n-1}=\sum\limits_{0}^{n-1}\dfrac{t^{mr}}{x-t^r}$.

56. Eliminate θ and ϕ from the equations

$$\tan\theta+\tan\phi=\tan\alpha,\quad \cot\theta+\cot\phi=\cot\beta,\quad c\sin(\theta+\phi)=1.$$

---

57. A fly, stationed at a point of the circular base, radius r, of a cylindrical tower, finds that he can just see a distant flagstaff by walking along the tangent line to the base either a distance p in one direction or a distance q in the other. Show that the distance of the flagstaff from the centre of the base of the tower is

$$r\sqrt{\{(p^2+r^2)(q^2+r^2)\}}/(pq-r^2).$$

58. Straight lines are drawn through the vertices of a triangle ABC making the same angle θ in the same sense with the opposite sides, prove that the area of the triangle formed by them is $4\Delta\cos^2\theta$.

59. Does $\sum\limits_{1}^{n}\sin^2 r\theta\div\sum\limits_{1}^{n}\cos^2 r\theta$ tend to a limit when $n\to\infty$?

60. Prove that $x^2>(1+x)\{\log(1+x)\}^2$, where $x>-1$, $x\neq 0$.

61. Solve the equation $x^6+2x^3+2=0$.

62. If $\log\sin(\theta+i\phi)=\alpha+i\beta$, prove that $2e^{2\alpha}=\text{ch}\,2\phi-\cos 2\theta$.

63. Use the result of No. 55 to show that if n and s are odd and $s<2n-1$, then

$$\sum_{0}^{n-1}\frac{\cos\dfrac{rs\pi}{n}}{\cos\dfrac{r\pi}{n}}=n\sin\frac{s\pi}{2}\quad\text{and}\quad\sum_{0}^{n-1}\frac{\sin\dfrac{rs\pi}{n}}{\cos\dfrac{r\pi}{n}}=0.$$

64. If θ_1, θ_2, θ_3 are roots of $\tan(\theta+\alpha)=k\tan 2\theta$, no two of which differ by $n\pi$, prove that $\theta_1+\theta_2+\theta_3=m\pi-\alpha$.

65. From a point on the ground of a square courtyard of area a^2 the angles of elevation of the buildings (of equal height) at three consecutive corners are 60°, 60°, and 45°. Show that the height is

$$a\sqrt{\{3(1-\tfrac{1}{2}\sqrt{2})\}}.$$

66. For a cyclic quadrilateral, prove that the square of the distance from the centre of the circumcircle to the point of intersection of the diagonals is

$$(ac+bd)^2\{bd(a^2-c^2)^2+ac(b^2-d^2)^2\}/\{\sigma(ab+cd)(ad+bc)\},$$

where σ is the product of expressions like $b+c+d-a$. Verify the result by applying it to a square, a triangle ($d=0$), and a rectangle ($a=c$, $b=d$).

67. Draw the graph of $\sin x+\cos 2x$, for $0<x<2\pi$, and use it to find, roughly, the values of x for which $5(\sin x+\cos 2x)=1$.

68. If θ is positive, prove that $\theta^3-\dfrac{\theta^5}{10}<3(\sin\theta-\theta\cos\theta)<\theta^3$.

69. Find the values of $(3+4i)^{\frac{1}{2}}+(3-4i)^{\frac{1}{2}}$.

70. If the point of the Argand Diagram which represents the complex number z describes the unit circle centre the origin, what curve is described by the point representing $4z+z^4$?

71. If $\alpha=\operatorname{cis}\dfrac{2\pi}{13}$, show that

$$\alpha+\alpha^3+\alpha^4+\alpha^9+\alpha^{10}+\alpha^{12} \quad\text{and}\quad \alpha^2+\alpha^5+\alpha^6+\alpha^7+\alpha^8+\alpha^{11}$$

are the roots of $x^2+x=3$.

Deduce that $\cos\dfrac{\pi}{13}\cos\dfrac{3\pi}{13}\cos\dfrac{4\pi}{13}=\frac{1}{16}(\sqrt{13}+3)$.

72. If $\dfrac{\tan(\alpha+\beta-\gamma)}{\tan(\alpha-\beta+\gamma)}=\dfrac{\tan\gamma}{\tan\beta}$, prove that

either $\sin(\beta-\gamma)=0$ or $\sin 2\alpha+\sin 2\beta+\sin 2\gamma=0$.

---

73. From two points A and B at the same level on a cliff, the angles of depression of a ship S are observed to be α and β and the difference of bearing to be γ. Prove that the plane ABS makes an angle ϕ with the horizontal given by

$$\sin^2\gamma\tan^2\phi=\tan^2\alpha+\tan^2\beta-2\tan\alpha\tan\beta\cos\gamma.$$

74. ABC is a triangle with B $=80°=$ C, and points S, T are taken in the productions of AC, AB, so that $\angle$BCT $=30°$ and $\angle$CBS $=60°$. Show that ST is inclined at 50° to BC.

75. Find an equation which shows how to divide a given circle into two parts of equal area by a circular arc whose centre is on the circumference of the given circle. Solve the equation graphically.

76. If $u_n = (n+1)u_{n-1} - (n-1)u_{n-2}$, prove that

$$\lim_{n\to\infty} \frac{u_n}{n!} = u_2(e-2) - u_1(2e-5).$$

77. Find the locus in the Argand Diagram of the point which represents $\alpha t + \beta t^2$, where t is a real variable and α and β are complex constants.

78. Find the sum to infinity of $1 - \frac{1}{5} + \frac{1}{7} - \frac{1}{11} + \frac{1}{13} - \frac{1}{17} + \dots$.

79. Show that $\sec\frac{\pi}{11} - \sec\frac{2\pi}{11} + \sec\frac{3\pi}{11} - \sec\frac{4\pi}{11} + \sec\frac{5\pi}{11} = 6$.

80. If $x < y < 0$, and $x^3 = 9(x+1)$, and $y^3 = 9(y+1)$, prove that

$$x + 7 = (y-1)^2.$$

ANSWERS

CHAPTER I

EXERCISE I. a. (p. 1.)

2. $97° 54'$. **3.** $111° 2'$. **4.** $38° 12', 48° 7', 93° 41'$.

5. $99{\cdot}7$. **6.** $c=7{\cdot}01, a=22{\cdot}4, \mathrm{A}=76° 18'$.

7. $-2° 21', \mathrm{B}=10° 30', \mathrm{C}=15° 12', a=40{\cdot}2$.

8. $33{\cdot}6$; $(\theta=32° 14')$. **9.** $\mathrm{C}=59° 6', a=35{\cdot}4, b=54{\cdot}8$.

10. $\mathrm{C}=66° 5'$, $\mathrm{A}=70° 55'$, $a=8{\cdot}76$ or $\mathrm{C}=113° 55'$, $\mathrm{A}=23° 5'$, $a=3{\cdot}63$.

11. $36° 54', 87° 4', 56° 2'$.

12. $\mathrm{A}=25° 25\frac{1}{2}', \mathrm{C}=34° 49\frac{1}{2}', b=13{\cdot}77$. **13.** Impossible.

14. $<5{\cdot}1$; $=5{\cdot}1$ or $>14{\cdot}5$; between $5{\cdot}1$ and $14{\cdot}5$; (i) $\mathrm{C}=37° 56'$, $\mathrm{B}=121° 28'$, $b=20{\cdot}13$, or $\mathrm{C}=142° 4'$, $\mathrm{B}=17° 20'$, $b=7{\cdot}03$; (ii) $\mathrm{C}=18° 21'$, $\mathrm{B}=141° 3'$, $b=28{\cdot}95$; (iii) impossible ; (iv) $\mathrm{C}=90°$, $b=13{\cdot}57$.

15. $a^2-2ac\cos\mathrm{B}+c^2-b^2=0$; $2c\cos\mathrm{B}$, c^2-b^2 ; $180°$; $180°-2\mathrm{B}$.

21. $40{\cdot}2$; $(\theta=77° 22')$. **24.** $-5° 5'$; $(\theta=53° 38')$.

27. $\dfrac{a^2+b^2+c^2}{2abc}$. 29. $\dfrac{a}{\sin\mathrm{A}}$. **36.** $\dfrac{a+b}{c}$.

EXERCISE I. b. (p. 7.)

1. $18{\cdot}1$. **2.** $\dfrac{7}{\sqrt{3}}, \dfrac{\sqrt{3}}{2}$. **3.** $10\frac{1}{2}, 12, 14$.

4. $13{\cdot}3$, $6{\cdot}65$. **7.** $2\mathrm{R}\sin\mathrm{A}\sin\mathrm{B}\sin\mathrm{C}$.

25. $7{\cdot}84$. 28. $13{\cdot}9$. 29. $2\sqrt{(\mathrm{R}r_1)}$.

37. $2\mathrm{R}\cos\mathrm{A}\cos\mathrm{B}\cos\mathrm{C}$, $-2\mathrm{R}\cos\mathrm{A}\sin\mathrm{B}\sin\mathrm{C}$.

39. $\mathrm{R}(\cos\mathrm{B}\cos\mathrm{C}+\cos\mathrm{C}\cos\mathrm{A}+\cos\mathrm{A}\cos\mathrm{B})$.

EXERCISE I. c. (p. 12.)

2. $\frac{1}{2}(\sqrt{3}-1)$.

4. $\frac{33}{56}, \frac{5}{12}, \frac{29}{336}$.

5. $\frac{2279}{1100}$.

6. $\frac{4873}{364}$.

11. $\cot^{-1}\left(\frac{17}{5\sqrt{87}}\right)$.

12. $\cot^{-1}\{\frac{1}{2}(\cot\alpha-\cot\beta)\}$.

13. $3\sqrt{2}$.

18. $AK \cdot AL = bc$.

19. $\frac{1}{2}aR\cos A$, $aR\cos B\cos C$; $\frac{1}{4}aR\cos(B-C)$; $\frac{c\cos(A-B)}{b\cos(C-A)}$.

21. I_1.

22. $\tan A : \tan B : \tan C$.

23. N; H.

39. (ii) G.

40. A circle, centre O.

EXERCISE I. d. (p. 17.)

26. $-4Rr_3(s-c)$.

EXERCISE I. e. (p. 19.)

1. 15° 43′.

2. 98° 56′ or 43° 39′.

3. 89° 36′, 12·0.

4. 78° 28′.

5. 95° 1′.

6. $7\sqrt{5}$, $8\sqrt{5}$.

8. $\frac{4R\Delta}{a}$, $a^2 \pm \frac{4\Delta}{a}\sqrt{(4R^2-a^2)}$.

9. $r_1/\sqrt{\{(r_1+r_2)(r_1+r_3)\}}$.

10. $\frac{b\sin A}{c-b\cos A}$.

11. $\sqrt{\{a^2+b^2\pm 2\sqrt{(a^2b^2-4\Delta^2)}\}}$.

13. 29° 44′.

14. 10·2.

15. 8, 113° 25′ or 1, 6° 35′.

16. $8\sqrt{3}$, $\frac{1}{4}$.

17. 2 or 17.

19. $2r^2\cot\frac{1}{2}A \div (p-2r)$

EXERCISE I. f. (p. 21.)

6. 0·05 ft.

EXERCISE I. g. (p. 22.)

9. $2R\sin\frac{1}{2}A$.

16. $\tan\frac{1}{2}B\sqrt{(b^2-a^2\sin^2 B)}$.

21. $\angle ACB = 45°$.

24. $\cos^{-1}\frac{1}{4}$.

CHAPTER II

EXERCISE II. a. (p. 26.)

1. 29·0. 2. 7·43, 7·94. 3. 7·95. 4. 110° 47′.

7. $\dfrac{s-c}{s-a}$. 10. $\sqrt{(abcd)}$; $\dfrac{cd}{ab}$. 11. 3·51.

15. $\dfrac{abcd(ac+bd)}{(ab+cd)(ad+bc)}$. 16. $\mathrm{QA}=\dfrac{a(bc+ad)}{c^2-a^2}$, $\mathrm{QD}=\dfrac{c(ab+cd)}{c^2-a^2}$, etc.

EXERCISE II. b. (p. 28.)

1. $\cos(\mathrm{B}+\mathrm{D})=-\frac{12}{13}$. 2. $\tan\theta=-\frac{46}{3}$.

3. $30\sqrt{85}$. 4. $\tan\theta=2{\cdot}4$.

5. $\dfrac{55\sqrt{3}}{4}$; 55. 6. $\frac{7}{13}$. 14. $\sin\dfrac{\mathrm{B}+\mathrm{D}}{2}\cdot\sqrt{(abcd)}$.

16. (i) $6\sqrt{10}$; (ii) $\frac{2}{3}\sqrt{10}$; (iii) $\cos^{-1}\frac{3}{13}$; (iv) 5·94, 6·40; (v) 3·29.

17. 7; 2. 18. 2358.

EXERCISE II. c. (p. 29.)

1. $\cos^{-1}(-\frac{7}{9})$. 2. 13·8, 14·3, 14·1. 3. 3528; no.

4. 7. 5. 1, 120° or 3, 60°. 6. Two solutions.

7. 15. 8. $7\pm\sqrt{22}$. 9. $\sqrt{3}$, $\sqrt{3}$. 10. $\frac{7}{3}\sqrt{3}$, $\frac{2}{3}\sqrt{3}$.

17. $\cos\dfrac{\pi}{n}:1$. 18. $\cos^2\dfrac{\pi}{n}:1$. 22. $\frac{1}{2}\sqrt{3}$.

EXERCISE II. d. (p. 30.)

3. 834, $\frac{12}{5}\sqrt{865}$, $\tan^{-1}\frac{5}{12}$. 4. 90° or 118° 28′.

6. $x^2=\dfrac{(ac-bd)(bc-ad)}{ab-cd}$, $y^2=\dfrac{(ac-bd)(ab-cd)}{bc-ad}$.

10. $\sqrt{(2p-qs+ts^2-s^4)}$.

17. The values of x, y are any pair selected from 6·53, 8·27, 10·9.

CHAPTER III

EXERCISE III. a. (p. 35.)

1. $n\pi$.
2. $2n\pi$.
3. $n\pi$.
4. $2n\pi + \frac{\pi}{2}$.
5. $(2n+1)\pi$.
6. $2n\pi - \frac{\pi}{2}$.
7. $2n\pi + \frac{\pi}{3}$ or $2n\pi + \frac{2\pi}{3}$.
8. $2n\pi \pm \frac{\pi}{4}$.
9. $n\pi + \frac{\pi}{3}$.
10. $2n\pi - \frac{\pi}{6}$ or $2n\pi - \frac{5\pi}{6}$.
11. $2n\pi \pm \frac{5\pi}{6}$.
12. $n\pi - \frac{\pi}{4}$.
13. $n\pi$.
14. $\frac{n\pi}{3}$.
15. $\frac{n\pi}{4}$.
16. $\frac{(4n-1)\pi}{6}$.
17. $\frac{(4n-1)\pi}{12}$.
18. $\frac{(2n+1)\pi}{8}$.
19. $n\pi + \frac{\pi}{4}$.
20. $n\pi - \frac{\pi}{4}$.
21. $2n\pi \pm \frac{\pi}{3}$.
22. $2n\pi + a$ or $(2n+1)\pi - a$.

EXERCISE III. b. (p. 37.)

1. $\frac{(2n+1)\pi}{4}$.
2. $\frac{n\pi}{6}$.
3. $n\pi$ or $\frac{(2n+1)\pi}{10}$.
4. $\frac{(4n+1)\pi}{10}$.
5. $\frac{(2n+1)\pi}{5}$ or $\frac{2n\pi}{7}$.
6. $\frac{(4n-1)\pi}{4}$ or $\frac{(4n-1)\pi}{16}$.
7. $\frac{n\pi}{3}$.
8. $\frac{(2n+1)\pi}{14}$.
9. $n\pi$ or $n\pi + \frac{\pi}{4}$.
10. $\frac{2n\pi}{3}$.
11. $2n\pi + \frac{\pi}{2}$ or $2n\pi - \frac{\pi}{6}$.
12. $2n\pi$ or $2n\pi - \frac{\pi}{2}$.
13. $n\pi$.
14. $n \,.\, 360° + 96° \; 52'$ or $n \,.\, 360° - 23° \; 8'$.
15. $n\pi + \frac{\pi}{12}$ or $n\pi + \frac{5\pi}{12}$.
16. $n \,.\, 360° + 92° \; 44'$ or $n \,.\, 360° - 110° \; 20'$.
17. $\frac{(2n+1)\pi}{4}$ or $2n\pi \pm \frac{2\pi}{3}$.
18. $\frac{n\pi}{3}$ or $\frac{(6n \pm 1)\pi}{12}$.
19. $\frac{n\pi}{2}$.
20. $\frac{n\pi}{3}$.
21. $n\pi \pm \frac{\pi}{3}$ or $\frac{(2n+1)\pi}{8}$.
22. $\frac{2n\pi}{3} + \frac{\pi}{4}$.

23. $2n\pi$ or $2n\pi+\frac{\pi}{2}$ or $n\pi-\frac{\pi}{4}$. **24.** $n\pi$ or $\frac{(4n-1)\pi}{8}$.

25. $2n\pi+\alpha\pm\frac{\pi}{3}$. 26. $\frac{n\pi}{12}$. **27.** $\frac{(4n+1)\pi}{8}$.

28. $n\pi$ or $n\pi+\frac{1}{2}(\alpha+\beta)$, provided that $\alpha+\beta\neq(2m+1)\pi$.

29. $2n\pi$ or $2n\pi-\frac{\pi}{2}$. **30.** $\frac{n\pi}{2}+\alpha$.

31. $n\pi+\frac{1}{2}\alpha$ or $n\pi+\frac{1}{2}(\pi-\alpha)$ where α is a value of $\sin^{-1}\left[\frac{4}{(2n+1)\pi}\right]$.

32. $\pm\frac{1}{3}\sqrt{3}$.

33. (i) $2n\pi$ or $2n\pi+\frac{\pi}{2}$; (ii) no solution; (iii) $2n\pi+\frac{\pi}{3}$ or $2n\pi+\frac{\pi}{6}$.

34. $2n\pi-\frac{2\pi}{3}$. **35.** $n\pi+\frac{\pi}{12}$. **37.** Yes.

EXERCISE III. c. (p. 40.)

1. $\pm 0{\cdot}824$. **2.** $0{\cdot}642$. **3.** 7. **4.** 5.

5. $0{\cdot}913$ or $-2{\cdot}517$. **6.** 3. **7.** $n\pi-\frac{\pi}{3}<\theta<n\pi+\frac{\pi}{3}$.

10. $a^2+b^2\leqslant 4$.

11. x, y are $n\pi+\frac{\pi}{2}$, $(n-2k)\pi-\frac{\pi}{3}$, or $n\pi-\frac{\pi}{3}$, $(n-2k)\pi+\frac{\pi}{2}$, or $n\pi+\frac{5\pi}{6}$, $(n-2k)\pi$, or $n\pi$, $(n-2k)\pi+\frac{5\pi}{6}$.

12. $x=k\pi+\alpha$, $y=(2s+k)\pi+\beta$ where α, β are $\frac{\pi}{4}$, $\frac{\pi}{6}$, or $-\frac{\pi}{4}$, $-\frac{\pi}{6}$, or $\frac{\pi}{6}$, $\frac{\pi}{4}$, or $-\frac{\pi}{6}$, $-\frac{\pi}{4}$.

13. $x=n\pi$, $y=\frac{m\pi}{3}$.

14. $x=m\pi$, $y=n\pi$, or any values which satisfy $\sin x-\sin y=1$.

15. $x=n\pi+\frac{\pi}{2}$, $y=k\pi$.

16. $\theta=126°\ 47'$, $\phi=357°\ 5'$ or $\theta=357°\ 5'$, $\phi=126°\ 47'$.

17. $x=36°\ 52'$, $y=90°$ or $x=351°\ 12'$, $y=298°\ 4'$.

18. $x=(5k-3n+\frac{1}{2})\frac{\pi}{8}$, $y=(5n-3k+\frac{1}{2})\frac{\pi}{8}$.

19. $x=\frac{p\pi}{7}$, $y=\frac{q\pi}{7}$, $z=\frac{r\pi}{7}$ where p, q, r in order are (4, 2, 1); (1, 4, 2); (2, 1, 4); (6, 3, 5); (5, 6, 3); (3, 5, 6).

20. Seven solutions: $x=y=z=\frac{2\pi}{3}$ or $x=\frac{6\pi}{7}$, $y=\frac{4\pi}{7}$, $z=\frac{5\pi}{7}$, etc., or $x=\frac{8\pi}{9}$, $y=\frac{5\pi}{9}$, $z=\frac{7\pi}{9}$, etc.

EXERCISE III. d. (p. 44.)

3. $\sin\frac{1}{2}\theta = +\sqrt{\{\frac{1}{2}(1-\cos\theta)\}}$ if $4n\pi < \theta < (4n+2)\pi$.

4. (i) $-$, $+$; (ii) $-$, $-$; (iii) $+$, $-$.

5. (i) $+$, $+$; (ii) $-$, $-$; (iii) $+$, $-$.

6. $-$, $-$. 7. $-$, $+$. 8. $+$, $-$.

10. $\sin\frac{\theta}{2}-\cos\frac{\theta}{2} = +\sqrt{(1-\sin\theta)}$ if $4n\pi+\frac{\pi}{2} < \theta < (4n+2)\pi+\frac{\pi}{2}$.

11 and 12. $4n\pi+\frac{\pi}{2} < \theta < 4n\pi+\frac{3\pi}{2}$.

13. $2n\pi+\frac{5\pi}{4} < \theta < 2n\pi+\frac{7\pi}{4}$. 14. $2n\pi+\frac{3\pi}{4} < \theta < 2n\pi+\frac{5\pi}{4}$.

15. (i) $\pm\frac{1}{2}\sqrt{3}$; (ii) $\pm\frac{1}{2}$. 16. (i) $\pm\frac{1}{2}$, $\pm\frac{1}{2}\sqrt{3}$; (ii) $\pm\frac{1}{2}$, $\pm\frac{1}{2}\sqrt{3}$.

17. $\frac{1}{2}\sqrt{\{2-\sqrt{(2+\sqrt{2})}\}}$. 18. $-\frac{2}{3}$. 19. $\frac{3}{5}$.

20. $\frac{1}{\tan\theta}\{-1\pm\sqrt{(1+\tan^2\theta)}\}$; (i) $-$; (ii) $-$; (iii) $+$.

21. $\frac{1}{\sin\theta}\{1\pm\sqrt{(1-\sin^2\theta)}\}$. 22. $+$ if $2n\pi<\theta<(2n+1)\pi$.

23. $\sin\frac{1}{3}a$, $\sin\frac{1}{3}(2\pi+a)$, $\sin\frac{1}{3}(4\pi+a)$.

24. (i) cos 20°, cos 100°, cos 140°; (ii) cos 50°, cos 70°, cos 170°.

26. (i) 4 cos 40°, 4 cos 80°, 4 cos 160°;
(ii) 4 cos 25° 10′, 4 cos 94° 50′, 4 cos 145° 10′.

27. (i) 6 cos 20°, 6 cos 100°, 6 cos 140°;
(ii) 6 cos 45° 56′, 6 cos 74° 4′, 6 cos 165° 56′.

28. $-1+4\cos 40°$, $-1+4\cos 80°$, $-1+4\cos 160°$.

29. $2\cos a$ where $a=24°$ or $48°$ or $96°$ or $120°$ or $168°$.

EXERCISE III. e. (p. 48.)

1. $p = \pm\frac{1}{x}\sqrt{(1-x^2)}$, $q = \pm\frac{1}{\sqrt{(1-x^2)}}$.

2. $p = \pm\frac{x}{\sqrt{(1+x^2)}}$, $q = \pm\frac{1}{\sqrt{(1+x^2)}}$, $r = \frac{1}{x}$.

3. $p = \frac{3}{5}$, $q = \frac{4}{3}$.

5. (i) $\pm\sqrt{(1-x^2)}$; (ii) $\pm\frac{x}{\sqrt{(1-x^2)}}$.

6. 0, $4\cos^{-1}x$.

7. $\pm\sin^{-1}\{2x\sqrt{(1-x^2)}\}$.

9. $\frac{\pi}{4}$.

10. $\frac{\pi}{4}$.

12. $1-2x^2$.

13. $x=3$.

14. $\sin^{-1}(3x-4x^3)$.

15. $\frac{\pi}{4}$.

16. $k\pi - 2x$.

17. $\frac{1-x^2-y^2-4xy+x^2y^2}{(1+x^2)(1+y^2)}$.

20. $p^2+q^2=1$.

23. $x=3$

24. $x=\frac{1}{3}$ or $\frac{2}{3}$.

25. $x=\tan\frac{\pi}{9}$.

26. $x=\cos\frac{\pi}{12}$.

27. $x=\frac{1}{3}\sqrt{3}$.

EXERCISE III. f. (p. 49.)

1. $\frac{n\pi}{3}$.

2. $\frac{(4n-1)\pi}{10}$.

3. $n\pi+\frac{\pi}{2}$ or $n\pi+\frac{\pi}{4}$.

4. $\frac{n\pi}{2}$.

5. $\frac{1}{3}(n\pi-\alpha-\beta)$.

6. $n\pi+\beta$ or $(2n+1)\frac{\pi}{2}-\alpha-\beta$.

7. $n\pi+\frac{\pi}{2}$ or $2n\pi+a+b\pm c$.

8. $n\pi-\frac{5\pi}{8}$, $n\pi-\frac{11\pi}{24}$, $n\pi-\frac{19\pi}{24}$.

9. $n\times 360^\circ + 41^\circ\ 4'$ or $n\times 360^\circ - 78^\circ\ 56'$.

10. $(2n+1)\pi$ or $n\pi+\frac{\pi}{2}$ or $\frac{2n\pi}{5}$.

11. $-\sin\frac{1}{2}(\alpha+\beta)\sec\frac{1}{2}(\alpha-\beta)$ or $-\sin\frac{1}{2}(\alpha-\beta)\sec\frac{1}{2}(\alpha+\beta)$.

12. $2n\pi<\theta<2n\pi+\frac{\pi}{3}$ or $2n\pi+\frac{\pi}{2}<\theta<(2n+1)\pi$ or

$$(2n+1)\pi+\frac{\pi}{2}<\theta<(2n+1)\pi+\frac{2\pi}{3}.$$

13. $x=a+n\pi$, $y=a-n\pi$, if $a\neq n\pi+\frac{\pi}{2}$;

indeterminate, $x+y=(2n+1)\pi$ if $a=n\pi+\frac{\pi}{2}$.

14. $\theta = 2m\pi + \alpha \pm \beta$, $\phi = 2n\pi + \alpha \mp \gamma$ where β, γ are positive angles, less than π, given by $2a^2c^2 \cos\beta = a^4 + c^4 - b^4$, $2b^2c^2 \cos\gamma = b^4 + c^4 - a^4$.

15. $x = m\pi$, $y = n\pi$ or $x = m\pi + \frac{\pi}{2}$, $y = n\pi + \frac{\pi}{2}$.

16. $x = 2m\pi \pm \frac{\pi}{3}$, $y = 2n\pi \pm \frac{2\pi}{3}$ or $x = 2m\pi \pm \frac{2\pi}{3}$, $y = 2n\pi \pm \frac{\pi}{3}$;

and $z = k\pi + \frac{\pi}{2}$.

19. One value if $\frac{5}{7} < c < \frac{5}{3}$; otherwise no values.

20. Two possible values of $\cos x$ if $0 < m < \frac{1}{3}$, one possible value if $-1 < m < 0$; otherwise no values.

22. $\sin\frac{1}{2}\theta = \frac{1}{2}\{-\sqrt{(1+\sin\theta)} + \sqrt{(1-\sin\theta)}\}$; $270° < \theta < 450°$.

23. $-\cot\frac{\theta}{4}$, $\tan\frac{\pi-\theta}{4}$, $-\cot\frac{\pi-\theta}{4}$.

25. (i) $4p$ values; (ii) $2p+1$ values.

26. $2k\cos\frac{2n\pi \pm \alpha}{5}$ for $n = 0, 1, 2$.

27. $\tan^{-1}p - \tan^{-1}r$.

29. $\tan^{-1}\frac{1}{2} = \tan^{-1}\frac{1}{3} + \tan^{-1}\frac{1}{7}$; $\tan^{-1}\frac{1}{3} = \tan^{-1}\frac{1}{4} + \tan^{-1}\frac{1}{13}$.

33. $\pm\frac{1}{2}$.

34. $\frac{3\pi}{10}$; $\frac{\pi}{2}$.

CHAPTER IV

EXERCISE IV. a. (p. 54.)

1. $\cdot 693$, $1\cdot 099$, $1\cdot 386$, $-\cdot 693$.

9. $\text{sq}\, 2t = 4t^2$; $\text{sq}(-t) = \text{sq}\, t$.

13. Yes. $\frac{d}{dt}(\text{hyp}\, t) = \frac{1}{t}$.

14. $\text{hyp}\, t = -\text{hyp}\frac{1}{t}$.

15. $\text{hyp}\, t$.

16. $1, \frac{1}{2}, \frac{1}{5}, \frac{1}{10}$.

EXERCISE IV. b. (p. 58.)

1. $\frac{1}{x}, \frac{1}{x}, \frac{1}{x}, 0.$

2. $\frac{2}{x}, \frac{3}{x}, \frac{-1}{x}, \frac{n}{x}.$

3. $\frac{na}{ax+b}, \frac{1}{x+1} - \frac{1}{x+2}.$

4. $\cot x$, $2 \operatorname{cosec} 2x$, $-2 \operatorname{cosec} 2x$.

5. $\text{hyp}\, x$, $\text{hyp}\,(x+1)$, $\text{hyp}\,(x-1)$, $-\text{hyp}\,(1-x)$, $\frac{1}{2}\text{hyp}\,(2x+3)$, $-\frac{1}{5}\text{hyp}\,(4-5x)$, $x - 2\,\text{hyp}\,(x+2)$, $x + 3\,\text{hyp}\,(x-3)$.

6. $\frac{1}{f(x)}\frac{d}{dx}f(x)$; $\text{hyp}\,(\sin x)$, $-\text{hyp}\,(\cos x)$, $\frac{1}{2}\text{hyp}\,(\sin 2x)$,

   $\frac{1}{2}\text{hyp}\,(ax^2+2bx+c)$, $\frac{1}{n}\text{hyp}\,(1+x^n)$, $\text{hyp}\,(1+\sin x)$.

7. $\text{hyp}\,(\tan x)$, $\frac{1}{2}\text{hyp}\,(\tan x)$.

8. $1 + \text{hyp}\, x$, $x\,\text{hyp}\, x - x$.

9. +.

10. +, −.

EXERCISE IV. c. (p. 61.)

1. 1·386, 1·792, 2·079, 2·198, −·693, −1·386, ·406, ·812, ·98; 2·718.

2. ·406; ·406; ·693; ·693.

3. $\text{hyp}\, b$; $\text{hyp}\,(ab) = \text{hyp}\, a + \text{hyp}\, b$.

7. $\text{hyp}\,(tb) - \text{hyp}\,(ta)$ is independent of t and equals $\text{hyp}\,\frac{b}{a}$.

EXERCISE IV. d. (p. 66.)

4. $ae, \frac{e}{b}, e, e^e.$

5. x^2, 2^x, $2 + \log x$.

8. $\log x$, $t \log t - t + 1$.

9. xe^x, $(t-1)e^t + 1$.

10. $x + 2x \log x$.

11. $\frac{1 - \log x}{x^2}$.

12. $-\frac{1}{x}$.

13. ae^{ax}.

14. $3x^2 e^{x^3}$.

15. 1.

16. $-\tan x$.

17. $\tan x$.

18. $\cos x \,.\, e^{\sin x}$.

19. $2 \tan x \sec^2 x \,.\, e^{\tan^2 x}$.

20. $\sec x\,(1 + x \tan x) \exp\,(x \sec x)$.

21. $-e^x \sin(e^x)$.

22. $-\frac{1}{x} \operatorname{cosec}\,(\log x) \cot\,(\log x)$.

23. $\frac{nb}{a+bx}$.

24. $\frac{e^{2x}-1}{e^{2x}+1}$.

25. $\operatorname{cosec} x$, $\sec x$, $\frac{1}{\sqrt{(a^2+x^2)}}$.

26. $\frac{1}{3}\log(3x+4)$. 27. $x+\frac{1}{2}\log(2x+3)$.

28. $\frac{1}{2}x-\frac{5}{4}\log(2x+3)$. 29. $\frac{1}{3}e^{3x}$. 30. e^x+e^{-x}.

31. $-\frac{1}{3}\log(2+3\cos x)$. 32. $\log(x^2+3x+4)$. 33. $\frac{1}{2}(\log x)^2$.

34 $2\log\tan\dfrac{x}{2}$. 35. $\frac{1}{2}e^{x^2}$. 36. $\frac{1}{3}\log\sin 3x$. 37. $-\log(1+e^{-x})$.

38. $\log\tan\left(\dfrac{\pi}{4}+\dfrac{x}{2}\right)$; $\log\tan\dfrac{x}{2}$; $\log\{x+\sqrt{(a^2+x^2)}\}$.

39. $\log(x+1)+2\log(x-1)$.

40. $2\log(x+1)-\log(x+2)-\log(x+3)$.

41. $-\frac{1}{4}\log x+\frac{1}{4}\log(x-2)-\dfrac{1}{2(x-2)}$.

42. $\log(x-3)-\frac{1}{2}\log(x^2+1)$. 43. $\frac{1}{3}x^3+\frac{1}{2}x^2+x+\log(x-1)$.

44. $x+\dfrac{a^2}{a-b}\log(x-a)+\dfrac{b^2}{b-a}\log(x-b)$.

45. $x\log x-x$. 46. $\dfrac{x^2}{2}\log x-\dfrac{x^2}{4}$. 47. $\dfrac{x^{n+1}}{n+1}\log x-\dfrac{x^{n+1}}{(n+1)^2}$.

48. xe^x-e^x. 49. $x^2e^x-2xe^x+2e^x$. 50. $\dfrac{x}{2}(\log x-1)$.

51. Differential coefficients, $e^{ax}(a\sin bx+b\cos bx)$;

$$e^{ax}(a\cos bx-b\sin bx).$$

Integrals, $\dfrac{1}{a^2+b^2}e^{ax}(a\sin bx-b\cos bx)$;

$$\frac{1}{a^2+b^2}e^{ax}(a\cos bx+b\sin bx).$$

52. $\dfrac{1}{\sqrt{e}}$.

53. $\dfrac{1}{e}$; 1 root for $A<0$, 2 roots for $0<A<\dfrac{1}{e}$, no roots for $A>\dfrac{1}{e}$.

EXERCISE IV. e. (p. 70.)

6. Untrue if $x>1$. 15. $\gamma+\log 2$. 16. $\log 2$. 17. $\log 2$.

EXERCISE IV. f. (p. 72.)

1. $(x^2-a^2)^{-\frac{1}{2}}$. 3. $x+3\log(x-1)-\dfrac{3}{(x-1)}-\dfrac{1}{(x-1)^2}$.

4. $I_{n+1}=x^{n+1}e^x-(n+1)I_n$. 5. $e^{\frac{1}{e}}$. 7. 2 or 3.

11. $\frac{1}{2}e^{ax}\left\{\frac{1}{a}-\frac{a\cos 2bx+2b\sin 2bx}{a^2+4b^2}\right\}$. 14. $x-\frac{1}{2}x^2+\frac{1}{2}x^3$.

17. (i) $\log t>\frac{1}{2}\left(1-\frac{1}{t^2}\right)$, (ii) $\log t>2\left(1-\frac{1}{\sqrt{t}}\right)$; the last.

23. $e^x(x^2-2x+2)$. 25. $\frac{y}{e^{ny}}\to 0$ when $y\to\infty$, if $n>0$.

EXERCISE IV. g. (p. 74.)

6. $\frac{a}{c}$. 21. $\int_0^\infty \frac{\log x}{1+c^2x^2}\,dx=\frac{\pi}{2c}\log\frac{1}{c}$. 24. $\frac{mn\,.\,B(m,n)}{(m+n)(m+n+1)}$.

CHAPTER V

EXERCISE V. a. (p. 82.)

1. sin 1. 2. cos 1. 3. 1 − sin 2. 4. sin 1 − cos 1.
5. 1. 6. sin 1 + cos 1. 7. 0·540.

8. 0·0523. 11. $1+\frac{x^2}{6}+\frac{7x^4}{360}$.

12. $(-1)^{[\frac{1}{2}r]}\cdot\frac{1}{2\sqrt{2}}\cdot\frac{(2x)^r}{r!}$; $(-1)^{[\frac{1}{2}(r-1)]}\cdot\frac{1}{2\sqrt{2}}\cdot\frac{2^r\left(\frac{\pi}{4}+x\right)^r}{r!}$.

13. $(-1)^n\frac{x^{2n}}{4\,.\,(2n)!}(3+3^{2n})$. 15. $\tan x-24\tan\frac{x}{2}$.

18. 0·4502. 19. 4·5.

EXERCISE V. b. (p. 86.)

1. (i) log 2; (ii) $\log\frac{3}{2}$; (iii) $\frac{1}{2}\log 2$.

2. (i) $-1<\frac{x}{a}\leqslant 1$; (ii) $0<\frac{y}{x}\leqslant 2$; (iii) $x^2>1$.

3. (i) $-\frac{1}{n}2^{-n}$, $-2\leqslant x<2$;

(ii) $-\frac{1}{n}[1+(-3)^n]$, $-\frac{1}{3}<x\leqslant\frac{1}{3}$;

(iii) $(-1)^{n-1} \cdot \frac{1}{n}(2^n+3^n)$, $-\frac{1}{3}<x\leqslant\frac{1}{3}$;

(iv) $(-1)^{n-1} \cdot \frac{2}{n}$, $-1<x\leqslant 1$;

(v) $(-1)^{n-1} \cdot \frac{1}{n}2^{-n}$, $-2<x\leqslant 2$;

(vi) $(-1)^{n-1} \cdot \frac{1}{n}(1+2^{-n})$, $-1<x\leqslant 1$;

(vii) n odd, $\frac{1}{n}$; n even, $\frac{1}{n}(1-2^{\frac{n}{2}+1})$, $x^2<\frac{1}{2}$;

(viii) $n=3p$, $-\frac{2}{n}$; $n=3p\pm 1$, $\frac{1}{n}$; $-1\leqslant x<1$.

4. (i) 0·6931; (ii) 0·4055, 0·2231, 0·1541; (iii) 1·0986, 1·3863, 1·6094, 1·7918, 1·9459, 2·0794, 2·1972, 2·3026.

5. (i) $\frac{1}{n(n-1)} \cdot 2^n$, $-\frac{1}{2}\leqslant x<\frac{1}{2}$;

(ii) $(-1)^{n-1} \cdot \frac{2}{n(n-1)(n-2)} \cdot 3^n$, $-\frac{1}{3}<x\leqslant\frac{1}{3}$;

(iii) n even, $(-1)^{\frac{n}{2}-1} \cdot \frac{1}{n} \cdot 2^{1-n}$; n odd, $(-1)^{\frac{n-1}{2}} \cdot \frac{1}{n-1} \cdot 2^{2-n}$; $x^2\leqslant 4$.

6. (i) $\log(1-x)+\frac{x}{1-x}$;

(ii) $\frac{1}{2}\{(1+x)\log(1+x)+(1-x)\log(1-x)\}$;

(iii) $1-\left(1+\frac{1}{x}\right)\log(1+x)$, for $x\neq 0$; if $x=0$, sum is 0.

7. $2\left\{\frac{1}{x}+\frac{1}{3x^3}+\frac{1}{5x^5}+\dots\right\}$.

8. $\Sigma r^{-1}n^{-2r}$.

9. $\Sigma\frac{2y^{2r-1}}{2r-1}$ where $y=\frac{2}{x^2-3x}$; $x^2>4$.

10. 1.

11. $-1+2\log 2$.

12. $2\log 2-1\frac{1}{4}$.

13. $\log 2-\frac{1}{2}$.

14. $2(1-\log 2)$.

15. $3-4\log 2$.

17. $-\frac{1}{3}$.

18. -1.

20. 0·175, 1·244

EXERCISE V. c. (p. 89.)

1. $2\tan^{-1}\frac{1}{2}$; $\dfrac{\pi}{12}$.

2. $x-\pi$; $x-2\pi$; $x-n\pi$.

5. $\dfrac{\pi}{4}$; $4\Sigma\left[(-1)^{r-1}\cdot\dfrac{1}{2r-1}(2^{1-2r}+3^{1-2r})\right]$.

6. $\dfrac{\pi}{4}$.

7. $\frac{1}{4}\log\dfrac{1+x}{1-x}+\frac{1}{2}\tan^{-1}x$, if $x^2<1$.

8. $(n+\frac{1}{4})\pi+\tan\theta-\frac{1}{3}\tan^3\theta+\frac{1}{5}\tan^5\theta-\dots$.

9. $\tan\theta\simeq\theta+\frac{1}{3}\theta^3+\frac{2}{15}\theta^5$.

10. $\frac{1}{2}\epsilon+\frac{1}{48}\epsilon^3-\frac{1}{1920}\epsilon^5$.

EXERCISE V. d. (p. 95.)

1. $\dfrac{1}{e}$.

2. $\frac{1}{2}\left(e-\dfrac{1}{e}\right)$.

3. $2e$.

4. 1.

5. $e-1$.

6. $5e$.

7. $\dfrac{1}{e}$.

8. $\dfrac{3e}{4}+\dfrac{1}{4e}$.

9. $\dfrac{3e}{2}$.

10. e^2-e.

11 $e+1$.

12. $\dfrac{e^2-3}{8e}$.

13. $\dfrac{e^2+8e-9}{4e}$.

14. $e+1$.

15. $-\dfrac{1}{e}$.

16. $1+\dfrac{1}{e}$.

17. 1.

18. $\dfrac{e-1}{e+1}$.

19. $4e-1$.

20. $\dfrac{17e}{6}$.

21. $\dfrac{7e}{4}$.

22. $\dfrac{1}{x}(e^x-1-x)$.

23. $\dfrac{1}{x^2}[(x-1)e^x+1-\frac{1}{2}x^2]$.

24. $\dfrac{1}{2\sqrt{x}}(e^{\sqrt{x}}-e^{-\sqrt{x}})-1$.

25. $\dfrac{1}{x^3}[(2-2x+x^2)e^x-2-\frac{1}{3}x^3]$.

26. $2+5x+\dots+(2+3n)\dfrac{x^n}{n!}+\dots$.

27. $1+x+\frac{3}{2}x^2+\dots+(-1)^n(1-5n+3n^2)\dfrac{x^n}{n!}+\dots$.

28. $e\left(1+x+\dots+\dfrac{x^n}{n!}+\dots\right)$.

29. $2\left[1+\dfrac{(2x)^2}{2!}+\dfrac{(2x)^4}{4!}+\dots\right]$; coefficient of x^n is 0 if n is odd; it is $2^{n+1}\cdot\dfrac{1}{n!}$ if n is even.

30. $1+\frac{5x}{2}-\frac{23}{8}x^2-\ldots+\frac{1+20n-16n^2}{2^n\,.\,n!}x^n+\ldots$. 31. $e^{-y^2}-e^{-x^2}$.

32. $\frac{1}{2}$. 33. $\frac{1}{3}n(3n-2)\,.\,2^{n-3}$. 34. $\frac{1}{2}$. 35. 3.

37. (i) $e^x<\frac{1}{1-x}$ if $x<0$; (ii) $e^x>\frac{2+x}{2-x}$ if $x<0$.

EXERCISE V. e. (p. 97.)

1. (i) 0; (ii) $\frac{1}{2}\log 2$; (iii) $\sqrt{2}\log(\sqrt{2}+1)$; (iv) $\frac{\pi\sqrt{3}}{6}$;

(v) $\frac{(e-1)^2}{2e}$; (vi) $\frac{1}{4}(e-\frac{1}{e}+2\sin 1)$.

4. $\pm 1{\cdot}033, 0$. 5. $a(1-2\cos\theta)+2a^2\sin 2\theta$. 9. $0{\cdot}26194$.

14. (i) $\log\frac{2}{3}+\Sigma\frac{1}{n}(3^{-n}-2^{-n})x^n$; $-2\leqslant x<2$;

(ii) $\Sigma(-1)^{p-1}\left[\frac{1}{3p-1}x^{6p-2}+\frac{2}{3p}x^{6p}+\frac{1}{3p+1}x^{6p+2}\right]$.

15. $\Sigma\left[\frac{2}{2r-1}\left(\frac{y}{x}\right)^{2r-1}\right]$; $x^2>y^2$.

16. $\Sigma\left[\frac{1}{r}\cdot\left(\frac{h}{x+h}\right)^{2r}\right]$; x, $(x+2h)>0$.

17. $\log 10=2{\cdot}30258$; $\log_{10}2=0{\cdot}30103$.

18. $\log 2=0{\cdot}693$. 19. Sum $=\log(-\cot\theta)$.

20. $\frac{1}{2}\log\frac{1+x+x^2}{1-x+x^2}$; all values of x; n even, coefficient of x^{3n} is zero; n odd, $-\frac{2}{3n}$.

21. (i) $\log 2$; (ii) $1-\log 2$; (iii) $\frac{1}{4}\log 2$; (iv) $3-2\log 2$.

22. $\frac{2}{3}\log 2-\frac{5}{12}$. 26. (i) $\frac{\pi}{8}$; (ii) $\frac{1}{4}\log\frac{1+x}{1-x}-\frac{1}{2}\tan^{-1}x$.

27. $-\frac{1}{4}$. 28. $a=3$, $b=-8$.

29. $\pm\Sigma\left[(-1)^{r-1}\,.\,\frac{1}{2r-1}\left(\tan\frac{a}{2}\tan\frac{\theta}{2}\right)^{2r-1}\right]$; $\tan^2\frac{a}{2}\tan^2\frac{\theta}{2}\leqslant 1$.

30. $\frac{1}{2}e$. 31. $\frac{1}{2}e$. 32. $4e+1$.

33. $e-1\frac{1}{2}$. 34. $\frac{e}{2}+\frac{2}{e}-1$. 35. $21e$.

42. $1-a\cos k+a^2\sin^2 k+\frac{3}{2}a^3\sin^2 k\cos k$.

EXERCISE V. f. (p. 100.)

1. $\dfrac{\theta^7}{1050}$.

10. $n=6p,\ -\dfrac{3}{n}$; $n=6p\pm1,\ \dfrac{2}{n}$; $n=6p\pm2,\ 0$; $n=6p+3,\ -\dfrac{1}{n}$.

12. $\dfrac{1}{3n}\left[1+(-1)^{n-1}\,.\,2\right]$.

16. $p^n-np^{n-2}q+\dfrac{n(n-3)}{2!}p^{n-4}q^2-\dfrac{n(n-4)(n-5)}{3!}p^{n-6}q^3+\dots$.

20. Order is (iii), (i), (v), (vi), (ii), (iv). 21. $\frac{3}{2}$. 25. $15e$.

28. (i) $(n-1)!$; (ii) $\frac{1}{2}(n+1)!$; (iii) $\frac{1}{24}(3n+1)(n+2)!$.

33. $\left(1-\dfrac{x}{n+1}\right)^{n+1}$ is the greater.

CHAPTER VI

EXERCISE VI. a. (p. 106.)

7. $2\,\text{sh}\,\frac{1}{2}(\theta+\phi)\,\text{ch}\,\frac{1}{2}(\theta-\phi)$; $2\,\text{ch}\,\frac{1}{2}(\theta+\phi)\,\text{ch}\,\frac{1}{2}(\theta-\phi)$.

9. $(3\,\text{th}\,\theta+\text{th}^3\theta)\div(1+3\,\text{th}^2\theta)$. 10. $\text{cosech}^2x=\coth^2x-1$.

11. $-\text{cosech}^2x$. 12. $\frac{1}{4}(\text{sh}\,3x-3\,\text{sh}\,x)$.

13. $\text{ch}^2x-\text{ch}^2y$. 14. $\frac{1}{2}\{\text{ch}\,(\theta+\phi)-\text{ch}\,(\theta-\phi)\}$.

15. $\frac{1}{2}\{\text{sh}\,(\theta+\phi)+\text{sh}(\theta-\phi)\}$. 16. $\frac{1}{2}\{\text{ch}\,(\theta+\phi)+\text{ch}\,(\theta-\phi)\}$.

17. $\text{ch}\,x+\text{sh}\,x$. 18. $\text{ch}\,nx+\text{sh}\,nx$. 19. $\text{ch}\,nx-\text{sh}\,nx$.

20. $\{\Sigma\,(\text{th}\,x)+\text{th}\,x\,\text{th}\,y\,\text{th}\,z\}\div\{1+\Sigma\,(\text{th}\,x\,\text{th}\,y)\}$.

23. $+\sqrt{(1+\text{sh}^2\theta)}$; $\text{sh}\,\theta\div\sqrt{(1+\text{sh}^2\theta)}$.

24. $1\div\sqrt{(1-\text{th}^2\theta)}$; $\text{th}\,\theta\div\sqrt{(1-\text{th}^2\theta)}$.

25. $\pm\sqrt{\{\frac{1}{2}(k-1)\}}$; $\pm\sqrt{\left\{\dfrac{k-1}{k+1}\right\}}$. 26. $\dfrac{2t}{1-t^2}$; $\dfrac{1+t^2}{1-t^2}$.

27. $x^2\,\text{cosec}^2u-y^2\sec^2u=1$; $x^2\,\text{sech}^2v+y^2\,\text{cosech}^2v=1$.

32. $-\coth\frac{1}{2}\phi$. 33. $-4\,\text{sh}\,(y+z)\,\text{sh}\,(z+x)\,\text{sh}\,(x+y)$.

35. $\frac{1}{2}\left(x-\dfrac{1}{x}\right)$; $\frac{1}{2}\left(x+\dfrac{1}{x}\right)$.

EXERCISE VI. b. (p. 108.)

1. $\text{ch } x + \text{sh } x$. 2. $\text{sh } 2x$. 3. $\text{sh } 2x$. 4. $\text{ch } 2x$.
5. $-\coth x \text{ cosech } x$. 6. $-\text{th } x \text{ sech } x$. 7. $-\text{cosech}^2 x$.
8. $\coth x$. 9. $\text{th } x$. 10. $\text{cosech } x$.
11. $-\text{sech } 2x$. 12. 1. 13. $\frac{1}{2}\text{ sh } 2x$.
14. $\frac{1}{3}\text{ ch } 3x$. 15. $\log(\text{ch } x)$. 16. $\log(\text{sh } x)$.
17. $\frac{1}{4}\text{ sh } 2x - \frac{1}{2}x$. 18. $-\coth x$. 19. $x - \text{th } x$.
20. $x - \coth x$. 21. $2\tan^{-1}\left(\text{th}\frac{x}{2}\right)$. 22. $\log\left(\text{th}\frac{x}{2}\right)$.
23. $\frac{1}{6}\text{ sh } 3x - \frac{1}{2}\text{ sh } x$. 24. $\frac{1}{12}\text{ sh } 3x + \frac{3}{4}\text{ sh } x$.
25. $2\text{ sh } x \cos x$. 26. $\frac{1}{2}\{\sin x \text{ sh } x - \cos x \text{ ch } x\}$.
28. (iii) 1; (iv) $\to\infty$; (v) $\to\infty$, $\to-\infty$; (vi) $\frac{1}{2}$; (vii) $\frac{1}{2}$, $-\frac{1}{2}$.
29. (iii) 1, -1; 1, -1; (iv) $\to 0$; (v) 1

EXERCISE VI. c. (p. 112.)

6. (i) $\log\dfrac{1+\sqrt{(1+y^2)}}{y}$; (ii) $\log\dfrac{1-\sqrt{(1+y^2)}}{y}$.

11. (i) $\dfrac{x^2}{a^2}-\dfrac{y^2}{b^2}=1$; (ii) $\dfrac{x^2}{a^2}-\dfrac{y^2}{b^2}-\dfrac{2xy}{ab}\text{ sh }(\alpha-\beta)=\text{ch}^2(\alpha-\beta)$.

15. $\dfrac{1}{a}\text{th}^{-1}\dfrac{x}{a}$; $\dfrac{1}{2a}\log\dfrac{a+x}{a-x}$.

16. (i) $\text{ch}^{-1}\left|\dfrac{x}{a}\right|$; (ii) $-\text{ch}^{-1}\left|\dfrac{x}{a}\right|$.

17. $\frac{1}{2}\left\{a^2\text{ sh}^{-1}\dfrac{x}{|a|}+x\sqrt{(x^2+a^2)}\right\}$. 18. $\frac{15}{8}-2\log 2$.

19. $\log 2$. 20. $\log(5+\sqrt{34})-2\log 3$.

21. $\frac{9}{2}\text{ sh}^{-1}\dfrac{x}{3}+\frac{1}{2}x\sqrt{(x^2+9)}$. 22. $\frac{1}{2}x\sqrt{(x^2-a^2)}-\frac{1}{2}a^2\text{ ch}^{-1}\left|\dfrac{x}{a}\right|$.

23. $\frac{1}{2}x\sqrt{(x^2+1)}-\frac{1}{2}\text{ sh}^{-1}x$. 24. $2\coth^{-1}\left(\tan\dfrac{x}{2}\right)$.

EXERCISE VI. d. (p. 113.)

2. $\frac{1}{8}(3\text{ ch } 4x+5)$. 3. $\text{th } x$. 5. $2\sqrt{(1+x^2)}$.

6. $2\sqrt{(x^2-a^2)}$ or $\dfrac{2x^2}{\sqrt{(x^2-a^2)}}$ according as $\text{ch}^{-1}\dfrac{x}{a}$ is + or −.

7. $\mp \dfrac{1}{x\sqrt{(1-x^2)}}$.

8. $-\dfrac{a}{|x|\,.\,\sqrt{(x^2+a^2)}}$.

9. $x^{\text{sh}\,x}\left(\dfrac{1}{x}\,\text{sh}\,x + \text{ch}\,x\log x\right)$.

10. $e^{ax}(a\,\text{sh}\,bx + b\,\text{ch}\,bx)$.

11. $e^x\,\text{th}\,x$.

12. $\frac{1}{48}(2\,\text{ch}\,6x - 6\,\text{ch}\,2x - 3\,\text{ch}\,4x)$.

13. $\dfrac{1}{a^2-b^2}\,.\,e^{ax}(a\,\text{sh}\,bx - b\,\text{ch}\,bx)$.

14. $\log\dfrac{1+\text{ch}\,x}{2+\text{ch}\,x}$.

15. $y = 1 + \frac{1}{2}x^2$; 1.

16. $\dfrac{\pi}{8}$.

18. 1; 1.

19. $-\frac{2}{3}$.

22. $1 - \dfrac{x^2}{6} + \dfrac{7x^4}{360}$.

26. 1·91(6).

29. $\text{th}^{-1}\left(\dfrac{x+y}{1+xy}\right)$.

31. $x = \dfrac{a(1+t^2)}{1-t^2}$, $y = \dfrac{2bt}{1-t^2}$.

33. The parabola, $y^2 = 4ax$.

EXERCISE VI. e. (p. 115.)

1. $\dfrac{2a}{1+a^2+b^2}$; $\dfrac{2b}{1-a^2-b^2}$.

4. $\text{sh}\,u = -\tan\theta$; $u = \log(\sec\theta - \tan\theta)$; $\text{th}\,\dfrac{u}{2} = -\tan\dfrac{\theta}{2}$.

5. $\dfrac{y_1}{y(x_1-x)}$.

6. $-\dfrac{1}{y_1}\,\text{th}^{-1}\dfrac{axx_1 + b(x+x_1) + c}{yy_1}$.

9. $\dfrac{x}{\text{sh}\,x} > \dfrac{\text{th}\,x}{x}$.

11. $x = 1$ or 6.

12. $\log 2 + \dfrac{a}{8} - \dfrac{a^2}{64}$.

16. $\text{ch}\,x = u$, $16u^5 - 20u^3 + 5u$.

17. $\text{sh}\,x = v$, $16v^5 + 20v^3 + 5v$.

18. $\text{sh}\,x = v$, $32v^5 + 32v^3 + 6v$.

19. $\frac{1}{64}(\text{sh}\,7x - 7\,\text{sh}\,5x + 21\,\text{sh}\,3x - 35\,\text{sh}\,x)$.

20. $\frac{1}{32}(\text{ch}\,6x - 6\,\text{ch}\,4x + 15\,\text{ch}\,2x - 10)$.

22. $2^n\,\text{sh}^n\dfrac{\theta}{2}\left(\text{ch}\,\dfrac{n\theta}{2} + \text{sh}\,\dfrac{n\theta}{2}\right)$.

24. $\text{sh}\,(n+1)\dfrac{\alpha}{2}\,\text{sh}\,\dfrac{n\alpha}{2}\,\text{cosech}\,\dfrac{\alpha}{2}$.

25. $\text{ch}\left[\alpha + (n-1)\dfrac{\beta}{2}\right]\text{sh}\,\dfrac{n\beta}{2}\,\text{cosech}\,\dfrac{\beta}{2}$.

26. $\frac{1}{2}\,\text{cosech}^2\,\frac{1}{2}\theta\left\{n\,\text{sh}\,\dfrac{\theta}{2}\,\text{sh}\,(n+\frac{1}{2})\theta - \text{sh}^2\frac{1}{2}n\theta\right\}$.

29 $\text{sh}\,(\text{sh}\,\theta)e^{\text{ch}\,\theta}$.

32. $e^{\sin\theta\,\text{ch}\,\theta}\,\text{ch}\,(\theta + \sin\theta\,\text{sh}\,\theta)$.

CHAPTER VII

EXERCISE VII. a. (p. 121.)

1. $\frac{2\pi}{3}+\theta$; $\frac{4\pi}{3}+\theta$; $\pi+\theta$; $-\frac{\pi}{3}-\theta$.

2. $\phi-\frac{\pi}{2}$; $\phi-\pi$; $\phi+\frac{\pi}{4}$; $\phi+\frac{3\pi}{4}$.

3. AC ; CA. 4. SP ; PS.

5. (i) $-4\cos\phi$, $-5\cos\theta$, 0 ; (ii) $-4\sin\phi$, $5\sin\theta$, 3.

6. (i) $-c\sin\phi$, $-c\cos\phi$, $-c\cos\left(\phi+\frac{\pi}{4}\right)$;

(ii) $c\cos\phi$, $-c\sin\phi$, $-c\sin\left(\phi+\frac{\pi}{4}\right)$.

7. $a\cos\alpha$, $a\cos\left(\alpha+\frac{\pi}{3}\right)$, $a\cos\alpha+a\cos\left(\alpha+\frac{\pi}{3}\right)$, $2a\cos\left(\alpha+\frac{\pi}{3}\right)$, $-2a\cos\alpha$.

8. $3\cos\alpha-2\sin\alpha$; $3\sin\alpha+2\cos\alpha$.

9. $4-2\cos\alpha+3\sin\alpha-2\sin(\alpha+\beta)$; $2\sin\alpha+3\cos\alpha-2\cos(\alpha+\beta)$.

10. $5\sin\alpha+\cos\alpha$. 11. $(h+r\cos\theta,\ k+r\sin\theta)$.

12. $(h-s\cos\phi,\ k-s\sin\phi)$. 13. $\sqrt{\{r_1^2+r_2^2+2r_1r_2\cos(\theta_1-\theta_2)\}}$.

14. $h\cos\alpha+k\sin\alpha-p$.

EXERCISE VII. b. (p. 126.)

2. $(x\cos A+y\sin A,\ -x\sin A+y\cos A)$.

3. As in No. 2. 9. Yes. 12. $\sin^2 n\theta \operatorname{cosec}\theta$.

EXERCISE VII. c. (p. 129.)

1. $\frac{1}{2}\sin\frac{(2n+1)\theta}{4}\operatorname{cosec}\frac{\theta}{4}-\frac{1}{2}$.

7. (i) $\sin[\alpha+(n-\frac{1}{2})\beta]\sin n\beta\sec\frac{1}{2}\beta$;

(ii) $\cos(\alpha+n\beta)\cos(n+\frac{1}{2})\beta\sec\frac{1}{2}\beta$;

(iii) $\cos[\alpha+\frac{1}{2}(m-1)(\beta+\pi)]\sin\frac{m}{2}(\beta+\pi)\sec\frac{1}{2}\beta$.

8. $\sin[\alpha+\frac{1}{2}(n-1)(\beta+\pi)]\sin\frac{n}{2}(\beta+\pi)\sec\frac{1}{2}\beta$.

9. $\frac{1}{2}+(-1)^{n+1}\frac{1}{2}\cos(n+\frac{1}{2})\theta \sec\frac{1}{2}\theta$.

10. $\frac{1}{4}(2n+1)-\frac{1}{4}\sin(2n+1)\theta \operatorname{cosec}\theta$.

11. $\frac{1}{2}\sin 2n\theta \operatorname{cosec}\theta$.

12. $\frac{1}{2}n\sin\theta+\frac{1}{2}\sin(n+2)\theta\sin n\theta \operatorname{cosec}\theta$.

13. $\cos\left[(n+1)\frac{\theta}{2}+(n-1)\frac{\pi}{4}\right]\sin\left(\frac{n\theta}{2}+\frac{n\pi}{4}\right)\operatorname{cosec}\left(\frac{\theta}{2}+\frac{\pi}{4}\right)$.

14. $\frac{1}{2}n+\frac{1}{2}\cos[2\theta+(n-1)\phi]\sin n\phi \operatorname{cosec}\phi$.

15. $\frac{1}{4}\sin\frac{n\theta}{2}\operatorname{cosec}\frac{\theta}{2}\left\{2\sin(n+3)\frac{\theta}{2}+\sin(n-1)\frac{\theta}{2}\right\}$

$-\frac{1}{4}\sin(3n+5)\frac{\theta}{2}\sin\frac{3n\theta}{2}\operatorname{cosec}\frac{3\theta}{2}$.

16. $\frac{1}{8}\left\{\sin(3n+\frac{3}{2})\theta\operatorname{cosec}\frac{3\theta}{2}+3\sin(n+\frac{1}{2})\theta\operatorname{cosec}\frac{\theta}{2}-4\right\}$.

17. $\frac{1}{16}\{6n-5+4\sin(2n+1)\theta\operatorname{cosec}\theta+\sin(4n+2)\theta\operatorname{cosec}2\theta\}$.

18. $\frac{1}{2}\left[\cot\frac{\theta}{2}-\cos(2n+1)\frac{\theta}{2}\operatorname{cosec}\frac{\theta}{2}\right]$;

$\frac{1}{4}\operatorname{cosec}^2\frac{\theta}{2}\{(n+1)\cos n\theta-n\cos(n+1)\theta-1\}$.

EXERCISE VII. d. (p. 132.)

1. $\tan(2^n\theta)-\tan\theta$.

2. $2^{1-n}\cot(2^{1-n}\theta)-2\cot 2\theta$.

3. $\tan\alpha\tan(\alpha+\beta)+\tan(\alpha+\beta)\tan(\alpha+2\beta)+\ldots$
$=\cot\beta\{\tan(\alpha+n\beta)-\tan\alpha\}-n$.

4. $\Sigma\, 2^{1-r}\tan^2(2^{r-1}\theta)\tan(2^r\theta)=2^{1-n}\tan(2^n\theta)-2\tan\theta$.

5. $\Sigma\, 2^{r-1}\cos(2^{1-r}\theta)\sin^2(2^{-r}\theta)=\frac{1}{2}\sin^2\theta-2^{n-1}\sin^2(2^{-n}\theta)$.

6. $\operatorname{cosec}^2\theta\sin n\theta\operatorname{cosec}(n+1)\theta$.

9. $2\operatorname{cosec}2\theta\sin n\theta\sec(n+1)\theta$.

10. $\tan(n+1)\theta\cot\theta-n-1$.

11. $\cot^2\theta-\cot\theta\cot(n+1)\theta-n$.

12. $\operatorname{cosec}2\theta\sin n\theta\sec(n+1)\theta$.

13. $\frac{1}{2}\{\tan(3^n\theta)-\tan\theta\}$.

14. $4\operatorname{cosec}^2 2\theta-4^{1-n}\operatorname{cosec}^2(2^{1-n}\theta)$.

15. $4\operatorname{cosec}^2 2\theta-4^{1-n}\operatorname{cosec}^2(2^{1-n}\theta)-\frac{4}{3}(1-4^{-n})$.

16. $\tan^{-1}\left(\frac{n}{n+2}\right)$.

17. $\tan^{-1}\left(\frac{n}{n+1}\right)$.

18. $\tan^{-1}n+\tan^{-1}(n+1)-\frac{\pi}{4}$.

19. $\tan^{-1}\frac{1}{2}(n+2)+\tan^{-1}\frac{1}{2}(n+1)+\tan^{-1}\frac{1}{2}n+\tan^{-1}\frac{1}{2}(n-1)-\frac{\pi}{4}$.

20. $\tan^{-1}\left(\frac{n}{n+1}\right)$.

EXERCISE VII. e. (p. 133.)

1. $\sqrt{2}$.
2. 25.

4. $+\sqrt{(a^2+b^2)},\quad \theta=\frac{\pi}{2}-\alpha;\qquad -\sqrt{(a^2+b^2)},\quad \theta=-\frac{\pi}{2}-\alpha,$ where $\sin\alpha:\cos\alpha:1=a:b:+\sqrt{(a^2+b^2)}$.

7. $\frac{1}{2}(n-1)\cos\theta+\frac{1}{4}\sin(2n+2)\theta\operatorname{cosec}\theta$.

8. $\frac{1}{2}n\cos\beta-\frac{1}{2}\cos[2\alpha+(n+2)\beta]\sin n\beta\operatorname{cosec}\beta$.

9. $\frac{1}{2}\sin\frac{1}{2}(n+3)\theta\sin\frac{1}{2}n\theta\operatorname{cosec}\frac{1}{2}\theta-\frac{1}{4}\sin(n+3)\theta\sin n\theta\operatorname{cosec}\theta$.

12. $\frac{1}{4}\operatorname{cosec}^2\frac{\theta}{2}\{(n+1)^2\cos n\theta-n^2\cos(n+1)\theta$
$-\sin(n+\frac{1}{2})\theta\operatorname{cosec}\frac{1}{2}\theta\}$.

13. $\log(\sin\theta)-\log(\sin 2^{-n}\theta)-n\log 2$; $2^{-n}\cot(2^{-n}\theta)-\cot\theta$.

14. $\log(\sin 2^n\theta)-\log\sin\theta-n\log 2$.
15. $\frac{1}{2}$.

16. 0.
17. $\cot\theta-2^n\cot(2^n\theta)$.

18. $4^n\operatorname{cosec}^2(2^n\theta)-\operatorname{cosec}^2\theta-\frac{1}{3}(4^n-1)$.

19. $\Sigma\sin^2(3^{-r}\theta)\cos(3^{-r}\theta)=\frac{1}{4}[\cos(3^{-n}\theta)-\cos\theta]$, for $r=1$ to n.

EXERCISE VII. f. (p. 135.)

5. $\frac{1}{4}\Sigma\{\sin\frac{1}{2}(n+1)\alpha\sin\frac{1}{2}n\alpha\operatorname{cosec}\frac{1}{2}\alpha\}$ where $\alpha=x+y-z$, $y+z-x,\quad z+x-y,\quad -x-y-z$.

6. $$\frac{1}{16}\left\{2\sin\tfrac{1}{2}(n+1)\theta\sin\tfrac{1}{2}n\theta\operatorname{cosec}\tfrac{1}{2}\theta+\sin\tfrac{3}{2}(n+1)\theta\sin\frac{3n\theta}{2}\operatorname{cosec}\frac{3\theta}{2}-\sin\tfrac{5}{2}(n+1)\theta\sin\frac{5n\theta}{2}\operatorname{cosec}\frac{5\theta}{2}\right\}.$$

7. $$\frac{1}{32}\left\{10\sin(n+\tfrac{1}{2})\theta\operatorname{cosec}\tfrac{1}{2}\theta+5\sin3(n+\tfrac{1}{2})\theta\operatorname{cosec}\frac{3\theta}{2}+\sin5(n+\tfrac{1}{2})\theta\operatorname{cosec}\frac{5\theta}{2}-16\right\}.$$

8. $\frac{1}{2}\left\{n+\frac{1-\cos n\theta}{1-\cos\theta}\right\}$. 9. $\frac{1}{2}\left\{n+\frac{1-\cos n\theta}{1-\cos\theta}\right\}$.

10. $\frac{1}{4}\{n^2+2n+\sin^2 n\theta\operatorname{cosec}^2\theta\}$. 12. $\frac{1}{4}\operatorname{cosec}^2\frac{\theta}{2}(n\sin\theta-\sin n\theta)$.

13. $1-x\cos\theta-x^{n+1}\cos(n+1)\theta+x^{n+2}\cos n\theta$.

14. $x\sin\theta-x^{n+1}\sin(n+1)\theta+x^{n+2}\sin n\theta$.

17. $\frac{1}{4}\sin 2\theta-2^{-n-2}\sin(2^{n+1}\theta)$.

18. $\frac{1}{2}\operatorname{cosec}\theta\{\sec(2n+2)\theta-\sec 2\theta\}$.

19. $\sin 3(n+\frac{1}{2})\theta\operatorname{cosec}\frac{3\theta}{2}+\sin(n+\frac{1}{2})\theta\operatorname{cosec}\frac{\theta}{2}-2$.

20. $\frac{1}{2}\{\cot\theta-\cot(3^n\theta)\}$. 21. $\frac{1}{3}+\operatorname{cosech}^2 x-\frac{1}{x^2}$.

CHAPTER VIII

EXERCISE VIII. a. (p. 139.)

1. $[3, 8]$. 2. $[-14, 23]$. 5. $4, 3$.

6. $[1, 13]$; $[-16, 1]$; $[-8, 5]$; $[-8, 5]$; $[18, 3]$; $[-3, 18]$; $[-42, -32]$; $[3, 1]$.

7. $[a, b]$; $[a, b]$.

8. $[a^2-b^2, 2ab]$; $[1, 0]$; $[1, 0]$; $[-1, 0]$; $[-1, 0]$.

9. $[\cos(\theta+\phi), \sin(\theta+\phi)]$.

EXERCISE VIII. b. (p. 144.)

1. $3+5i$; 6; $7i$; 0; $-i$; $a-\beta i$.

2. $[1, 2]$; $[5, 0]$; $[3, -2]$; $[0, 7]$; $[0, -2]$; $[-2, -1]$; $[\beta, a]$.

3. $[4, -3]$; $[-2, 0]$; $[-5, 10]$; $[\frac{2}{3}, 0]$; $7i-1$; 2; -12; $\frac{3}{4}$; $[a, \beta]$; $[0, 0]$; $[a, \beta]$; $-1-8i$.

4. $3+6i$. 5. $5+5i$. 6. 2. 7. $8-i$.

8. $7+24i$. 9. a^2-b^2+2abi. 10. $-b+ai$. 11. $-ab$.

12. $-ai$. 13. $\cos(\theta+\phi)+i\sin(\theta+\phi)$. 14. 1.

15. $\cos(\theta-\phi)+i\sin(\theta-\phi)$. 16. $\cos 2\theta+i\sin 2\theta$.

17. $rs\{\cos(\theta+\phi)+i\sin(\theta+\phi)\}$. 18. $\cos(\theta-\phi)+i\sin(\theta-\phi)$.

19. $x^2-2x\cos\theta+1$. 20. $x^2+2x\sin\phi+1$.

21. $2i$. 22. $-2-2i$. 23. i. 24. $-\frac{1}{2}+\frac{3}{2}i$.

25. $-\frac{4}{5}-\frac{7}{5}i$. 26. $-1+i$. 27. x^2-y^2-2ixy.

28. $\dfrac{x+yi}{x^2+y^2}$. 29. $\dfrac{x^2-y^2+2ixy}{x^2+y^2}$.

30. $\dfrac{1-x^2-y^2+2iy}{(1-x)^2+y^2}$. 31. $0, -\frac{1}{2}$. 32. 325.

34. x^2+y^2 ; $(x-1+yi-2i)(x-1-yi+2i)$. 35. $(x^2+y^2)^n$.

36. $0+0+0+4+0+0+0+4+\ldots$. 37. $-\frac{6}{7}, -\frac{2}{7}$.

38. $[3, -4]$. 39. $3+2i$; $\dfrac{1}{\sqrt{2}}+\dfrac{i}{\sqrt{2}}$.

40. Either $d^2-4abd+4b^2c=0$ and $b\neq 0$, or $b=0$, $d=0$, $a^2 \geqslant c$.

EXERCISE VIII. c. (p. 147.)

1. $1, 0$. 2. $1, \pi$. 3. $1, \dfrac{\pi}{2}$. 4. $1, -\dfrac{\pi}{2}$.

5. $2, \dfrac{\pi}{3}$. 6. $2, -\dfrac{\pi}{3}$. 7. $2, \dfrac{2\pi}{3}$. 8. $2, -\dfrac{2\pi}{3}$.

9. $2, \dfrac{5\pi}{6}$. 10. $\sqrt{2}, \dfrac{\pi}{4}$. 11. $\sqrt{2}, \dfrac{3\pi}{4}$. 12. $\sqrt{2}, -\dfrac{3\pi}{4}$.

13. $\sqrt{2}, -\dfrac{\pi}{4}$. 14. $\sqrt{2}, \dfrac{3\pi}{4}$. 15. $2, -\dfrac{\pi}{6}$. 16. $2, -\dfrac{5\pi}{6}$.

17. $5(\frac{3}{5}+i\frac{4}{5})$. 18. $\sqrt{(4+2\sqrt{2})}\left\{\cos\left(-\dfrac{\pi}{8}\right)+i\sin\left(-\dfrac{\pi}{8}\right)\right\}$.

19. $5(-\frac{3}{5}+i\frac{4}{5})$. 20. $5(-\frac{3}{5}-i\frac{4}{5})$.

21. $(\sqrt{6}-\sqrt{2})\left\{\cos\left(-\dfrac{7\pi}{12}\right)+i\sin\left(-\dfrac{7\pi}{12}\right)\right\}$.

22. $\cos(-a)+i\sin(-a)$. 23. $\cos\left(a-\dfrac{\pi}{2}\right)+i\sin\left(a-\dfrac{\pi}{2}\right)$.

24. $\cos\left(\dfrac{\pi}{2}-a\right)+i\sin\left(\dfrac{\pi}{2}-a\right)$.

25. (i) $\cos a, a$; (ii) $-\cos a, a-\pi$; (iii) $-\cos a, a+\pi$.

26. $\sec a(\cos a+i\sin a)$ or $(-\sec a)\{\cos(a+\pi)+i\sin(a+\pi)\}$.

27. $\operatorname{cosec} \alpha \left\{\cos\left(\frac{\pi}{2} - \alpha\right) + i \sin\left(\frac{\pi}{2} - \alpha\right)\right\}$ or

$$(-\operatorname{cosec} \alpha) \left\{\cos\left(\frac{3\pi}{2} - \alpha\right) + i \sin\left(\frac{3\pi}{2} - \alpha\right)\right\}.$$

28. $\sec\beta \left\{\cos\left(\beta - \frac{\pi}{2}\right) + i \sin\left(\beta - \frac{\pi}{2}\right)\right\}$ or

$$(-\sec\beta) \left\{\cos\left(\beta + \frac{\pi}{2}\right) + i \sin\left(\beta + \frac{\pi}{2}\right)\right\}.$$

29. $\cos\left(-\beta - \frac{\pi}{2}\right) + i \sin\left(-\beta - \frac{\pi}{2}\right).$

30. $2 \cos \frac{\theta}{2} \left(\cos \frac{\theta}{2} + i \sin \frac{\theta}{2}\right)$ or

$$\left(-2 \cos \frac{\theta}{2}\right)\left\{\cos \frac{\theta}{2} + \pi\right) + i \sin\left(\frac{\theta}{2} + \pi\right)\right\}.$$

31. $2 \cos \frac{\theta}{2} \left\{\cos\left(-\frac{\theta}{2}\right) + i \sin\left(-\frac{\theta}{2}\right)\right\}$ or

$$-2 \cos \frac{\theta}{2} \left\{\cos\left(\pi - \frac{\theta}{2}\right) + i \sin\left(\pi - \frac{\theta}{2}\right)\right\}.$$

32. $2 \cos\left(\frac{\pi}{4} - \frac{\theta}{2}\right)\left\{\cos\left(\frac{\pi}{4} - \frac{\theta}{2}\right) + i \sin\left(\frac{\pi}{4} - \frac{\theta}{2}\right)\right\}$ or

$$-2 \cos\left(\frac{\pi}{4} - \frac{\theta}{2}\right)\left\{\cos\left(\frac{5\pi}{4} - \frac{\theta}{2}\right) + i \sin\left(\frac{5\pi}{4} - \frac{\theta}{2}\right)\right\}.$$

33. $2 \cos \frac{1}{2}(\alpha - \beta)\{\cos \frac{1}{2}(\alpha + \beta) + i \sin \frac{1}{2}(\alpha + \beta)\}$ or

$$-2 \cos \tfrac{1}{2}(\alpha - \beta)\{\cos \tfrac{1}{2}(\alpha + \beta + 2\pi) + i \sin \tfrac{1}{2}(\alpha + \beta + 2\pi)\}.$$

34. $2 \sin \frac{1}{2}(\alpha - \beta)\{\cos \frac{1}{2}(\alpha + \beta + \pi) + i \sin \frac{1}{2}(\alpha + \beta + \pi)\}$ or

$$2 \sin \tfrac{1}{2}(\beta - \alpha)\{\cos \tfrac{1}{2}(\alpha + \beta - \pi) + i \sin \tfrac{1}{2}(\alpha + \beta - \pi)\}.$$

35. $s(\cos\psi + i \sin\psi)$ where $s = +\sqrt{(1 + 2r\cos\phi + r^2)}$ and

$$\cos\psi : \sin\psi : 1 = 1 + r\cos\phi : r\sin\phi : s.$$

36. $(\cos\theta + i\sin\theta) + \left\{\cos\left(\theta + \frac{2\pi}{3}\right) + i\sin\left(\theta + \frac{2\pi}{3}\right)\right\}$

$$+\left\{\cos\left(\theta + \frac{4\pi}{3}\right) + i\sin\left(\theta + \frac{4\pi}{3}\right)\right\} = 0.$$

38. $-\cos\theta_2, \cos\theta_1.$

EXERCISE VIII. d. (p. 149.)

2. $2\beta - a$. 3. $2\gamma + a = 3\beta$.

6. Q in (i) OP′, (ii) OP where P′ represents $-z_1$.

7. Circles. Centre O rad. 3; centre (3, 0), rad. 1; centre (9, 0), rad. 4.

8. On circles (i) centre O, rad. 5; (ii) centre (1, 0), rad. 2; (iii) centre (−2, 0), rad. 3; (iv) centre ($\frac{1}{2}$, 0), rad. $\frac{3}{2}$; (v) centre (2, 3), rad. 4; (vi) on positive x-axis.

9. $|z-8-9i|<7$, $|z-a-bi|=c$, $|z+1|>1$. 10. 4, 2.

11. 3, 1. 12. 7, 3. 13. 2, 8. 14. 6, 4.

15. Between concentric circles, centre (−2, 3), radii 1, 2.

16. Within circle centre (1, 0), radius 1. 18. $-\frac{\pi}{6}$ to $+\frac{\pi}{6}$.

20. $\frac{\Sigma mz}{\Sigma m}$ is centre of mass of $m_1, m_2, m_3, \ldots$ at points representing $z_1, z_2, z_3, \ldots$.

EXERCISE VIII. e. (p. 153.)

1. $\text{cis}\, a$. 2. $\text{cis}\, 2\beta$. 3. $\text{cis}(-2\theta)$. 4. $\text{cis}(-3\phi)$.

5. $\text{cis}\, 4a$. 6. $\text{cis}(-2\theta)$. 7. $\text{cis}(a+\beta-\gamma)$.

8. $\text{cis}(2\theta+\phi)$. 9. $\text{cis}(-3\theta)$. 10. $\text{cis}\, 6\theta$. 11. $\text{cis}(-5\theta)$.

12. $\text{cis}(-18\theta)$. 13. -1. 14. -1.

15. $-\sin 3\theta - i\cos 3\theta$. 16. $\sin 5\theta - i\cos 5\theta$.

17. $8\cos^3\frac{\theta}{2}\,\text{cis}\,\frac{3\theta}{2}$. 18. $16\sin^4\frac{\theta}{2}\,\text{cis}(-2\theta)$.

19. $2\cos\theta\,\text{cis}(-\theta)$. 20. $i\tan\theta$. 21. $\text{cis}(-8\theta)$.

22. $\frac{1}{8}\operatorname{cosec}^3\left(\frac{\pi}{4}-\frac{\theta}{2}\right)\text{cis}\,\frac{3\pi+6\theta}{4}$. 23. $\text{cis}(-4\theta)$.

24. $\tan^3\frac{\theta}{2}(-\sin 3\theta + i\cos 3\theta)$.

25. $\text{cis}\,\theta$; $\text{cis}\,\theta$ has 3 cube roots.

26. $\text{cis}\,4a$; $\text{cis}\,4a$ has 4 fourth roots. 27. -1. 28. $2\cos n\theta$.

29. $2\cos\theta$; $2i\sin\theta$; $2\cos n\theta$; $2i\sin n\theta$. 30. $2\cos(\theta-\phi)$.

34. From $(a-1, b)$ to $(a+1, b)$; from $(-a, 0)$ to $(+a, 0)$; from $(0, -1)$ to $(0, +1)$; from $(-a, -b)$ to (a, b); in each case in a straight line.

35. (i) From (1, 0) to the origin and back with constant acceleration; (ii) from $(-1, 0)$ away from the origin with increasing acceleration and back along the positive x-axis to (1, 0).

36. (i) See No. 35; (ii) z_2 moves from $(\frac{1}{2}, 0)$ negatively along the x-axis; (iii) z_3 moves from (2, 0) along the axis to the origin.

37. (i) Describes $|z|=2$, anti-clockwise; (ii) describes $|z|=1$, clockwise; (iii) describes $|z+\frac{2}{3}|=\frac{1}{3}$, anti-clockwise; (iv) describes cardioid $r=2(1+\cos\theta)$, anti-clockwise.

38. (i) Circle, $|z+1|=1$, anticlockwise; (ii) $x=-\frac{1}{2}$, upwards; (iii) $x=-1$, upwards; (iv) $x=0$, upwards.

39. (i) Circle, $|z-2|=1$, anti-clockwise; (ii) circle, $|z-2|=1$, anti-clockwise; (iii) circle, $|z|=1$, clockwise.

40. (i) Circle, $|z|=1$, anti-clockwise; (ii) circle, $|z-i|=1$, anti-clockwise; (iii) $y=\frac{1}{2}$, from left to right; (iv) $x=\frac{1}{2}$, downwards.

41. From $(-1, 0)$ to (1, 0) along the upper half of the circle $|z|=1$.

42. For the motion of z_2 see No. 35. z_3 describes the lower half of the circle $|z|=1$. $z_4 \equiv z$.

43. The lemniscate, $r^2=2\cos 2\theta$.

EXERCISE VIII. f. (p. 156.)

6. (i) $\frac{\pi}{2}$; (ii) $\frac{\pi}{2}$ (x pos.); $-\frac{\pi}{2}$ (x neg.).

7. (i) $0 \leqslant x \leqslant 1$; (ii) $x=1$; (iii) $-1 \leqslant x \leqslant 0$.

9. (i) 0 (x pos.), π (x neg.); (ii) $\frac{\pi}{2}$ (y pos.), $-\frac{\pi}{2}$ (y neg.).

10. (i) $\frac{\pi}{4}$, $x>0$; $-\frac{3\pi}{4}$, $x<0$; (ii) $-\frac{\pi}{4}$, $x>0$; $\frac{3\pi}{4}$, $x<0$.

11. $n>\frac{1}{m}$. None. $n<\frac{1}{m}<0$.

14. $-\pi$ for $x<-1$; 0 for $-1<x<1$; π for $x>1$.

15. 0 for $1 \geqslant x \geqslant 0$; $-4\sin^{-1}x$ for $0>x\geqslant -1$.

16. (i) $2n\pi - \frac{\pi}{2} < \mathrm{Sin}^{-1} x < 2n\pi + \frac{\pi}{2}$;

(ii) $2n\pi < \mathrm{Cos}^{-1} x < 2n\pi + \pi$;

(iii) $n\pi - \frac{\pi}{2} < \mathrm{Tan}^{-1} x < n\pi + \frac{\pi}{2}$.

EXERCISE VIII. g. (p. 157.)

1. $\mathrm{cis}\,\frac{\pi}{4},\ 2^{\frac{n}{2}}\,\mathrm{cis}\,\frac{n\pi}{4}$. **2.** -256.

5. Mid-point of BC; centroid of ABC; point dividing AB as $1-k : k$.

6. $\frac{1}{2}(\alpha+\beta),\ \frac{1}{2}(\gamma+\delta),\ \frac{1}{4}(\alpha+\beta+\gamma+\delta)$; EG, FH, PQ have a common mid-point.

7. Points form a parallelogram.

8. Circle centre A; perp. bisector of AB; circle of Apollonius.

9. $(x^2 - x\sqrt{2}+1)(x^2+x\sqrt{2}+1)$.

10. (i) $2\cos(\alpha-\beta)\,\mathrm{cis}\,(\alpha+\beta)$ or $\{-2\cos(\alpha-\beta)\}\,\mathrm{cis}(\alpha+\beta+\pi)$;

(ii) $2\sin(\alpha-\beta)\,\mathrm{cis}\left(\alpha+\beta+\frac{\pi}{2}\right)$ or

$$\{-2\sin(\alpha-\beta)\}\,\mathrm{cis}\left(\alpha+\beta-\frac{\pi}{2}\right);$$

(iii) $4\sin(\alpha-\gamma)\cos(\beta-\delta)\,\mathrm{cis}\left(\alpha+\beta+\gamma+\delta+\frac{\pi}{2}\right)$ or

$$\{-4\sin(\alpha-\gamma)\cos(\beta-\delta)\}\,\mathrm{cis}\left(\alpha+\beta+\gamma+\delta-\frac{\pi}{2}\right).$$

13. (i) $\mathrm{cosec}\,\theta$ for $0<\theta<\pi$, $-\mathrm{cosec}\,\theta$ for $-\pi<\theta<0$;

(ii) $\frac{\pi}{2}-\theta$ for $0<\theta<\pi$, $-\frac{\pi}{2}-\theta$ for $-\pi<\theta<0$.

15. (i) Half-line; (ii) arc of circle.

16. (i) Displaced position of Σ; (ii) magnified and rotated position of Σ; (iii) magnification of the reflection in OX of the inverse of Σ.

17. $-1, 2, 1, 1$; or $-1, 0, -1, 1$.

18. $z=(1-Zi)/(Z-i)$. **19.** $p^2>25c^2$.

21. -1 ; 3 ; 0 if $n=3p$, 1 if $n=3p+1$, $-\omega^2$ if $n=3p-1$.

22. a^3-b^3, $a^3+b^3+c^3-3abc$.

EXERCISE VIII. h. (p. 159.)

1. (i) $2^{n+1}(-1)^{\frac{n}{2}}\sin^n\dfrac{\alpha-\beta}{2}\cos\dfrac{n(\alpha+\beta)}{2}$;

(ii) $2^{n+1}(-1)^{\frac{n+1}{2}}\sin^n\dfrac{\alpha-\beta}{2}\sin\dfrac{n(\alpha+\beta)}{2}$.

3. $x=\dfrac{3(2+\cos\theta)}{5+4\cos\theta}$, $y=\dfrac{-3\sin\theta}{5+4\cos\theta}$.

4. $p^2+q^2=x^2+y^2+2xy\cos(\alpha-\beta)$.

6. (i) $2\sin 2(\alpha-\beta)\operatorname{cis}\left(2\alpha+2\beta+\frac{\pi}{2}\right)$ or

$2\sin 2(\beta-\alpha)\operatorname{cis}\left(2\alpha+2\beta-\frac{\pi}{2}\right)$;

(ii) $2\sin(\alpha+\beta-\gamma-\delta)\operatorname{cis}\left(\alpha+\beta+\gamma+\delta+\frac{\pi}{2}\right)$ or

$2\sin(\gamma+\delta-\alpha-\beta)\operatorname{cis}\left(\alpha+\beta+\gamma+\delta-\frac{\pi}{2}\right)$;

(iii) $2\sin\theta\operatorname{cis}\frac{\pi}{2}$ or $-2\sin\theta\operatorname{cis}\left(-\frac{\pi}{2}\right)$.

9. A congruent curve. **10.** A circle. **12.** A cardioid.

15. Arcs of circles of radius $\dfrac{AB}{\sqrt{3}}$. **17.** $\dfrac{-a\alpha+b\beta+c\gamma}{-a+b+c}$ etc.

18. Equilateral triangle. **19.** Similar triangles.

23. Isolated points $(2m\pi+\pi,\ 2n\pi+\pi)$ together with oval curves inscribed in squares bounded by

$$x=2m\pi\pm\cos^{-1}(-\tfrac{1}{3}),\ y=2n\pi\pm\cos^{-1}(-\tfrac{1}{3}).$$

25. The upper half of $|z|=1$. The part of the x-axis outside $(\pm1,\ 0)$. The lower half of $|z|=1$.

30. $x^2+y^2+z^2-yz-zx-xy$; $(x+y+z)(x+\omega y+\omega^2 z)(x+\omega^2 y+\omega z)$.

31. (i) $3(a_3x^3+a_6x^6+\dots)$; (ii) $3(a_2x^2+a_5x^5+\dots)$.

CHAPTER IX

EXERCISE IX. a. (p. 168.)

1. $\pm\text{cis}\,\theta$; $\pm\text{cis}\left(-\frac{3\theta}{2}\right)$; $\pm\text{cis}\left(\frac{\pi}{4}-\frac{\theta}{2}\right)$; $\pm\frac{1+i}{\sqrt{2}}$; $\pm\frac{1-i}{\sqrt{2}}$.

2. $\text{cis}\left(\theta+\frac{2r\pi}{3}\right)$, $r=0, 1, 2$; $1, -\frac{1}{2}(1\pm i\sqrt{3})$; $-i, \frac{1}{2}(\pm\sqrt{3}+i)$; $i, \frac{1}{2}(\pm\sqrt{3}-i)$; $\text{cis}\,\frac{2r\pi-\theta}{3}$, $r=0, 1, 2$; $\text{cis}\left[\frac{\theta}{3}+(4r-1)\frac{\pi}{6}\right]$, $r=0, 1, 2$.

3. $1, -\frac{1}{2}(1\pm i\sqrt{3})$; $\text{cis}\left[(4r-3)\frac{\pi}{8}\right]$ $r=0, 1, 2, 3$; $\pm\sqrt[4]{2}\,.\,\text{cis}\,\frac{\pi}{8}$; $\sqrt[5]{4}\,.\,\text{cis}\left[\frac{2\pi}{15}(3r-1)\right]$, $r=0$ to 4; $2\,\text{cis}\,\frac{2r\pi}{7}$, $r=0$ to 6.

4. $\text{cis}\,\frac{2r\pi}{5}$, $r=0$ to 4; $\pm\frac{1}{\sqrt{2}}(1\pm i)$. **6.** -1. **7.** -1.

8. (i) $\pm\,\text{cis}\,\frac{\pi}{3}$; p.v. $\text{cis}\,\frac{\pi}{3}$; (ii) $\text{cis}\left(\frac{\pi}{3}+\frac{r\pi}{4}\right)$, $r=0$ to 7, p.v. $\text{cis}\,\frac{\pi}{3}$.

9. $\frac{1}{\sqrt{2}}(1-i)$; $\frac{1}{\sqrt{2}}(1-i)$. **10.** $\text{cis}\left(-\frac{\pi}{12}\right)$; $\text{cis}\,\frac{\pi}{6}$.

11. (i) 1; (ii) 0; $z^5=1$. **13.** $\text{cis}\,\frac{2r\pi}{n}$, $r=0$ to $n-1$; 0.

14. $\text{cis}\,\frac{r\pi}{3}$, $r=0, 1, 3, 5$. **15.** $\cot\frac{(2r-1)\pi}{12}$, $r=1$ to 6.

16. $i\tan\frac{r\pi}{n}$, $r=0$ to $n-1$. **17.** $\text{cis}\left(\frac{2r\pi}{n}\pm a\right)$, $r=0$ to $n-1$.

18. $\text{cis}\,(2r-1)\frac{\pi}{4}$, $r=1$ to 4, and $\text{cis}\,\frac{2s\pi}{5}$, $s=0$ to 4.

19. (i) $\sqrt{\left(\frac{1}{2}\cot\frac{\theta}{2}\right)}\,.\,(1+i)$; (ii) $\sqrt{\left(-\frac{1}{2}\cot\frac{\theta}{2}\right)}\,.\,(1-i)$. **20.** 7; 3.

21. (i) $\left(2\cos\frac{\theta}{2}\right)^{\frac{3}{4}}\text{cis}\,\frac{3\theta}{8}$; (ii) $\left(-2\cos\frac{\theta}{2}\right)^{\frac{3}{4}}\text{cis}\,\frac{3}{8}(\theta-2\pi)$.

22. $\text{cis}\left(a+\frac{r\pi}{3}\right)$, $r=1$ to 5.

23. (i) Circle, centre (3, 0), rad. 2; (ii) circle, centre (2, 0), rad. 1; (iii) circle, $|z|=3$, twice.

24. (i) Circle $|z|=1$, 3 times ; (ii) circle, centre (1, 0), rad. 1, 3 times ; (iii) two semicircles of $|z|=1$.

25. (i) Right half circle of $|z|=1$; (ii) cardioid, $r=2(1+\cos\theta)$; (iii) cardioid in (ii) displaced 1 unit to left.

26. (i) Right loop of lemniscate, $r^2=2\cos 2\theta$; (ii) right branch of rect. hyperbola, $2r^2\cos 2\theta=1$; (iii) both loops of (i) simultaneously.

EXERCISE IX. b. (p. 171.)

1. $2\cos 3\theta$; $2i\sin 4\theta$.
2. $\frac{1}{2}\{(\text{cis}\,\theta)^7+(\text{cis}\,\theta)^{-7}\}$; $\dfrac{1}{2i}\{(\text{cis}\,\theta)^6-(\text{cis}\,\theta)^{-6}\}$.
3. $\frac{1}{4}(\cos 3\theta+3\cos\theta)$.
4. $\frac{1}{8}(\cos 4\theta+4\cos 2\theta+3)$.
5. $\frac{1}{64}(\cos 7\theta+7\cos 5\theta+21\cos 3\theta+35\cos\theta)$.
6. $\frac{1}{8}(\cos 4\theta-4\cos 2\theta+3)$.
7. $-\frac{1}{64}(\sin 7\theta-7\sin 5\theta+21\sin 3\theta-35\sin\theta)$.
8. $-\frac{1}{8}(\sin 4\theta-2\sin 2\theta)$.
9. $-\frac{1}{64}(\sin 7\theta+\sin 5\theta-3\sin 3\theta-3\sin\theta)$.
10. $\frac{1}{256}(\cos 9\theta+\cos 7\theta-4\cos 5\theta-4\cos 3\theta+6\cos\theta)$.

12. (i) $-\frac{1}{16}(\frac{1}{5}\cos 5\theta-\frac{5}{3}\cos 3\theta+10\cos\theta+a)$;
 (ii) $\frac{1}{64}(\frac{1}{7}\sin 7\theta+\frac{7}{5}\sin 5\theta+7\sin 3\theta+35\sin\theta+a)$;
 (iii) $\frac{1}{256}(\frac{1}{9}\sin 9\theta+\frac{1}{7}\sin 7\theta-\frac{4}{5}\sin 5\theta-\frac{4}{3}\sin 3\theta+6\sin\theta+a)$.

13. $\frac{1}{512}(\frac{1}{10}\sin 10\theta+\frac{1}{4}\sin 8\theta-\frac{1}{2}\sin 6\theta-2\sin 4\theta+\sin 2\theta+6\theta+a)$.

19. 24.

EXERCISE IX. c. (p. 173.)

1. $5s-20s^3+16s^5$.
2. $16c^5-20c^3+5c$.
3. $32c^5-32c^3+6c$.
4. $1-18s^2+48s^4-32s^6$.
5. $\dfrac{4t-4t^3}{1-6t^2+t^4}$; $\dfrac{5t-10t^3+t^5}{1-10t^2+5t^4}$.
6. $t(t^2-3)(3t^2-1)=0$.
7. $1-21t^2+35t^4-7t^6=0$.
8. $\dfrac{t_1+t_2-t_3+t_1t_2t_3}{1-t_1t_2+t_2t_3+t_3t_1}$.
9. $t_1+t_2+t_3=t_1t_2t_3$; $\Sigma t_1=\Sigma(t_1t_2t_3)$; $t_1t_2+t_2t_3+t_3t_1=1$.
10. $s^n-\binom{n}{2}s^{n-2}c^2+\ldots=\cos\left(\dfrac{n\pi}{2}-n\theta\right)$;

$$ns^{n-1}c-\binom{n}{3}s^{n-3}c^3+\ldots=\sin\left(\frac{n\pi}{2}-n\theta\right).$$

11. (i) $(-1)^{\frac{n}{2}} . s^n$; $(-1)^{\frac{n}{2}-1} . n\,cs^{n-1}$;

(ii) $(-1)^{\frac{n-1}{2}} n\,cs^{n-1}$; $(-1)^{\frac{n-1}{2}} . s^n$.

12. (i) $(-1)^{\frac{n}{2}-1} . nt^{n-1}$; $(-1)^{\frac{n}{2}} . t^n$;

(ii) $(-1)^{\frac{n-1}{2}} . t^n$; $(-1)^{\frac{n-1}{2}} . nt^{n-1}$.

14. n even, $(-1)^{\frac{n}{2}}$; n odd, $(-1)^{\frac{n-1}{2}} . n$.

17. $t^6 - 21t^4 + 35t^2 - 7 = 0$. 18. $\dfrac{e-b}{1-c+f}$.

EXERCISE IX. d. (p. 176.)

1. $\dfrac{4\cos\theta - 2 + 2^{1-n}\cos n\theta - 2^{2-n}\cos(n+1)\theta}{5 - 4\cos\theta}$; $\dfrac{4\cos\theta - 2}{5 - 4\cos\theta}$.

2. (i) If $\theta \neq r\pi$, $\cot\theta\cos^n\theta\sin n\theta$; if $\theta = r\pi$, n; (ii) 0.

3. (i) $\dfrac{\sin^2\theta - \sin^{n+1}\theta\sin(n+1)\theta + \sin^{n+2}\theta\sin n\theta}{1 - \sin 2\theta + \sin^2\theta}$;

(ii) $\dfrac{\sin^2\theta}{1 - \sin 2\theta + \sin^2\theta}$.

4. $\operatorname{cosec}\frac{1}{2}\beta\sin\frac{1}{2}n\beta \operatorname{cis}\{\alpha + \frac{1}{2}(n-1)\beta\}$; equations (11) and (12), Ch. VII, pp. 125, 126.

6. $(-1)^{\frac{1}{2}n+1}\sin n\theta\,(2\sin\theta)^n$. 7. $\cos n\theta$. 8. $\sin^n\theta\cos n\phi$.

10. $\cos^n(\theta-\phi) - \binom{n}{1}\cos^{n-1}(\theta-\phi)\cos\phi\cos\theta + \dots$

$+(-1)^r\binom{n}{r}\cos^{n-r}(\theta-\phi)\cos^r\phi\cos r\theta \dots$.

12. (i) $\cos(n+1)\theta$; (ii) $2\cos n\theta$. 13. $\operatorname{cosec}\theta\cos(\alpha-\theta)$.

14. $(1 + 2x\cos\theta + x^2)^{\frac{n}{2}}\cos n\alpha$, where $\cos\alpha : \sin\alpha : 1$

$= (1 + x\cos\theta) : x\sin\theta : +\sqrt{(1 + 2x\cos\theta + x^2)}$.

15. $(1 + 2x^2\cos 2\theta + x^4)^{\frac{n}{2}}\cos n\alpha$, where $\cos\alpha : \sin\alpha : 1$

$= \cos\theta(1 + x^2) : \sin\theta(1 - x^2) : +\sqrt{(1 + 2x^2\cos 2\theta + x^4)}$.

16. For $r = 0$ to $n-1$, $\Sigma(r+1)\cos r\theta =$

$-\frac{1}{4}\operatorname{cosec}^2\frac{1}{2}\theta\,\{\cos\theta - (n+1)\cos(n-1)\theta + n\cos n\theta\}$,

$\Sigma(r+1)\sin r\theta =$

$\frac{1}{4}\operatorname{cosec}^2\frac{1}{2}\theta\,\{\sin\theta + (n+1)\sin(n-1)\theta - n\sin n\theta\}$.

17. $2^{2n-1}\cos n\theta\,\{\cos^{2n}\frac{1}{2}\theta - (-1)^n\sin^{2n}\frac{1}{2}\theta\}$.

EXERCISE IX. e. (p. 182.)

1. In $\sin\theta$; $n\sin\theta$; $(-1)^{\frac{1}{2}(n-1)}\,2^{n-1}\sin^n\theta$.

2. (i) in $\cos\theta$; $(-1)^{\frac{1}{2}n}$; $2^{n-1}\cos^n\theta$;
(ii) in $\sin\theta$; 1; $(-1)^{\frac{1}{2}n}\,2^{n-1}\sin^n\theta$.

3. In $\sin\theta$; $n\sin\theta$; $(-1)^{\frac{1}{2}n-1}\,2^{n-1}\sin^{n-1}\theta$.

4. (i) In $\cos\theta$; $n(-1)^{\frac{1}{2}(n-1)}$; $2^{n-1}\cos^{n-1}\theta$;
(ii) in $\sin\theta$; 1; $(-1)^{\frac{1}{2}(n-1)}\,2^{n-1}\sin^{n-1}\theta$.

5. (i) In $\sin\theta$; n; $(-1)^{\frac{1}{2}(n-1)}\,2^{n-1}\sin^{n-1}\theta$;
(ii) in $\cos\theta$; $(-1)^{\frac{1}{2}(n-1)}$; $2^{n-1}\cos^{n-1}\theta$.

6. (i) In $\sin\theta$; n; $(-1)^{\frac{1}{2}n-1}\,2^{n-1}\sin^{n-2}\theta$;
(ii) in $\cos\theta$; $n(-1)^{\frac{1}{2}n-1}$; $2^{n-1}\cos^{n-2}\theta$.

7. 1; see No. 22.

8. n; see No. 25.

26. $\dfrac{n(n-5)(n-6)(n-7)}{4!}$; $-\dfrac{n(n-6)(n-7)(n-8)(n-9)}{5!}$.

27. $$\sin\frac{p\pi}{2}=p\left\{1-\frac{p^2-1^2}{3!}\,.\,2+\frac{(p^2-1^2)(p^2-2^2)}{5!}\,.\,2^2-\dots,\ p\text{ terms}\right\}.$$

31. $$\left(x+\frac{1}{x}\right)^n-n\left(x+\frac{1}{x}\right)^{n-2}+\frac{n(n-3)}{2!}\left(x+\frac{1}{x}\right)^{n-4}-\frac{n(n-4)(n-5)}{3!}\left(x+\frac{1}{x}\right)^{n-6}+\dots$$

32. $y^9-9y^7+27y^5-30y^3+9y$, where $y=x+\dfrac{1}{x}$.

33. $y^3+7y^2+14y+7$, where $y=\left(x-\dfrac{1}{x}\right)^2$.

EXERCISE IX. f. (p. 185.)

1. $\operatorname{cis}\dfrac{2r\pi}{7}$, $r=1$ to 6.

2. $\operatorname{cis}\dfrac{(2r+1)\pi}{18}$, $r=0$ to 17, excluding $r=3k+1$.

3. $\left\{(a^2+b^2)\cos\frac{2r\pi}{n}+2ab+i(a^2-b^2)\sin\frac{2r\pi}{n}\right\}$

$\div\left\{a^2+b^2+2ab\cos\frac{2r\pi}{n}\right\}$, $r=0$ to $n-1$.

4. $\tan\frac{(4r+1)\pi}{4n}$, $r=0$ to $n-1$. 5. $\operatorname{cis}\frac{k\pi}{5}$, $k=2, 4, 6, 8$.

7. $\frac{2}{11}$; $-2-4i$, $1+2\sqrt{3}+i(2-\sqrt{3})$, $1-2\sqrt{3}+i(2+\sqrt{3})$.

8. $1-nx-\binom{n}{2}x^2+\binom{n}{3}x^3+\binom{n}{4}x^4-\binom{n}{5}x^5-\dots$.

9. (i) $3\left\{1+\binom{3n}{3}x^3+\binom{3n}{6}x^6+\dots\right\}$;
(ii) $3\left\{\binom{3n}{2}x^2+\binom{3n}{5}x^5+\dots\right\}$;
(iii) $3\left\{\binom{3n}{1}x+\binom{3n}{4}x^4+\dots\right\}$.

11. $\frac{2k\pi}{r-s}$. 13. The line $x=-\frac{1}{2}$. 14. The lines $y=\pm x$.

16. (i) $r^4+6r^2\cos 2\theta+5=0$; (ii) $r=8\cos^3\frac{\theta}{3}$.

18. $(2\cos\frac{1}{2}\theta)^n\cos\frac{1}{2}n\theta$. 19. 1; $2n+1$.

21. $2n\sin n\theta(1-\cos\theta)$.

EXERCISE IX. g. (p. 186.)

1. $(x+1)\left(x^2-2x\cos\frac{\pi}{7}+1\right)$

$\left(x^2-2x\cos\frac{3\pi}{7}+1\right)\left(x^2-2x\cos\frac{5\pi}{7}+1\right)$.

2. $(x^2+x+1)\left(x^2-2x\cos\frac{2\pi}{9}+1\right)$

$\left(x^2-2x\cos\frac{4\pi}{9}+1\right)\left(x^2-2x\cos\frac{8\pi}{9}+1\right)$.

7. $\Sigma\sin^4\frac{1}{2}(\beta-\gamma)\,\frac{\cos}{\sin}\,2(\beta+\gamma)$

$=2\Sigma\sin^2\frac{1}{2}(\alpha-\beta)\sin^2\frac{1}{2}(\alpha-\gamma)\,\frac{\cos}{\sin}(2\alpha+\beta+\gamma)$.

21. $n!\left\{\frac{(2\cos\theta)^n}{n!}+\dots+\frac{(2\cos\theta)^{n-2r}}{(r!)^2(n-2r)!}+\dots\right\}$.

CHAPTER X

EXERCISE X. a. (p. 196.)

1. $-e$.
2. $2\cos 1$.
3. $\dfrac{1+i\sqrt{3}}{2e}$.
4. $e^{\cos\theta}\operatorname{cis}(\sin\theta)$.
5. e^{2a}.
6 $r\operatorname{cis}\theta$.
7. $e\operatorname{cis}(\tan\alpha)$.
8. $e^{x\cos\theta+y\cos\phi}\operatorname{cis}(x\sin\theta+y\sin\phi)$.
9. -1.
10. -1.
11. $2e^{\cos\theta}\cos(\sin\theta)$.
12. $e^{\sin\theta}\operatorname{cis}(\sin^2\theta\sec\theta)$.
13. $2ie^{-\sin\theta}\sin(\cos\theta)$.
14. $e^{e^{\cos\theta}\cos(\sin\theta)}\cdot\operatorname{cis}\{e^{\cos\theta}\sin(\sin\theta)\}$.
16. $X^2+Y^2=e^{2c}$; $X\sin m=Y\cos m$.
17. $u=\exp\left\{\dfrac{x^2-a^2+y^2}{(x+a)^2+y^2}\right\}\cos\left\{\dfrac{2ay}{(x+a)^2+y^2}\right\}$.
18. $(1-a^2)\cos\theta\,.\,D^{-1}$, $(1+a^2)\sin\theta\,.\,D^{-1}$, where
$$D=1-2a^2\cos 2\theta+a^4.$$
19. (i) Down the y-axis; (ii) from left to right along $y=2n\pi-\dfrac{\pi}{2}$.
20. $a=2b=4m\pi$.
22. $e^{\cos\alpha}\cos(\sin\alpha)=1+\dfrac{\cos\alpha}{1!}+\dfrac{\cos 2\alpha}{2!}+\dots$;

$$e^{\cos\alpha}\sin(\sin\alpha)=\frac{\sin\alpha}{1!}+\frac{\sin 2\alpha}{2!}+\dots;$$

$$e\cos(\tan\beta)=1+\Sigma\left(\frac{1}{n!}\sec^n\beta\cos n\beta\right);$$

$$e\sin(\tan\beta)=\Sigma\left(\frac{1}{n!}\sec^n\beta\sin n\beta\right).$$

23. $\dfrac{\sin n\theta}{n!}$.
24. $u=v=\theta$; $\dfrac{2^{\frac{1}{2}n}}{n!}\cos\dfrac{n\pi}{4}$.
26. $\Sigma\dfrac{x^n}{n!}\sin\left(\dfrac{n\pi}{2}-n\alpha\right)$.
27. $\dfrac{x\cos\theta-x^2}{1-2x\cos\theta+x^2}$.
28. $\dfrac{x\sin\alpha-x^2\sin(\alpha-\beta)}{1-2x\cos\beta+x^2}$.
29. $e^{\cos\theta}\cos(\sin\theta)$.
30. $e^{-\cos\theta}\sin(\sin\theta)$.
31. $e^{-x\cos\beta}\cos(\alpha-x\sin\beta)$.
32. $e^{\sin\theta}\cos(\sin^2\theta\sec\theta)$.
33. $e^{\cos^2\beta}\cos(\alpha+\cos\beta\sin\beta)$.
34. $\cos(\sin\alpha)\operatorname{sh}(\cos\alpha)$.

EXERCISE X. b. (p. 200.)

1. $\frac{1}{2}(e^{-\frac{\pi}{2}} \pm e^{\frac{\pi}{2}})$.

5. $\sin x \operatorname{ch} y - i \cos x \operatorname{sh} y$.

6. $\frac{1}{2} + \frac{1}{2}\cos 2x \operatorname{ch} 2y - \frac{i}{2}\sin 2x \operatorname{sh} 2y$.

7. $\dfrac{\sin 2x - i \operatorname{sh} 2y}{\operatorname{ch} 2y - \cos 2x}$.

8. $\operatorname{ch} x \cos y + i \operatorname{sh} x \sin y$.

9. $\dfrac{\operatorname{sh} 2x - i \sin 2y}{\operatorname{ch} 2x + \cos 2y}$.

10. $\dfrac{2(\sin x \operatorname{ch} y - i \cos x \operatorname{sh} y)}{\operatorname{ch} 2y - \cos 2x}$.

11. $e^{\sin x \operatorname{ch} y} \operatorname{cis}(\cos x \operatorname{sh} y)$.

12. $e^{\operatorname{sh} x \cos y} \operatorname{cis}(-\operatorname{ch} x \sin y)$.

13. $\frac{1}{2}(\operatorname{sh} 2x \cos 2y - i \operatorname{ch} 2x \sin 2y)$.

14. $i \operatorname{sh} x$, $i \operatorname{ch} x$, $\coth x$, $-\operatorname{ch} x$, $-\operatorname{sh} x$, $\operatorname{th} x$.

15. $\dfrac{2 + 2\cos 2x \operatorname{ch} 2y}{\cos 2x + \operatorname{ch} 2y}$.

21. Ellipse. Hyperbola.

22. Confocal parabolas.

24. $\cos(\cos\theta)\operatorname{sh}(\sin\theta)$.

25. $1 - \cos\left(\cos\frac{\theta}{2}\right)\operatorname{ch}\left(\sin\frac{\theta}{2}\right)$.

26. $\operatorname{sh}(\cos\theta)\sin(\sin\theta)$.

27. $\operatorname{ch}(\sin\theta\cos\theta)\sin(\sin^2\theta)$.

28. $\frac{1}{2}\{\operatorname{ch}(\cos\theta)\cos(\sin\theta) + \cos(\cos\theta)\operatorname{ch}(\sin\theta)\}$.

30. (i) $1 - \dfrac{2^2 z^4}{4!} + \dfrac{2^4 z^8}{8!} - \dots$; (ii) $z + \dfrac{2z^3}{3!} - \dfrac{2^2 z^5}{5!} - \dfrac{2^3 z^7}{7!} + \dfrac{2^4 z^9}{9!} + \dots$.

EXERCISE X. c. (p. 202.)

1. $(-1)^n i$.

2. $e^{x^2 - y^2} \operatorname{cis}(2xy)$.

3. $\dfrac{\sin x + i \operatorname{sh} y}{\cos x + \operatorname{ch} y}$.

4. $\dfrac{2(\cos x \operatorname{ch} y + i \sin x \operatorname{sh} y)}{\cos 2x + \operatorname{ch} 2y}$.

5. $\dfrac{2(\sin x \operatorname{ch} y + i \cos x \operatorname{sh} y)}{\operatorname{ch} 2y - \cos 2x}$.

6. $\dfrac{2(\operatorname{sh} x \cos y + i \operatorname{ch} x \sin y)}{\operatorname{ch} 2x - \cos 2y}$.

9. $\pm\sec(a - ib)$.

11. $\dfrac{x \operatorname{sh} a}{1 - 2x \operatorname{ch} a + x^2}$ if $|x| e^{|a|} < 1$.

12. $1 - \dfrac{x\cos\alpha + x^2\cos(\alpha - \beta)}{1 + 2x\cos\beta + x^2}$ if $|x| < 1$.

13. $e^{x\cos\beta}\sin(\alpha + x\sin\beta)$.

14. $e^{x\cos\theta}\sin(x\sin\theta)$.

15. $\operatorname{ch}(\cos\theta)\cos(\sin\theta)$.

16. $\operatorname{ch}(\cos\theta)\sin(\sin\theta)$.

19. $\Sigma \dfrac{\cos n\theta}{n!}(a^2+b^2)^{\frac{n}{2}} x^n$, where $\cos\theta : \sin\theta : 1 = a : b : \sqrt{(a^2+b^2)}$.

21. $\text{ch}^2 \dfrac{\pi}{2}$. 22. $2 \text{ sh}(\cos\theta) \text{ cis}(\sin\theta)$.

23. $\sin\alpha + \Sigma \dfrac{x^n}{n!} \sin(\alpha + n\beta)$.

24. $2\sin\theta(1-\sin\theta\cos\theta)/(1-\sin 2\theta+\sin^2\theta)^2$.

25. $\cos(\alpha-\beta)\text{ sh}(\cos\beta)\cos(\sin\beta) - \sin(\alpha-\beta)\text{ ch}(\cos\beta)\sin(\sin\beta)$.

26. $e^{e^{\cos\theta}\cos(\sin\theta)} \cos\{e^{\cos\theta}\sin(\sin\theta)\}$.

27. $\frac{1}{2}\{\text{ch}(\cos\theta)\sin(\sin\theta) + \cos(\cos\theta)\text{ sh}(\sin\theta)\}$.

28. $\frac{1}{2}\{\text{sh}(\cos\theta)\cos(\sin\theta) - \sin(\cos\theta)\text{ ch}(\sin\theta)\}$.

CHAPTER XI

EXERCISE XI. a. (p. 207.)

2. $8c^3 - 4c^2 - 4c + 1 = 0$. 4. $64x^3 - 112x^2 + 56x - 7 = 0$.

7. 8 ; 416. 10. $x^2 + x + 2 = 0$. 17. $2\frac{1}{4}$.

18. $2\sqrt{2}(4x^3 - 3x) + \sqrt{3} + 1 = 0$.

EXERCISE XI. b. (p. 210.)

4. $3\tan 3\theta$. 17. $\dfrac{\pi^2}{12}$. 18. $\dfrac{4\pi^2}{27}$.

EXERCISE XI. c. (p. 214.)

2. $q = 1$. 15. The normals at the four points are concurrent.

20. $t^2(a^2+b^2+2b) - 4at + a^2 + b^2 - 2b = 0$;

$4a\{16a^2 + 12b^2 - 3(a^2+b^2)^2\}/(a^2+b^2-2b)^3$; $\dfrac{2ab}{a^2+b^2}$.

EXERCISE XI. d. (p. 216.)

1. $16x^5 - 20x^3 + 5x = \frac{1}{2}\sqrt{2}$;

$$-\tfrac{1}{2}\sqrt{2} \text{ and } \sin\frac{k\pi}{20} \text{ for } k = 1,\ 3,\ 9, \text{ and } -7.$$

2. $2\frac{1}{2}$. 5. 5. 6. 9. 11. $4x^2 + 2x = 3$. 14. $x^2 + x + 3 = 0$.

17. $16n^4 \operatorname{cosec}^4 2n\theta - \dfrac{16n^2(2n^2+1)}{3}\operatorname{cosec}^2 2n\theta + 2n$.

25. $\dfrac{\pi^4}{45} + \dfrac{10\pi^2}{3} - 35$; $126 - \dfrac{35\pi^2}{3} - \dfrac{\pi^4}{9}$.

26. (i) $\dfrac{\pi^6}{945}$; (ii) $\dfrac{2\pi^6}{945} + \dfrac{7\pi^4}{15} + 42\pi^2 - 462$;

(iii) $1716 - 154\pi^2 - \dfrac{28\pi^4}{15} - \dfrac{2\pi^6}{135}$.

27. $4ab(a^2 - b^2)/\{4c^4 - 4c^2(a^2+b^2) + (a^2-b^2)^2\}$. 28. $\dfrac{6bc}{c^2 - a^2}$.

CHAPTER XII

EXERCISE XII. a. (p. 220.)

1. $(x-1)\left(x - \operatorname{cis}\dfrac{2\pi}{3}\right)\left(x - \operatorname{cis}\dfrac{-2\pi}{3}\right)$.

2. $(x-1)(x^2+x+1)$. 3. $(x^2 + x\sqrt{2} + 1)(x^2 - x\sqrt{2} + 1)$.

4. $(x-1)\,\Pi\left(x - \operatorname{cis}\dfrac{2r\pi}{5}\right)$, $r = \pm 1,\ \pm 2$;

$$\{x^2 - \tfrac{1}{2}(\sqrt{5}-1)x + 1\}\{x^2 + \tfrac{1}{2}(\sqrt{5}+1)x + 1\}.$$

5. (i) $(x^2+1)(x^2 + x\sqrt{3} + 1)(x^2 - x\sqrt{3} + 1)$;

(ii) $\Pi\left(x^2 - 2xy\cos\dfrac{k\pi}{8} + y^2\right)$, $k = 1,\ 3,\ 5,\ 7$;

(iii) $(x^2 - a^2)(x^2 - ax + a^2)(x^2 + ax + a^2)$;

(iv) $(x^2 - 4)(x^2 + 4)(x^2 - 2x\sqrt{2} + 4)(x^2 + 2x\sqrt{2} + 4)$.

6. $\displaystyle\prod_0^4 \left\{x^2 - 2x\cos(6r+1)\frac{\pi}{15} + 1\right\}$.

7. (i) $\cos(2r+1)\dfrac{\pi}{2n}$, for $r = 0$ to $n - 1$;

(ii) $\cos\dfrac{2r\pi}{n}$, for $r = 0$ to $[\frac{1}{2}n]$;

(iii) $\cos\dfrac{(2r-1)\pi}{n}$, for $r=1$ to $[\frac{1}{2}(n+1)]$;

(iv) $\cos\left(\alpha+\dfrac{2r\pi}{n}\right)$, for $r=0$ to $n-1$;

(v) $\cos\dfrac{r\pi}{n}$, for $r=1$ to $n-1$.

8. (i) $\sin\dfrac{r\pi}{2n}$, for $r=\pm1,\ \pm3,\ \ldots\ \pm(n-1)$;

(ii) $\sin\dfrac{r\pi}{n}$, for $r=\pm1,\ \pm2,\ \ldots\ \pm\left(\dfrac{n}{2}-1\right)$.

9. (i) $\sin\dfrac{r\pi}{n}$, for $r=0,\ \pm1,\ \pm2,\ \ldots\ \pm\frac{1}{2}(n-1)$;

(ii) $\sin\dfrac{r\pi}{2n}$, for $r=\pm1,\ \pm3,\ \ldots\ \pm(n-2)$.

10. $\operatorname{cis}\left\{\pm\left(\alpha+\dfrac{2r\pi}{n}\right)\right\}$, for $r=0$ to $n-1$.

11. $\prod_{1}^{n}\left\{x^2-2x\cos\dfrac{(2r-1)\pi}{2n}+1\right\}$; $2\prod_{1}^{n}\left\{x^2+\tan^2\dfrac{(2r-1)\pi}{4n}\right\}$.

14. $(x^2-1)\prod_{1}^{\frac{1}{2}n-1}\left(x^2-2x\cos\dfrac{2r\pi}{n}+1\right)$.

EXERCISE XII. b. (p. 223.)

1. (i) $5\sin\theta\left(1-\sin^2\theta\operatorname{cosec}^2\dfrac{\pi}{5}\right)\left(1-\sin^2\theta\operatorname{cosec}^2\dfrac{2\pi}{5}\right)$;

(ii) $32\sin\theta(\sin^2\theta-\frac{1}{4})(\sin^2\theta-\frac{3}{4})$;

(iii) $16(\sin\theta-\sin\alpha)\left\{\sin\theta+\sin\left(\alpha-\dfrac{\pi}{5}\right)\right\}\left\{\sin\theta-\sin\left(\alpha+\dfrac{2\pi}{5}\right)\right\}$
$\left\{\sin\theta+\sin\left(\alpha+\dfrac{\pi}{5}\right)\right\}\left\{\sin\theta-\sin\left(\alpha-\dfrac{2\pi}{5}\right)\right\}$.

2. See Nos. 4 and 3.

14. $8\left(\sin\theta-\sin\dfrac{\pi}{14}\right)\left(\sin\theta+\sin\dfrac{3\pi}{14}\right)\left(\sin\theta-\sin\dfrac{5\pi}{14}\right)$.

15. $(-4)^{\frac{1}{2}(n-1)}\prod\left\{\sin\theta-\sin\left[\dfrac{r\pi}{n}+(-1)^r\alpha\right]\right\}$
for $r=0,\ \pm1,\ \ldots\ \pm\frac{1}{2}(n-1)$ if n is odd.

17. (i) n odd, $(-1)^{\frac{1}{2}(n-1)}\dfrac{\sin n\theta}{\sin\theta}$; n even, $(-1)^{\frac{1}{2}n}\cos n\theta$;

(ii) n odd, $(-1)^{\frac{1}{2}(n-1)}\dfrac{\cos n\theta}{n\cos\theta}$;

n even, $(-1)^{\frac{1}{2}n+1}\dfrac{\sin n\theta}{n\sin\theta\cos\theta}$.

24. (i) for n odd and (ii) for n even : $2^{\frac{1}{2}(1-n)}$;
(ii) for n odd and (i) for n even : $2^{\frac{1}{2}(1-n)}\sqrt{n}$.

30. 1. **32.** $16\left(\cos\theta-\cos\dfrac{\pi}{5}\right)^2\left(\cos\theta+\cos\dfrac{2\pi}{5}\right)^2$.

33. $16\left(\cos\theta+\cos\dfrac{\pi}{15}\right)\left(\cos\theta-\cos\dfrac{2\pi}{15}\right)\left(\cos\theta-\cos\dfrac{4\pi}{15}\right)\left(\cos\theta+\cos\dfrac{7\pi}{15}\right)$.

34. $\prod_{0}^{n-1}\left\{x^2-2x\cos\left(a+\dfrac{2r\pi}{n}\right)+1\right\}$.

EXERCISE XII. c. (p. 229.)

7. $-n\tan n\beta=\sum_{0}^{n-1}\cot\left\{\beta+\dfrac{(2r+1)\pi}{2n}\right\}$. **8.** $\sin 3\theta$; $\frac{1}{8}\sin 4\theta$.

9. $\frac{1}{4}(\cos 3\theta-\cos 3a)$. **24.** $\frac{1}{90}(n^2-4)(n^2+14)$.

25. $\frac{1}{6}n^2(n^2+2)$; $\frac{1}{6}(n^2-1)(n^2+3)$.

EXERCISE XII. d. (p. 234.)

1. $$\frac{1}{5(x-1)}+\frac{2\left(x\cos\frac{2\pi}{5}-1\right)}{5\left(x^2-2x\cos\frac{2\pi}{5}+1\right)}+\frac{2\left(x\cos\frac{4\pi}{5}-1\right)}{5\left(x^2-2x\cos\frac{4\pi}{5}+1\right)}.$$

2. $$\sum_{1}^{4}\frac{1-x\cos\frac{(2k-1)\pi}{8}}{4\left(x^2-2x\cos\frac{(2k-1)\pi}{8}+1\right)}.$$

4. $$\sum_{1}^{3}\frac{(-1)^{r-1}\sin^2\frac{r\pi}{7}\cos^3\frac{r\pi}{7}}{7\left(x^2+\tan^2\frac{r\pi}{7}\right)}.$$

5. $$\sum \frac{(-1)^{r-1}\sin(2r-1)\frac{\pi}{2n}\cos^2(2r-1)\frac{\pi}{2n}}{n\left\{\cos\theta-\cos(2r-1)\frac{\pi}{2n}\right\}},$$
for $r=1$ to n, omitting $\frac{1}{2}(n+1)$ if n is odd.

12. $$\sum_0^{n-1} \frac{\operatorname{cosec} n\theta \sin\left(\theta+\frac{2r\pi}{n}\right)}{n\left\{x^2-2x\cos\left(\theta+\frac{2r\pi}{n}\right)+1\right\}}.$$

13. $n\cos(n-1)\theta \operatorname{cosec} n\theta \operatorname{cosec}\theta - \operatorname{cosec}^2\theta$.

19. $\Sigma\{\sin^{n-1}a_1 \operatorname{cosec}(x-a_1)\operatorname{cosec}(a_1-a_2)\operatorname{cosec}(a_1-a_3)\ldots \operatorname{cosec}(a_1-a_n)\}$.

EXERCISE XII. e. (p. 236.)

1. $(x^2-a^2)\prod_{k=1}^{n-1}\left(x^2-2xa\cos\frac{k\pi}{n}+a^2\right)$.

2. $x-\operatorname{cis}\left\{\pm\frac{(6r+1)\pi}{9}\right\}$, $r=0, 1, 2$.

5. $(3^{2n}-1)/8n$.

6. $16\left(\sin\theta-\sin\frac{\pi}{14}\right)\left(\sin\theta-\sin\frac{5\pi}{14}\right)\left(\sin\theta+\sin\frac{3\pi}{14}\right)\left(\sin\theta+\frac{1}{2}\right)$.

17. $A=\operatorname{cosec}(a-b)\operatorname{cosec}(a-c)$, etc. $\prod_1^{2n+1}\operatorname{cosec}(x-a_r)$
$=\Sigma\operatorname{cosec}(x-a_r)\operatorname{cosec}(a_r-a_1)\operatorname{cosec}(a_r-a_2)\ldots \operatorname{cosec}(a_r-a_{r-1})\operatorname{cosec}(a_r-a_{r+1})\ldots\operatorname{cosec}(a_r-a_{2n+1})$.

18. $A=\operatorname{cosec}(a-b)=-B$. $\prod_1^{2n}\operatorname{cosec}(x-a_r)$
$=\Sigma\cot(x-a_r)\operatorname{cosec}(a_r-a_1)\operatorname{cosec}(a_r-a_2)\ldots \operatorname{cosec}(a_r-a_{r-1})\operatorname{cosec}(a_r-a_{r+1})\ldots\operatorname{cosec}(a_r-a_{2n})$.

19. $A=\cos^3 a\operatorname{cosec}(a-b)\operatorname{cosec}(a-c)$,
$D=\Sigma\sin a\cos^2 a\operatorname{cosec}(c-a)\operatorname{cosec}(a-b)$.

EXERCISE XII. f. (p. 238.)

1. $\cos 2\theta-\cos 3\theta$.

2. $16\,.\prod_1^4\left[\sin\theta+\sin\frac{(12r+5)\pi}{30}\right]$.

5. $\operatorname{ch} nx=\operatorname{ch}^n x\,\Pi\left\{1+\operatorname{th}^2 x\cot^2(2r-1)\frac{\pi}{2n}\right\}$, for $r=1$ to $[\frac{1}{2}n]$.

6. $n\operatorname{sh}x\operatorname{ch}^{n-1}x\prod_1^{\frac{1}{2}(n-1)}\left(1+\operatorname{th}^2 x\cot^2\frac{r\pi}{n}\right)$.

22. $(-1)^{[\frac{1}{2}n]}$.

CHAPTER XIII

EXERCISE XIII. a. (p. 243.)

1. (i) $2n\pi i$, $(4n+1)\dfrac{\pi i}{2}$, $\log 2+(2n+1)\pi i$, $\log 2+(4n-1)\dfrac{\pi i}{2}$;

(ii) as (i) with $n=0$.

2. (i) $\frac{1}{2}\log 2+i\left(\dfrac{\pi}{4}+2n\pi\right)$, $(8n+1)\dfrac{\pi i}{4}$,

$\log 5+i(\pi-\tan^{-1}\frac{4}{3}+2n\pi)$, $\log 5+i(-\pi+\tan^{-1}\frac{4}{3}+2n\pi)$;

(ii) as (i) with $n=0$.

3. $1+(2n+1)\pi i$.

4. (i) $i(a+2k\pi)$; (ii) $i(a-2n\pi)$.

5. $-\dfrac{e^2}{\sqrt{2}}$, $-\dfrac{e^2}{\sqrt{2}}$.

6. $\log(-2\cos\theta)+i(\theta-\pi)$.

7. $-(4n+1)\dfrac{\pi}{2}$, $\log 2+i\left(\dfrac{\pi}{3}+2n\pi\right)$.

8. $\log(-\sec a)+i(a-\pi)$.

9. $\frac{1}{2}e^{-\frac{\pi}{3}}$, $\dfrac{\sqrt{3}}{2}e^{-\frac{\pi}{3}}$.

10. $2i(a+\pi)$, $2ia$.

11. For $4m\pi<\theta<(4m+2)\pi$, $\log\left(2\sin\dfrac{\theta}{2}\right)+i\left(\dfrac{\theta-\pi}{2}+2k\pi\right)$;

p.v., $k=-m$.

For $(4m+2)\pi<\theta<(4m+4)\pi$,

$\log\left(-2\sin\dfrac{\theta}{2}\right)+i\left(\dfrac{\theta+\pi}{2}+2k\pi\right)$; p.v., $k=-m-1$.

12. $c=e^{\mathsf{A}}\cos\mathsf{B}$, $d=e^{\mathsf{A}}\sin\mathsf{B}$, where $\mathsf{A}=\dfrac{af+bg}{a^2+b^2}$, $\mathsf{B}=\dfrac{ag-bf}{a^2+b^2}$.

13. $3\operatorname{Log}\omega^2$.

15. $\pm a$.

16. $ie^{-(4n+1)\frac{\pi}{2}}$, $\sqrt{2}\cdot e^{-(8n+1)\frac{\pi}{4}}\operatorname{cis}\left(\frac{1}{2}\log 2+\dfrac{\pi}{4}\right)$.

17. $e^{-a-2n\pi}$.

18. $\log(\theta+2n\pi)+i\left(\dfrac{\pi}{2}+2m\pi\right)$.

20. $\frac{1}{2}\log\{(x+1)^2+y^2\}+i\tan^{-1}\dfrac{y}{x+1}+ik\pi$.

21. $\log\dfrac{\sqrt{\{(x^2+y^2-1)^2+4y^2\}}}{(x+1)^2+y^2}+i\tan^{-1}\dfrac{2y}{x^2+y^2-1}+ik\pi$.

22. $\log\tan\frac{\theta}{2}+\frac{\pi}{2}i$ or $\log\left(-\tan\frac{\theta}{2}\right)-\frac{\pi}{2}i$.

23. Loci are circles of two orthogonal coaxal systems.

24. Constant except at $(\pm a, 0)$.

EXERCISE XIII. b. (p. 250.)

1. $-\Sigma\frac{z^n}{n}$; $2\Sigma\frac{z^{2n-1}}{2n-1}$; unless $y=0$ and $x>1$.

2. $-\frac{1}{2}\log(1-2r\cos a+r^2)$. 3. $\frac{1}{2}\log(1-2\cos a\cos\beta+\cos^2 a)$.

6. $\frac{\pi}{2}-\theta$; $\frac{3\pi}{2}-\theta$. 7. $\log(-2\cos\theta)$.

8. $|x|<\frac{\pi}{2}$; $|x-\pi|<\frac{\pi}{2}$. 9. $\frac{1}{2}(2n+1)\pi-\theta$.

11. $\frac{1}{4}\log\{4(\cos a+\cos\beta)^2\}$. 12. $\frac{1}{4}\log\operatorname{cosec}^2 x$.

15. $\frac{1}{2}\tan^{-1}\left(\frac{2x\sin\theta}{1-x^2}\right)$.

17. $k\pi+\phi-x\sin 2\phi+\frac{1}{2}x^2\sin 4\phi-\frac{1}{3}x^3\sin 6\phi+\ldots$.

18. $k\pi+a+\tan^2\omega\sin 2a+\frac{1}{2}\tan^4\omega\sin 4a+\frac{1}{3}\tan^6\omega\sin 6a+\ldots$.

19. $-\frac{\pi}{4}$. 20. $2\Sigma(-1)^{r-1}\frac{1}{2r-1}x^{2r-1}\cos(2r-1)\theta$.

EXERCISE XIII. c. (p. 254.)

1. $e^{2n\pi}\operatorname{cis}(\log 2)$. 2. $e^{2n\pi}$. 3. $e^{\frac{1}{2}(4n-1)\pi}$.

4. $e^{2n\pi+\frac{\pi}{4}}\operatorname{cis}(\frac{1}{2}\log 2)$. 5. $e^{2n\pi+\frac{\pi}{4}}\operatorname{cis}(-\frac{1}{2}\log 2)$.

6. $e^{\frac{1}{2}(4n-1)\pi}$. 7. $-e^{2n\pi^2}$. 8. $ie^{n\pi^2}$.

9. $(-1)^n e^{2nk\pi^2}$. 10. $e^{x+2n\pi y}\operatorname{cis}(2n\pi x-y)$.

11. $\operatorname{cis}[\frac{1}{2}(4n-1)\pi^2]$. 12. $e^{2n\pi}\operatorname{cis}(-\log\pi)$.

13. $\frac{\log 2+2k\pi i}{\log 10+2n\pi i}$. 14. $\frac{2k\pi-i\log 3}{2n\pi}$. 15. $\frac{4k+1}{4n+1}$.

16. $\frac{\frac{1}{2}\log 3+k\pi i}{\log 3+n\pi i}$. 17. $u-2n\pi v+i(v+2n\pi u+2k\pi)$.

18. $e^{\theta+2n\pi}$. 19. -1. 20. $v=n\pi$. 22. $(-x)^x$; πx.

23. $e^{2\cos\theta+2m\pi\sin\theta}\operatorname{cis}\{(2m+4n)\pi\cos\theta\}$.

24. $\frac{1}{2}b\log(x^2+y^2)+a\tan^{-1}\frac{y}{x}-a\pi$ is a multiple of π.

26. $\frac{2(a+2k)}{4n+1}$. 27. No; yes; yes.

31. $\theta \neq (2r+1)\frac{\pi}{2}$, $n > -1$;

(i) (a) $(2\cos\theta)^n\cos n(\theta-2k\pi)$;

(b) $(-2\cos\theta)^n\cos n(\theta-\pi-2k\pi)$;

(ii) (a) $(2\cos\theta)^n\sin n(\theta-2k\pi)$;

(b) $(-2\cos\theta)^n\sin n(\theta-\pi-2k\pi)$.

$\theta=(2r+1)\frac{\pi}{2}$, (i) is zero for $n>0$ and unity for $n=0$, and divergent for $n<0$; (ii) is zero.

32. (i) $-\left(-2\sin\frac{a}{2}\right)^{-n}\sin\frac{n}{2}(a+\pi-4k\pi)$;

(ii) $-\left(2\sin\frac{a}{2}\right)^{-n}\sin\frac{n}{2}(a-\pi-4k\pi)$; (iii) 0.

33. $\frac{1}{2}\sin\frac{\theta}{4}\sqrt{\left(-2\sec\frac{\theta}{2}\right)}$. 34. $(-\cos\theta)^n\cos n(\theta-\pi)$.

35. $(\sin 2\theta-2x\sin\theta)(1-2x\cos\theta+x^2)^{-2}$.

EXERCISE XIII. d. (p. 258.)

2. $\text{Sh}^{-1}z=i\{n\pi-(-1)^n\sin^{-1}(iz)\}$; $\text{Th}^{-1}z=i\{n\pi-\tan^{-1}(iz)\}$.

10. $2n\pi\pm i\log 2$. 11. $2n\pi+\frac{\pi}{2}\pm i\log 3$.

12. $(2n+1)\pi\pm i\log 5$. 13. $k\pi+i(-1)^k\log(1+\sqrt{2})$.

14. (i) $\frac{\pi}{4}+\frac{i}{4}\log\frac{1+\sin\theta}{1-\sin\theta}$; (ii) $-\frac{\pi}{4}+\frac{i}{4}\log\frac{1+\sin\theta}{1-\sin\theta}$.

17. $2a=(2n-1)\pi+\frac{\pi}{2}+\phi$; $\beta=\frac{1}{2}\log\left(-\cot\frac{\phi}{2}\right)$.

EXERCISE XIII. e. (p. 260.)

1. $(-1)^n e^{-k\pi^2(2n+1)} \,.\, i.$

2. $e^{x^2-y^2-4k\pi xy} \operatorname{cis}(2xy+2k\pi x^2-2k\pi y^2).$

3. $e^{\theta-2n\pi-\frac{\pi}{2}}.$ 4. $\log(2n\pi+\theta)+i\left(2k\pi+\dfrac{\pi}{2}\right).$

5. $\dfrac{\sin^3\theta+i}{\cos\theta(1+\sin^2\theta)}.$ 6. $\sqrt{2}\,e^{-2n\pi-\frac{\pi}{4}} \operatorname{cis}\left(\log\sqrt{2}+\dfrac{\pi}{4}\right).$

7. Add $-\pi i$ to r.h.s. 9. $\cot^{-1}(1+\tan\theta+\tan^2\theta).$

12. $k\pi+c\sin 2x-\frac{1}{2}c^2\sin 4x+\frac{1}{3}c^3\sin 6x-\ldots.$ 13. $\frac{1}{2}(3\pi-\theta).$

14. $+\sqrt{\left(\frac{1}{2}\operatorname{cosec}\dfrac{\theta}{2}\right)}\sin\dfrac{\pi-\theta}{4}.$

15. $\cos n\left(\pi-\dfrac{\theta}{2}\right).\left(-2\cos\dfrac{\theta}{2}\right)^{-n}.$

16. $\theta\neq r\pi,\ \frac{1}{2}\tan^{-1}(2\cot^2\theta);\ \theta=r\pi,\ \dfrac{\pi}{4}.$

17. $x=2n\pi\pm\sin^{-1}(\sqrt{\tfrac{3}{5}}),\ y=\mp\operatorname{sh}^{-1}(\sqrt{\tfrac{3}{5}}).$ 18. $\dfrac{2a+4k\pi-2i}{(4n+1)\pi}.$

19. $-\dfrac{\pi}{2}<\theta<0,\ \sqrt[3]{\left(\frac{1}{2}\operatorname{cosec}\dfrac{3\theta}{2}\right)}\sin\left(\dfrac{\theta}{2}+\dfrac{\pi}{6}\right);$

$\theta=0,0;\ 0<\theta<\dfrac{\pi}{6},\ \sqrt[3]{\left(\frac{1}{2}\operatorname{cosec}\dfrac{3\theta}{2}\right)}\sin\left(\dfrac{\pi}{6}-\dfrac{\theta}{2}\right).$

EXERCISE XIII. f. (p. 261.)

2. $2(h\cos\theta-\frac{1}{2}h^2\cos 2\theta+\frac{1}{3}h^3\cos 3\theta-\ldots).$ 4. $-\operatorname{ch}2k\pi.$

10. $+\sqrt{\{\cos\theta+\cos^2\theta\}}.$ 11. $\frac{1}{2}\log\cot\dfrac{\theta}{2}.$

13. $e^{-\tan^{-1}(\cot x\coth y)}\operatorname{cis}\{\frac{1}{2}\log(\sin^2 x+\operatorname{sh}^2 y)\}.$

14. $y=\frac{1}{4}\log\dfrac{1-\sin\theta}{1+\sin\theta};$ (i) $x=\frac{1}{2}(\pi-\theta);$ (ii) $x=-\frac{1}{2}(\pi+\theta).$

16. $\pm[\operatorname{ch}^{-1}(t_1+t_2)+i\cos^{-1}(t_1-t_2)+2k\pi i]$, with the notation of Ex. XIII. d., No. 15.

17. $2^{\frac{n}{2}}\cos\dfrac{n\pi}{4}.$

18 (i) $\cos^2\frac{\theta}{2}\log\left(4\cos^2\frac{\theta}{2}\right)+\frac{1}{2}\theta\sin\theta-1$;

(ii) $\cos^2\frac{\theta}{2}\log\left(4\cos^2\frac{\theta}{2}\right)+\frac{1}{2}(\theta-2\pi)\sin\theta-1$.

19. $x=a_1y-a_3y^3+a_5y^5-\ldots+2k\pi$.

21. (i) $2n\pi+\frac{\pi}{3}<\theta<2(n+1)\pi-\frac{\pi}{3}$; $-\frac{\pi}{3}<\theta<\frac{\pi}{3}$.

CHAPTER XIV

EXERCISE XIV. d. (p. 270.)

1. $\tan^2\beta\cos^2\alpha=\sin^2\alpha+\tan^2\gamma$.
2. $a^2\sin^2\alpha=b^2+c^2-2bc\cos\alpha$.
3. $b-a=d$.
4. $(c^2-d^2)(a^2d-abc-2b^2d)=0$.
5. $(a^2+b^2)(a^2+b^2-3)=2b$.
6. $a^{\frac{2}{3}}+b^{\frac{2}{3}}=1$; $(a^2+b^2-1)^3+27a^2b^2=0$.
7. $13(a^2+b^2)=24ab+25$.
8. $(4a-3b)^2\{49-(a+b)^2\}=49(a+b)^2$.
9. $x^2y^2(x^2+y^2+3)=1$.
10. $\frac{x^2}{a^2}+\frac{y^2}{b^2}=1$.
11. $(a^2+2b^2-9)^2=a^2(b^2-3)(a^2+b^2-6)$.
12. $x^{\frac{2}{3}}+y^{\frac{2}{3}}=(2a)^{\frac{2}{3}}$; $(x^2+y^2-4a^2)^3+108a^2x^2y^2=0$.
13. $x^2=4a(a-y)$.
14. $x^2+y^2=c^2\sec^2\frac{1}{2}\alpha$.
15. $ac+b^2=c^2$.
16. $(ax)^{\frac{2}{3}}+(by)^{\frac{2}{3}}=c^{\frac{4}{3}}$; $(a^2x^2+b^2y^2-c^4)^3+27a^2b^2c^4x^2y^2=0$.
17. $\frac{x^2}{a}+\frac{y^2}{b}=a+b$.
18. $(a^2+b^2-c^2)(a^2+b^2-cd)^2=a^2b^2\{d^2-4(cd-a^2-b^2)\}$.
19. $cd^2(c+2d)=(c+3d)^2$.
20. $(1+c^2)\tan\beta=(1-c^2)\tan^2\alpha$.
21. $b^2=a^2+c^2-2ac\cos 2\alpha$.
22. $x^2+x\cos\alpha=2$.
23. $x^2=2abe\sin\alpha$.
24. $2\sin y\cot x=a^2+\cos^2 y$.

EXERCISE XIV. e. (p. 272.)

1. $a^2 + b^2 = \sin^2 \gamma$.
2. $x^2 + y^2 - 2ay \cos a = a^2 \sin^2 a$.
3. $(a^2 + b^2 - 2)^2 + 4c^2 = 4$.
4. $(x^2 + y^2)(1 - z) = 2y(1 + z)$.
5. $4t(2k - st) = s(k^2 - t^2)^2$.
6. $c = a(3 + 3b - 2a^2)$.
7. $x + a = 0$.
8. $x^2 + y^2 = a^2 + b^2$.
9. $(\sec a - 1)(\sec \beta - 1) = \sin^2 \gamma$.
10. $y^2 = 4a(x + a)$.
11. $(a - q)(b - q)(c - r) = p^2(a + b - c - q)$.
12. $a^3 + b^3 + c^3 = 3abc$.
13. $c^2 = a^2 + b^2 + 2abd$.
14. $\cot^2 a = \cot^2 \beta + \cot^2 \gamma$.
15. $(a^2 + b^2)(a^2 + b^2 - 4) + c^2 + d^2 = 2c(a^2 - b^2) + 4abd$.
16. $\cos^2 \theta_1 + \cos^2 \theta_2 + \cos^2 \theta_3 = 1$.
17. $\sin a + \sin \beta + \sin \gamma = 0$.
19. $\cos \frac{1}{2}(a + \beta + \gamma) \cdot \cos \frac{1}{2}(\beta + \gamma - a)$
$\cdot \cos \frac{1}{2}(\gamma + a - \beta) \cdot \cos \frac{1}{2}(a + \beta - \gamma) = 0$.
20. $\dfrac{x^2 - y^2}{a - b} = \dfrac{xy}{h}$.

EXERCISE XIV. f. (p. 274.)

1. Min. $4\sqrt{3}$ for $\tan x = \dfrac{\sqrt{3}}{2}$; max. $-4\sqrt{3}$ for $\tan x = -\dfrac{\sqrt{3}}{2}$.
2. Min. $\frac{3}{4}$ for $\sin x = \frac{1}{2}$; max. 3 for $\sin x = -1$.
3. Max. 10 for $\sin x = -1$; min. 2 for $\sin x = 1$.
4. Min. 4 for $\cos \theta = \frac{4}{5}$; max. -4 for $\cos \theta = -\frac{4}{5}$.
5. Max. and min. $14 \pm \frac{17}{2}$ for $2\theta + a = 2n\pi \pm \dfrac{\pi}{2}$, $\sin a = \frac{8}{17}$, $\cos a = \frac{15}{17}$.
7. $\sqrt{(a^2 - b^2)}$.
8. Max. $17 - 12\sqrt{2}$; min. $17 + 12\sqrt{2}$.
10. $\sqrt{(a^2 + 2ab \cos a + b^2)}$.
11. $\tan^2 \dfrac{a}{2}$.
12. $\sin^3 a$.
13. $\cos^3 a$.
14. 1.
15. 3 tan D, 3 cot D, 3 cosec D.

EXERCISE XIV. g. (p. 275.)

4. $\dfrac{2sc}{s^2 + c^2}$; $\dfrac{c^2 - s^2}{c^2 + s^2}$.
17. $a^2 + b^2 = c^2 + d^2$.
18. $(a + b - 2)^2 + \dfrac{(a - b)^2}{(2a + 2b - 5)^2} = 2$.
19. $(x^2 + y^2)(x^2 + y^2 + 18) + 8x(x^2 - 3y^2) = 27$.

20. $\Sigma a\{a\cos(\beta+\alpha)+b\cos 2\beta+c\cos(\beta+\gamma)\}$
$\{a\cos(\gamma+\alpha)+b\cos(\gamma+\beta)+c\cos 2\gamma\}=0.$

21. $b^2c^2(4a+3)^3=\{(1+a)^2-(b^2+c^2)\}\{(1+a)(1+2a)^2-(b^2+c^2)\}^2.$

22. $(z-9)(z-3)^2+128x^2y^2=0$, where $z=x^2+y^2$.

23. $pq(p\cos^2\alpha+q\sin^2\alpha-\sin\alpha\cos\alpha)^2$
$=\sin^2\alpha\cos^2\alpha(p\cos^2\alpha+q\sin^2\alpha)^2.$

24. $x\sin(\beta-\alpha)=\sin 2\alpha.$ 25. $\frac{5}{3}$.

MISCELLANEOUS EXAMPLES

EXERCISE XV. (p. 278.)

3. $2n\pi+\frac{\pi}{2}$ or $n\pi\pm\frac{\pi}{6}$. 5. Confocal ellipses, foci (± 1, 0).

8. $4\sin s\sin(s-x)\sin(s-y)\sin(s-z)$ where $s=\frac{1}{2}(x+y+z)$.

11. $x=y=z=\frac{\pi}{3}$, $\frac{x}{5}=\frac{y}{1}=\frac{z}{3}=\frac{\pi}{7}$, $\frac{x}{3}=\frac{y}{5}=\frac{z}{1}=\frac{\pi}{7}$, $\frac{x}{1}=\frac{y}{3}=\frac{z}{5}=\frac{\pi}{7}$,
$\frac{x}{7}=\frac{y}{1}=\frac{z}{5}=\frac{\pi}{9}$, $\frac{x}{5}=\frac{y}{7}=\frac{z}{1}=\frac{\pi}{9}$, or $\frac{x}{1}=\frac{y}{5}=\frac{z}{7}=\frac{\pi}{9}$.

12. $\frac{1}{1+x}-\log(1+x)$.

13. $(-1)^{\frac{1}{2}(n-1)}n\sin^{n-1}\theta\cos^{n-1}\theta$, n odd;
$(-1)^{\frac{1}{2}n}\,2\sin^n\theta\cos^n\theta$, n even.

15. $\frac{1}{4}(1+\sqrt{21})$, $\frac{1}{4}(1-\sqrt{21})$.

16. $(2b^2-a^2-ac):b(c-a):(c^2-2b^2+ac)$.

17. $\cot^{-1}(\sqrt{3}-\frac{2}{3})=43°\ 11'$. 19. $b^2-1<c<\frac{1}{8}b^4+1<3$.

20. $x+x^2+\frac{x^3}{3}-\frac{x^5}{30}-\ldots+\frac{x^n}{n!}2^{\frac{1}{2}n}\sin\frac{n\pi}{4}+\ldots$.

28. $\frac{1}{2}(\operatorname{sh}x-\sin x)$. 31. $x^2+x-1=0$; $\frac{1}{2}(\sqrt{5}-1)$, $-\frac{1}{2}(\sqrt{5}+1)$.

33. Interpolation 7·16099, calculation 7·16070(5).

35. $\tan^{-1}\frac{a_nx-y}{a_ny+x}$; $\pi=4\sum_1^\infty\tan^{-1}\frac{1}{2r^2}$.

37. $\Sigma\{\cos(\alpha+\beta)\cos(\alpha+\gamma)\sin(\beta-\gamma)\}$
$=-\sin(\beta-\gamma)\sin(\gamma-\alpha)\sin(\alpha-\beta).$

41. $\frac{1}{2}(2c+d)\sec\alpha-\frac{2c(c+d)}{2c+d}\cos\alpha$.

43. $\frac{1}{4}[3^{n+1}\sin(3^{-n}\theta) - 3\sin\theta]$; $\frac{3}{4}(\theta - \sin\theta)$.

44. $(\Sigma\alpha)^4 - 5(\Sigma\alpha)^2(\Sigma\alpha\beta) + 5(\Sigma\alpha)(\Sigma\alpha\beta\gamma) + 5(\Sigma\alpha\beta)^2 - 5\alpha\beta\gamma\delta$.

46. (i) $\dfrac{\pi^2}{16} - \frac{1}{2}$; (ii) $\frac{1}{2} - \dfrac{3\pi^2}{64}$.

51. $\operatorname{cosec}\beta\,[\cos^{n+1}\beta\sin(\alpha + n\beta) - \sin(\alpha - \beta)]$.

56. $(\cot\alpha - \tan\beta)^2 = c^2 - 1$.

59. $\theta = k\pi$, limit 0 ; $\theta \neq k\pi$, limit 1.

61. $2^{\frac{1}{6}}\operatorname{cis}\left(\dfrac{2r\pi}{3} \pm \dfrac{\pi}{4}\right)$, $r = 0, 1, 2$.

67. 68° 30′, 111° 30′, 205° 30′, 334° 30′ ; in radians, 1·20, 1·95, 3·59, 5·84.

69. ± 4, $\pm 2i$.

70. An epicycloid, the locus of a point on the circumference of a circle of unit radius rolling on the circle $|z| = 3$.

75. Ratio of radii = 1·16 : 1.

78. $\dfrac{\pi}{2\sqrt{3}}$.

INDEX

[*The numbers refer to the pages*]

Absolute convergence, 189, 190.
Absolute value, 78, 145.
Addition theorems, 64, 105, 123, 193, 198, 199.
Algebraic factors, 219.
Ambiguous case, 2.
Amplitude, 145, *see* principal value.
Angle-bisectors in triangle, 11.
Angle measurement, 119.
Approximation, 82, 95.
Area, hyperbolic, 52.
 of quadrilateral, 24, 27, 28.
Argand diagram, defined, 145.
 $z_1 \pm z_2$, 148.
 $z_1 z_2$, $z_1 \div z_2$, 151.
 z^2, $1/z$, 152, 153.
 z^n, 165 to 167.
Argument, 145.

Base of logarithms, natural, 63.
 generalised, 241, 253.
Binomial Series, 142 ii. and *Ex.* 3, 253.

Centroid, 10.
ch, sh, behaviour of, 109.
 calculus applications, 107, 110.
 formulae for, 105.
 generalised, 198.
 inverse, 110, 256.
Circular functions, *see* cos.
Circumcentre, 4.
cis θ, 152.
Complex number, amplitude, 145, 146.
 conjugate, 143.
 definitions, 137 to 139.
 difference, 139, 148.
 first and second parts, 143.
 geometrical representation, *see* Argand.
 i, 140, 141.
Complex number, manipulation, 141.
 modulus, 145.
 modulus-amplitude, 145.
 nomenclature, 143.
 notation, 140, 152.
 product, quotient, 139, 151, 152.
 sum, 138, 148.
Complex variable, functions of
 geometrical representation, *see* Argand.
 principal value, 146, 164, 165.
 exp, 191.
 log, 241, 253.
 sh, ch, etc., 198.
 sin, cos, etc., 197, 199.
 $\sin^{-1}$, sh^{-1}, etc., 256.
Compound Interest law, 93.
Convergence, 77.
 absolute, 189, 190.
 circle of, 247.
 conditional, 190.
cos (A + B), 123.
cosh, *see* ch.
cos θ, sin θ, differential of, 79.
 expansion of, 80, 81.
 exponential form, 194.
 factors of, 223.
 generalised, 197.
cos $n\theta$, in factors, 223.
 in terms of *c* and *s*, 172.
 in terms of *c* or *s*, 178.
$\cos^n\theta$, $\cos^p\theta \sin^q\theta$, in multiple angles 169.
Cotes' properties, 228.
Cubic equations, 44.
Cyclic quadrilateral, 24.

De Moivre, property of, 228.
 theorem of, 151, 162 ;
  and applications to
 expansions, 169, 172.
 factors, 219, 226.

De Moivre, powers and roots, 165.
solution of equations, 167.
summation of series, 174.
Distances between points in triangle, 15.

e, defined, 63.
evaluated, 91.
irrationality, 92.
series, 91.
E-centre, 4.
Elimination, 270.
Equations, approx. solution, 82, 95.
construction of, 204.
cubic, 44.
functions of roots of, 206, 208.
graphical solution, 38.
trigonometrical, 34.
Errors, 20.
Essentially distinct roots, 212.
Euler's constant, 70.
Expansions, polynomials, etc.,
$\cos n\theta$, $\sin n\theta$, 172, 178.
$\cos^n\theta$, $\sin^n\theta$, 169.
$\tan n\theta$, $\tan \Sigma\theta$, 172, 173.
Expansions, power series
$(1+x)^{-1}$, 77.
$(1+z)^{-1}$, $(1+z)^m$, 191, 253.
ch, sh, 104, 198.
cos, sin, 79, 198.
e^x, $\exp z$, 90, 191.
log, 84, 85, 245.
tan, 82.
$\tan^{-1}$, 88.
Exponential function,
defined, 64, 192.
differential of, 64.
expansion of, 90.
limit form, 93, 75.

Factors, algebraic, 219.
$\cos\theta$, $\sin\theta$, 223.
Cotes and de Moivre, 228.
fundamental theorem, 142, 219.
series and products, 228.
$\sin n\theta$, etc., 222, 227.
trigonometrical, 221.
$x^n \pm 1$, 219, 220.
$x^{2n} - 2x^n \cos n\alpha + 1$, 226.
Feuerbach's theorem, 16.
Functions, circular, 79, 197.
exponential, 64, 192.
'hyp,' 52.
Functions, hyperbolic, 104, 198.
inverse, 46, 64, 88, 110, 256.
log, 63, 84, 241, 247, 253.
many-valued, 241.

Geometric progression, 77, 191.
Gregory's expansion, 88.

Hyperbolic functions, 52, and *see* ch.

i, 140, 141.
Identities, 263.
IH^2, I_1H^2, IN, I_1N, 16.
In-centre, 4.
Indices, z^n, 140.
$z^{p/q}$, 162.
z^w, $e^{i\alpha}$, 252, 253.
Inequalities, cos, sin, 80.
exp, 71, 74, 97.
log, 67, 71, 73, 74.
miscellaneous, 57.
trigonometrical, 274.
Infinite Integrals, 53, 54, 57 (No.17), 75, 76.
Infinite Products, 223, 240.
Infinity, Sum to, 77, 190.
Integration, 63, 64, 79.
Inverse Functions (*see also* Functions),
differentiation of, 157 (No. 16).
principal values, 155.
$\tan^{-1}m \pm \tan^{-1}m'$, 47, 156 (No. 12).

Limits for $n\to\infty$,
x^n, $x^n/n!$, 78.
$(1\pm x/n)^n$, 93.
$(\cos x/n)^n$, 70.
$n(\sqrt[n]{a}-1)$, 69.
$\sum_1^n \frac{1}{r} - \log n$, 69.
Limits for, $x\to\infty$,
hyp x, 53.
$(\log x)/x^p$, 68, 71 (No. 11), 73 (No. 25).
Limits for $y\to 0$,
$(e^y-1)/y$, 69.
hyp $|y|$, 54.
$\{\log(1+y)\}/y$, 68.
$y \log |y|$, 68.
Logarithms, base e, 63.
differentiation, 63.
inequalities, 67.

Logarithms, integration, 67 (No. 45).
  log w, $\log_z\zeta$, 241, **253.**

Machin's formula, 89.
Many-valued functions, 47, **155, 241.**
Mass-centre theorem, 10.
Maxima and Minima, **274.**
Median, 11.
Modulus, 78, 145, 148.
Modulus-amplitude form, **145.**

Nine-point circle, 6, 16.
Number, 137, and *see* Complex.

OH^2, OI^2, OI_1^2, **15.**
Ordered pairs, 138.
Orthocentre, 5.
Osborn's rule, 105.

π, evaluation, 89.
  product, 223.
$\pi^2/6$, 209, 228.
Partial fractions, **231.**
Pedal triangle, 5.
Polar circle, 6.
Power series, 77.
Powers, *see* Indices.
Powers of cos θ, sin θ, **169.**
Principal values,
  amplitude, 146, 156 (No. 8).
  $(\text{cis}\ \theta)^{p/q}$, **164.**
  $\cos^{-1}x$, etc., 155.
  Log w, Log$(1+w)$, $\text{Log}_z\zeta$, 241, 247, 254.
  range of, 155.
  $\text{Tan}^{-1}z$, 258.
  z^w, $(1+z)^m$, 252, 254.
Projection, cos, sin (A + B), **123.**
  points and lines, 118.
  summation, 125.
Ptolemy's theorem, 25, 27.

Quadrilateral, circumscribable, 27.
  cyclic, 24.
  general, 27.

R, r, r_1, ρ, 4, 5, 6, 25.
Roots, essentially distinct, **212.**
Roots of equations, 204.
Rutherford's formula, 89.

Series (*see also* Expansions)
  Σx^r, Σz^r, 77, 191.
  $\Sigma x^r/r!$, $\Sigma z^r/r!$, 90, **191.**
  Σr^{-2}, 209, 228.
  Σr^{-4}, 212 (No. 21).
  $\Sigma \cos(\alpha + r\beta)$, $\Sigma \sin(\alpha + r\beta)$, **125,** 127.
  $\Sigma x^r \cos r\theta$, $\Sigma x^r \sin r\theta$, **174.**
  $\Sigma(-1)^{n-1}\rho^n n^{-1} \cos(\text{or}\ \sin) n\phi$ **245.**
  $\Sigma\ \text{cosec}^2 r\pi/n$, 209, 228.
  binomial, 253.
  calculus method, 128, 132 (ii).
  definitions, 77, 189.
  difference method, 127, **130.**
  products and, 228.
sh, sin, sinh, *see* ch, cos.
sin (A + B), 123.
Solution of Triangles, **1, 2, 19.**
Submultiple angles,
  cos $\frac{1}{2}\theta$, sin $\frac{1}{2}\theta$ in terms of cos θ, **41.**
  cos $\frac{1}{2}\theta$, sin $\frac{1}{2}\theta$ in terms of sin θ, 41-43.
  cos $\frac{1}{3}\theta$ in terms of cos θ, **43.**
Subsidiary angle, 1.
Successive approximation, 82, **95.**
Sum to infinity, 77, 190.

tan $n\theta$, tan $\Sigma\theta$, **172, 173.**
tan x, 82.
$\tan^{-1}x$, 88, 57.
Trigonometrical factors, **221.**

Wallis' limit for π, 223.

EXPLANATION OF SYMBOLS

r! denotes factorial-r, that is $1 . 2 . 3 . 4 . \ldots (r-1) . r$.

$\binom{n}{r}$ denotes the binomial coefficient $\dfrac{n(n-1)(n-2)\ldots(n-r+1)}{1 . 2 . 3 . \ldots r}$.

[x] denotes the integral part of x; more precisely the largest whole number that is not algebraically greater than x; e.g. $[\frac{7}{2}]=3$ and $[-\frac{7}{2}]=-4$.

a = b (mod c) denotes that $a-b$ is a multiple of c.

$\simeq$ denotes "is approximately equal to."

a < b = c < d denotes: $a<b$, and $b=c$; $\therefore a<c$, but $c<d$; $\therefore a<d$.

$\alpha<(\theta, \phi, \ldots)<\beta$ denotes $\alpha<\theta<\beta$, $\alpha<\phi<\beta, \ldots$.

$\sum_{r=1}^{r=k} f(r)$ or $\sum_{1}^{k} f(r)$ denotes $f(1)+f(2)+f(3)+\ldots+f(k)$.

$\prod_{r=1}^{r=k} f(r)$ or $\prod_{1}^{k} f(r)$ denotes $f(1) . f(2) . f(3) . \ldots f(k)$.

increases steadily is taken to mean "always increases or at any rate does not decrease."

E.T. *Elementary Trigonometry*, by C. V. Durell and R. M. Wright.

M.G. *Modern Geometry*, by C. V. Durell.

A CATALOG OF SELECTED

DOVER BOOKS

IN SCIENCE AND MATHEMATICS

A CATALOG OF SELECTED
DOVER BOOKS
IN SCIENCE AND MATHEMATICS

Astronomy

BURNHAM'S CELESTIAL HANDBOOK, Robert Burnham, Jr. Thorough guide to the stars beyond our solar system. Exhaustive treatment. Alphabetical by constellation: Andromeda to Cetus in Vol. 1; Chamaeleon to Orion in Vol. 2; and Pavo to Vulpecula in Vol. 3. Hundreds of illustrations. Index in Vol. 3. 2,000pp. 6⅛ x 9¼.
23567-X, 23568-8, 23673-0 Three-vol. set

THE EXTRATERRESTRIAL LIFE DEBATE, 1750–1900, Michael J. Crowe. First detailed, scholarly study in English of the many ideas that developed from 1750 to 1900 regarding the existence of intelligent extraterrestrial life. Examines ideas of Kant, Herschel, Voltaire, Percival Lowell, many other scientists and thinkers. 16 illustrations. 704pp. 5⅜ x 8½. 40675-X

A HISTORY OF ASTRONOMY, A. Pannekoek. Well-balanced, carefully reasoned study covers such topics as Ptolemaic theory, work of Copernicus, Kepler, Newton, Eddington's work on stars, much more. Illustrated. References. 521pp. 5⅜ x 8½.
65994-1

AMATEUR ASTRONOMER'S HANDBOOK, J. B. Sidgwick. Timeless, comprehensive coverage of telescopes, mirrors, lenses, mountings, telescope drives, micrometers, spectroscopes, more. 189 illustrations. 576pp. 5⅝ x 8¼. (Available in U.S. only.)
24034-7

STARS AND RELATIVITY, Ya. B. Zel'dovich and I. D. Novikov. Vol. 1 of *Relativistic Astrophysics* by famed Russian scientists. General relativity, properties of matter under astrophysical conditions, stars, and stellar systems. Deep physical insights, clear presentation. 1971 edition. References. 544pp. 5⅝ x 8¼. 69424-0

Chemistry

CHEMICAL MAGIC, Leonard A. Ford. Second Edition, Revised by E. Winston Grundmeier. Over 100 unusual stunts demonstrating cold fire, dust explosions, much more. Text explains scientific principles and stresses safety precautions. 128pp. 5⅜ x 8½. 67628-5

THE DEVELOPMENT OF MODERN CHEMISTRY, Aaron J. Ihde. Authoritative history of chemistry from ancient Greek theory to 20th-century innovation. Covers major chemists and their discoveries. 209 illustrations. 14 tables. Bibliographies. Indices. Appendices. 851pp. 5⅜ x 8½. 64235-6

CATALYSIS IN CHEMISTRY AND ENZYMOLOGY, William P. Jencks. Exceptionally clear coverage of mechanisms for catalysis, forces in aqueous solution, carbonyl- and acyl-group reactions, practical kinetics, more. 864pp. 5⅜ x 8½.
65460-5

Math–Geometry and Topology

ELEMENTARY CONCEPTS OF TOPOLOGY, Paul Alexandroff. Elegant, intuitive approach to topology from set-theoretic topology to Betti groups; how concepts of topology are useful in math and physics. 25 figures. 57pp. 5⅜ x 8½. 60747-X

COMBINATORIAL TOPOLOGY, P. S. Alexandrov. Clearly written, well-organized, three-part text begins by dealing with certain classic problems without using the formal techniques of homology theory and advances to the central concept, the Betti groups. Numerous detailed examples. 654pp. 5⅜ x 8½. 40179-0

EXPERIMENTS IN TOPOLOGY, Stephen Barr. Classic, lively explanation of one of the byways of mathematics. Klein bottles, Moebius strips, projective planes, map coloring, problem of the Koenigsberg bridges, much more, described with clarity and wit. 43 figures. 210pp. 5⅜ x 8½. 25933-1

CONFORMAL MAPPING ON RIEMANN SURFACES, Harvey Cohn. Lucid, insightful book presents ideal coverage of subject. 334 exercises make book perfect for self-study. 55 figures. 352pp. 5⅝ x 8¼. 64025-6

THE GEOMETRY OF RENÉ DESCARTES, René Descartes. The great work founded analytical geometry. Original French text, Descartes's own diagrams, together with definitive Smith-Latham translation. 244pp. 5⅜ x 8½. 60068-8

THE THIRTEEN BOOKS OF EUCLID'S ELEMENTS, translated with introduction and commentary by Sir Thomas L. Heath. Definitive edition. Textual and linguistic notes, mathematical analysis. 2,500 years of critical commentary. Unabridged. 1,414pp. 5⅜ x 8½. Three-vol. set.
Vol. I: 60088-2 Vol. II: 60089-0 Vol. III: 60090-4

GEOMETRY OF COMPLEX NUMBERS, Hans Schwerdtfeger. Illuminating, widely praised book on analytic geometry of circles, the Moebius transformation, and two-dimensional non-Euclidean geometries. 200pp. 5⅝ x 8¼. 63830-8

DIFFERENTIAL GEOMETRY, Heinrich W. Guggenheimer. Local differential geometry as an application of advanced calculus and linear algebra. Curvature, transformation groups, surfaces, more. Exercises. 62 figures. 378pp. 5⅜ x 8½. 63433-7

CURVATURE AND HOMOLOGY: Enlarged Edition, Samuel I. Goldberg. Revised edition examines topology of differentiable manifolds; curvature, homology of Riemannian manifolds; compact Lie groups; complex manifolds; curvature, homology of Kaehler manifolds. New Preface. Four new appendixes. 416pp. 5⅜ x 8½. 40207-X

TOPOLOGY, John G. Hocking and Gail S. Young. Superb one-year course in classical topology. Topological spaces and functions, point-set topology, much more. Examples and problems. Bibliography. Index. 384pp. 5⅝ x 8¼. 65676-4

Physics

OPTICAL RESONANCE AND TWO-LEVEL ATOMS, L. Allen and J. H. Eberly. Clear, comprehensive introduction to basic principles behind all quantum optical resonance phenomena. 53 illustrations. Preface. Index. 256pp. 5⅜ x 8½. 65533-4

ULTRASONIC ABSORPTION: An Introduction to the Theory of Sound Absorption and Dispersion in Gases, Liquids and Solids, A. B. Bhatia. Standard reference in the field provides a clear, systematically organized introductory review of fundamental concepts for advanced graduate students, research workers. Numerous diagrams. Bibliography. 440pp. 5⅜ x 8½. 64917-2

QUANTUM THEORY, David Bohm. This advanced undergraduate-level text presents the quantum theory in terms of qualitative and imaginative concepts, followed by specific applications worked out in mathematical detail. Preface. Index. 655pp. 5⅜ x 8½. 65969-0

ATOMIC PHYSICS (8th edition), Max Born. Nobel laureate's lucid treatment of kinetic theory of gases, elementary particles, nuclear atom, wave-corpuscles, atomic structure and spectral lines, much more. Over 40 appendices, bibliography. 495pp. 5⅜ x 8½. 65984-4

AN INTRODUCTION TO HAMILTONIAN OPTICS, H. A. Buchdahl. Detailed account of the Hamiltonian treatment of aberration theory in geometrical optics. Many classes of optical systems defined in terms of the symmetries they possess. Problems with detailed solutions. 1970 edition. xv + 360pp. 5⅜ x 8½. 67597-1

THIRTY YEARS THAT SHOOK PHYSICS: The Story of Quantum Theory, George Gamow. Lucid, accessible introduction to influential theory of energy and matter. Careful explanations of Dirac's anti-particles, Bohr's model of the atom, much more. 12 plates. Numerous drawings. 240pp. 5⅜ x 8½. 24895-X

ELECTRONIC STRUCTURE AND THE PROPERTIES OF SOLIDS: The Physics of the Chemical Bond, Walter A. Harrison. Innovative text offers basic understanding of the electronic structure of covalent and ionic solids, simple metals, transition metals and their compounds. Problems. 1980 edition. 582pp. 6⅛ x 9¼. 66021-4

HYDRODYNAMIC AND HYDROMAGNETIC STABILITY, S. Chandrasekhar. Lucid examination of the Rayleigh-Benard problem; clear coverage of the theory of instabilities causing convection. 704pp. 5⅜ x 8¼. 64071-X

INVESTIGATIONS ON THE THEORY OF THE BROWNIAN MOVEMENT, Albert Einstein. Five papers (1905–8) investigating dynamics of Brownian motion and evolving elementary theory. Notes by R. Fürth. 122pp. 5⅜ x 8½. 60304-0

THE PHYSICS OF WAVES, William C. Elmore and Mark A. Heald. Unique overview of classical wave theory. Acoustics, optics, electromagnetic radiation, more. Ideal as classroom text or for self-study. Problems. 477pp. 5⅜ x 8½. 64926-1

METHODS OF THERMODYNAMICS, Howard Reiss. Outstanding text focuses on physical technique of thermodynamics, typical problem areas of understanding, and significance and use of thermodynamic potential. 1965 edition. 238pp. 5⅜ x 8½. 69445-3

TENSOR ANALYSIS FOR PHYSICISTS, J. A. Schouten. Concise exposition of the mathematical basis of tensor analysis, integrated with well-chosen physical examples of the theory. Exercises. Index. Bibliography. 289pp. 5⅜ x 8½. 65582-2

RELATIVITY IN ILLUSTRATIONS, Jacob T. Schwartz. Clear nontechnical treatment makes relativity more accessible than ever before. Over 60 drawings illustrate concepts more clearly than text alone. Only high school geometry needed. Bibliography. 128pp. 6⅛ x 9¼. 25965-X

THE ELECTROMAGNETIC FIELD, Albert Shadowitz. Comprehensive undergraduate text covers basics of electric and magnetic fields, builds up to electromagnetic theory. Also related topics, including relativity. Over 900 problems. 768pp. 5⅝ x 8¼. 65660-8

GREAT EXPERIMENTS IN PHYSICS: Firsthand Accounts from Galileo to Einstein, edited by Morris H. Shamos. 25 crucial discoveries: Newton's laws of motion, Chadwick's study of the neutron, Hertz on electromagnetic waves, more. Original accounts clearly annotated. 370pp. 5⅜ x 8½. 25346-5

RELATIVITY, THERMODYNAMICS AND COSMOLOGY, Richard C. Tolman. Landmark study extends thermodynamics to special, general relativity; also applications of relativistic mechanics, thermodynamics to cosmological models. 501pp. 5⅜ x 8½. 65383-8

LIGHT SCATTERING BY SMALL PARTICLES, H. C. van de Hulst. Comprehensive treatment including full range of useful approximation methods for researchers in chemistry, meteorology and astronomy. 44 illustrations. 470pp. 5⅜ x 8½. 64228-3

STATISTICAL PHYSICS, Gregory H. Wannier. Classic text combines thermodynamics, statistical mechanics and kinetic theory in one unified presentation of thermal physics. Problems with solutions. Bibliography. 532pp. 5⅜ x 8½. 65401-X

Paperbound unless otherwise indicated. Available at your book dealer, online at **www.doverpublications.com**, or by writing to Dept. GI, Dover Publications, Inc., 31 East 2nd Street, Mineola, NY 11501. For current price information or for free catalogues (please indicate field of interest), write to Dover Publications or log on to **www.doverpublications.com** and see every Dover book in print. Dover publishes more than 500 books each year on science, elementary and advanced mathematics, biology, music, art, literary history, social sciences, and other areas.